TROPICAL BOTANY

Proceedings of a Symposium held at the University of Aarhus on 10–12 August 1978, organized on the occasion of the 50th anniversary of this university.

TROPICAL BOTANY

Edited by

Kai Larsen and Lauritz B. Holm-Nielsen

Botanical Institute,
University of Aarhus, Denmark

1979

ACADEMIC PRESS
LONDON   NEW YORK   SAN FRANCISCO
A Subsidiary of Harcourt Brace Jovanovich, Publishers

ACADEMIC PRESS INC. (LONDON) LTD.
24/28 Oval Road
London NW1

United States Edition published by
ACADEMIC PRESS INC.
111 Fifth Avenue
New York, New York 10003

British Library Cataloguing in Publication Data

Tropical botany.
  1. Tropical plants—Congresses
  I. Larsen, Kai   II. Holm–Nielsen, Lauritz B
  581.9'09'3   QK936   79–41003
  ISBN 0–12–437350–X

Printed in Great Britain
by W & J Mackay Limited, Chatham

Contributors

M. Acosta-Solis Instituto Ecuatoriano de Ciencias Naturales, Apartado 408, Quito, Ecuador, SA

M. T. Kalin Arroyo Department de Biología, University of Chile, Casilla 653, Santiago, Chile

P. S. Ashton Arnold Arboretum of Harvard University, 22, Divinity Avenue, Cambridge, MA 02138, USA

L. Bernardi Conservatoire et Jardin Botaniques, Chemin de l'Impératrice 1, Case postale 60, CH-1292 Chambésy, Switzerland

M. M. Bhandari Department of Botany, University of Jodhpur, Jodhpur–342 001, India

J. P. M. Brenan Royal Botanic Gardens, Kew, Richmond, Surrey TW9 3AE, England

A. M. Cleef Institute for Systematic Botany, Heidelberglaan 2, Utrecht, Netherlands

Th. B. Croat Missouri Botanical Garden, 2345 Tower Grove Avenue, St Louis, MO 63110, USA

J. Cuatrecasas National Museum of Natural History, Smithsonian Institution, Washington, DC 20560, USA

F. R. Fosberg National Museum of Natural History, Smithsonian Institution, Washington, DC 20560, USA

A. Gentry Missouri Botanical Garden, 2345 Tower Grove Avenue, St Louis, MO 63110, USA

T. van der Hammen University of Amsterdam, Sarphatistraat 221, 1018 BX Amsterdam, Netherlands

G. Harling Botanical Institute, Carl Skottsbergs Gata 22, 413 19 Göteborg, Sweden

L. B. Holm-Nielsen Botanical Institute, Nordlandsvej 68, DK-8240 Risskov, Denmark

R. A. Howard The Arnold Arboretum of Harvard University, 22, Divinity Avenue, Cambridge, MA 02138, USA

A. T. Hunziker Casilla de Correo 495, 5000 Cordoba, Argentina

K. Iwatsuki Department of Botany, Faculty of Science, Kyoto University, Kyoto 606, Japan

C. Kalkman Rijksherbarium, Schelpenkade 6, Leiden-2404, Netherlands

H. Walter Lack Botanischer Garten und Botanisches Museum, Berlin-Dahlem, Königin Luise Strasse 6-8, 1000 Berlin, BRD

K. Larsen Botanical Institute, Nordlandsvej 68, DK-8240 Risskov, Denmark

P. J. M. Maas Institute for Systematic Botany, Heidelberglaan 2, Utrecht, Netherlands

B. Maguire The New York Botanical Garden, Bronx, NY 10458, USA

G. T. Prance The New York Botanical Garden, Bronx, NY 10458, USA

B. Øllgaard Botanical Institute, Nordlandsvej 68, DK-8240 Risskov, Denmark

Rolla S. Rao Department of Botany, Andhra University, Waltair, Visakhapatnam-53003, India

P. H. Raven Missouri Botanical Garden, 2345 Tower Grove Avenue, St Louis, MO 63110, USA

A. K. Sharma Department of Botany, University of Calcutta, 35 Ballygunj Circular Road, Calcutta 700019, India

J. A. Steyermark Instituto Botanico, Apartado 2156, Caracas, Venezuela

A. L. Stoffers Institute for Systematic Botany, Heidelberglaan 2, Utrecht, Netherlands

J. E. Vidal Laboratoire de Phanerogamie, 16, rue Buffon, 75005 Paris, France

Preface

The study of tropical plant life has been one of the major concerns of the Botanical Institute, University of Aarhus, since its foundation 15 years ago. The Botanical Institute is deeply involved in projects in Thailand and in Ecuador. This symposium was organized as an attempt to stimulate the exchange of ideas and experiences between botanists working in the tropics, and thus to encourage new efforts in this fascinating field.

The tropics hold the most diverse plant resources of any region in the world. Studies of tropical plant life is of the utmost importance and urgently needed since large areas of natural vegetation, particularly forests, are being destroyed. The accelerated exploitation of the tropical areas, called "the lowland invasion" (A. L. Luge in Ceres 65, FAO I, 1978) causes known and unknown problems and hampers optimal use due to lack of knowledge about the natural vegetation. In general the tropical floras are very poorly known and vast tropical areas are botanically unknown. Biological and ecological studies are even more scarce.

Throughout the meeting the speakers expressed their alarm about the rate of extinction of whole vegetations before their plants have been studied or recorded. Great emphasis was given to the need of giving higher priority to field work in the tropics, including simple biological and ecological observations, such as the dispersal of diaspores, the sexual behaviour and community studies, to mention just a few of many important subjects. Concludingly a resolution was adopted by the symposium plenum.

The symposium was open for contributions within the loose circumscription given in the invitation: history of tropical floras, present distribution of vegetation types, present distribution of taxa, relation between distribution of taxa and vegetation, and theories of distribution types. The meetings were attended by 96 scientists from 17 countries and followed by an excursion to western Jutland.

All papers printed in this volume were presented during the meetings, except the papers of Drs A. Cleef and B. Maguire, who were prevented from being present. The papers presented during the meetings by Drs E. Köhler, C. C. Berg and J. C. Lindeman are under publication elsewhere. In the present volume the sequence of papers has been rearranged, and they do not follow the geographically determined arrangement as they did during the meetings.

We are most grateful to all our colleagues who contributed to the symposium, and especially we wish to thank the chairmen of the sessions, viz. Prof. J. P. M. Brenan, Prof. C. Kalkman, Dr B. Sparre, and Dr B. Øllgaard. Also Dr J. Luteyn who read the summary of Maguire's paper. We are specially indebted to Dr G. T. Prance for summing up the symposium and for his and Mrs Anne Prance's assistance to the editors as linguistic consultants.

Last but not least we wish to express our thanks to the Danish Ministry of Education, the Board of Directors of the Danish Natural Science Research Council, and the University of Aarhus Research Foundation who all gave substantial aid to the symposium. Without these grants it would not have been possible to carry out this international meeting.

Kai Larsen and Lauritz B. Holm-Nielsen
September 1979

Contents

IV. Taxonomic and Biological Examples

Neotropics, Andes

V. Conclusion

I. History of Tropical Floras

Plate Tectonics and Southern Hemisphere Biogeography

P. H. RAVEN

Missouri Botanical Garden, Saint Louis, USA

Introduction

Evidence that has accumulated during the past decade for a wide separation of Australia from Asia and of North America from South America during Late Cretaceous and Paleogene time, coupled with the traditional acceptance of a broad Tethys Sea separating Africa from Europe during the same period, poses a fundamental dilemma concerning the route and method of dispersal of terrestrial organisms between the northern and southern hemispheres during this time. The problem is particularly acute for the angiosperms, known with certainty for the first time from the Barremian (127 m.y. BP; Raven and Axelrod, 1975; Doyle, 1977; Hickey and Doyle, 1977) and diverse by the Albian (110–113 m.y. BP), with the origin of many modern orders, a number of families, and some genera during the remaining 35 m.y. of the Cretaceous. The spread of successive waves of evolutionary novelties amongst the angiosperms between the northern and southern hemispheres appears to have been rapid throughout the Cretaceous (cf. Doyle, 1977, p. 528), so that migration between at least one of the three pairs of continents mentioned above must have been a relatively simple matter.

The remainder of this paper will be devoted to a consideration of where such dispersal occurred, and of the paths of dispersal within the southern hemisphere that were available at various times during this hundred-million-year period during which the stage was set for the remaining 27 million years of angiosperm history. The three major present-day pathways of dispersal between the northern and southern hemispheres will be reviewed in the context of Cretaceous and

Paleogene geology, in order to estimate the likelihood of migration between the two hemispheres by each of these routes at these times, so critical in the history of contemporary groups of plants and animals. Papers reviewed earlier by Raven and Axelrod (1972, 1975) will for the most part not be repeated here.

Asia–Australia

New Zealand and New Caledonia, both lands of great antiquity (cf. Paris and Lille, 1977; Paris and Bradshaw, 1977), separated from Australia–Antarctica about 80 m.y. BP and moved north-eastward into the Pacific, reaching their present position about 60 m.y. BP (Austin, 1975, 1977; Griffiths, 1975, 1977). Normal oceanic crust began to form between Australia and Antarctica about 55 m.y. BP, with the separation of their continental margins at about 49 m.y. BP (Veevers and McElhinny, 1976). More or less direct migration through the Tasmanian area via the South Tasman Rise may have been possible until about 38 m.y. BP, at the close of the Eocene (Kennett, 1978). Paleontological and geological evidence clearly contradicts the notion that any part of the South-East Asia or Indonesia was a part of Gondwanaland and moved north. Not until the Middle Miocene (about 15 m.y. BP) was the possibility for more or less direct migration of plants and animals between Asia and Australasia established.

As Australia moved northward, much of present-day New Guinea was elevated above the sea for the first time (Jenkins, 1974). Southern New Guinea is the leading edge of the Australian Plate, whereas northern New Guinea seems to consist of an island arc that was beached on the Australian Plate in the mid-Miocene and then uplifted (Hamilton, 1973). Relict genera that occurred widely in Australia when it was at southern latitudes and characterized by a cool-temperate climate (cf. Kemp, 1978) are now confined to the mountains of eastern Australia and Tasmania, and have migrated during the Late Pliocene and Pleistocene, probably mostly overland, to New Guinea (e.g. *Dawsonia*: van Zanten, 1973; marsupials: Ziegler, 1977). In New Guinea, they have become established along with immigrants from the northern hemisphere in the cooler mountains. They have also survived in the oceanic islands of New Zealand, and to an even greater extent in New Caledonia, which represents a virtual museum of the flora and fauna that was widespread in Australia in the Late Cretaceous. Relationships between the floras and faunas of New Guinea and New Caledonia are therefore held to be indirect; *Nothofagus* and Winteraceae in New Caledonia have been there for most of the history of the groups concerned—perhaps 75 m.y. or more—but in New Guinea apparently for less than a sixth of this time; probably no more than 12 m.y. The New Hebrides (Luyendyk *et al.*, 1974; Ravenne *et al.*, 1977) and other islands between New Guinea and New Caledonia are no more than 10–15 m.y. old and therefore were not available as a pathway for ancient migration. The Solomon Islands–New Ireland region was a volcanic arc from the Oligocene onward but reached its present extent and conformation only after the collision of the Australian and Asian plates 10–12 m.y. BP (de Broin *et al.*, 1977).

Among the elements represented in the biota of New Zealand are many groups that do not disperse well across water barriers, such as *Nothofagus*, austral gymnosperms, and ratite birds. Land between New Zealand and Australia–Antarctica was almost certainly continuous at the time of the Early Cretaceous Rangitata orogeny (Stevens, 1977), and perhaps subsequently also, until 80 m.y. BP.

Summarizing for relationships between Australasia and Asia, there was only indirect migration between them from the start of the Cretaceous (135 m.y. BP) to the Miocene ($\approx$10–15 m.y. BP). Owen's (1976) reconstructions are based on the now discredited expanding earth hypothesis (McElhinny *et al.*, 1978), and therefore the arguments of Stevens (1977; cf. especially Fig. 5, p. 325) are inappropriate, as well as contradicted by the biological evidence. Although I readily agree with Stevens (1977) that there may have been migration into Australasia via a warm-temperate pathway, this would seem to have been across the present area of the Indian Ocean (Raven and Axelrod, 1975; see also below), very ancient, and certainly not directly from the north.

Antarctica itself has now been shown to have been in a polar position by the start of the Tertiary (McElhinny, 1973; Lowrie and Hayes, 1975; Holden, 1976; Kemp, 1978; Kennett, 1978). This implies that the paleoposition of Australia was 15° farther south than we postulated earlier (Raven and Axelrod, 1972); at the time of separation of Australia from Antarctica, Australia would have been approximately 30° south of its present position. Northern Australia, consisting of present northern Queensland and the now submerged Queensland Plateau, was mainly south of latitude 38°S, approximately the latitude of Bass Straits which separate Tasmania from the mainland of Australia. Since the latitude of South-East Asia has not shifted significantly in latitude during the Cenozoic (McElhinny *et al.*, 1974; Haile *et al.*, 1977), a gap of at least 3000 km would have separated the lands on the Australian Plate from those on the Asian Plate. In addition,

South-East Asia would have had a full tropical climate (the mountains in the area are Pliocene or more recent) and Australasia a cool-temperate one, thus further lessening the chances of any direct dispersal between these areas. Even though there may have been scattered islands between, opportunities for direct dispersal across this gap would have been very limited indeed. The large islands east of Java have, like New Guinea, developed to their present size as a consequence of the collision of the eastern Indonesian area with the Australasian continental block (Milsom, 1977).

As Australia moved northward, it entered a zone of lowered precipitation, and various scleromorphic plant communities evolved, perhaps starting as early as the Oligocene (38–27 m.y. BP), but greatly accentuated subsequently and of wide extent by the close of the Miocene (10 m.y. BP; Kemp, 1978). In other words, temperate rain forest vegetation appears to have dominated the continent of Australia throughout the entire history of the angiosperms until approximately 10 m.y. ago (Kemp, 1978). Scleromorphic plant communities may have had their origin from the Late Cretaceous onward on ancient soils of reduced fertility where they may have existed locally even when the continent was in the far south (Johnson and Briggs, 1978). The distinctive families and genera that form such communities may, however, have existed much earlier on infertile soils, as suggested by Johnson and Briggs (1978). Analogous situations involving the survival of unusual or relict elements on infertile soils have been postulated by Axelrod (1972) for sites of edaphic aridity in tropical to temperate climates worldwide, and by Raven and Axelrod (1978) for serpentine soils in California.

North America–South America

About 180 m.y. BP, the Caribbean Sea and Gulf of Mexico began to form as North America started to move away from Africa and South America, then still linked (Barr, 1974; Raven and Axelrod, 1975; Donnelly, 1975; Van der Voo *et al.*, 1976; Holden, 1976; Bowin, 1976). The maps of Smith and Briden (1977) likewise greatly exaggerate the ease of migration between North and South America by showing present-day shorelines for ease of reference; those of Jurdy and Van der Voo (1975) and Irving (1977) show the relative positions very well. From then until about 127 m.y. BP North America moved north-west relative to South America, and then westward until about 84 m.y. BP (Ladd, 1976). During the interval 84 m.y. BP to the close of the Eocene,

38 m.y. BP, North America moved north-west again, relative to South America (Ladd, 1976). Although South America has been moving away from Africa since the mid-Cretaceous (about 125 m.y. BP; Barrett, 1977), with final separation about 90 m.y. BP (see also Kumar and Embley, 1977; Forster, 1978, Petters, 1978), it has been closer to Africa, and more accessible to immigration from Africa than from North America, until the Eocene (38–54 m.y. BP).

These relationships explain the existence of a mid-Cretaceous African–South American floral province which extended from South America through Central and North Africa to the Middle East (Herngreen, 1974; Sultan, 1978). They also indicate that a common fauna of mammals (Tedford, 1975) and other groups (cf. Cracraft, 1975) must have occurred in both continents when they were joined. They also underlie the many ancient relationships that link the floras of the Old World with South America, although such relationships are often obscured by the pattern of widespread extinction that has decimated the flora of Africa (Raven and Axelrod, 1975; Axelrod and Raven, 1978). Recently described examples include Dipterocarpaceae, discovered in the ancient Guyana Highlands of northern South America (Maguire *et al.*, 1977); Myristicaceae (Armstrong and Wilson, 1978); Cecropiaceae (Berg, 1978); and meliponine bees (Michener, 1979).

Extensive lands in North, Central, and South America that are now elevated were subsea until the Eocene. Although there were scattered volcanic islands between Oaxaca and the Guyana Highlands during the Late Cretaceous, some 3000 km of essentially open water seems to have separated North and South America at this time, a relationship not taken into account by such authors as Rosen (1975) and Schuster (1976) in their discussions of biogeography. Cuba appears to be the only one of the Antilles that contains rocks old enough (Jurassic) to indicate the possibility of direct overland connections with continental areas. Otherwise, the Antilles are no older than Late Cretaceous (≈ 75 m.y. BP) and have grown subsequently to their present extent. Consequently, the opportunities for dispersal to and between them, and the diversification of habitats within these islands, have been greater during the Quaternary than at any time in the past; Caribbean biogeography must be judged in this light. The Bahamas, on the other hand, appear to be largely foundered continental fragments derived from Africa in a manner analogous to the postulated separation of the Seychelles from North-West India (Mullins and Lynts, 1977).

Increasing opportunities for migration between North and South America during the past 38 m.y. culminated in the appearance of a terrestrial connection, the Isthmus of Panama, some 5·7 m.y. BP.

Increased elevation of lands through middle America during the Late Pliocene and more recently has also provided increasingly great opportunities for the dispersal of the plants and animals of middle and high elevations between North and South America, and for the first time in this region created rain shadows in which the biota of semiarid and desert habitats have dispersed.

In relation to these geological facts, the pattern of faunistic and floristic interchange between North and South America may be interpreted as follows. North and South America were moving farther apart throughout the Paleocene and Eocene (54–36 m.y. BP; Ladd, 1976), but teiid and iguanid lizards, arctostylopid mammals, and the genus *Bufo*, in addition to a number of plant groups derived from South America (Raven and Axelrod, 1975, p. 557–558, 621–630) seem to have dispersed between these continents prior to the close of the Eocene. To these may be added Zingiberaceae (Hickey and Peterson, 1978), reported from the Late Cretaceous to Early Eocene of western interior North America and *Tapirira* (Anacardiaceae), a mainly South American genus reported earlier from the Miocene of Chiapas (Langenheim, 1964) but now described from the Paleogene of Oregon (Manchester, 1977). Studies of pollen floras from Veracruz, Mexico have provided indications that the modern tropical rain forest in northern Central America and Mexico may be a community of postglacial derivation (Graham, 1976, 1977a,b). Indeed, the entire tropical rain forest as we know it today might be interpreted as a postglacial phenomenon derived in relation to the worldwide loss in equability following the mid-Pliocene and the development of torrid climates in equatorial regions. Paleogene and earlier tropical forests existed in an equable climate that was disrupted with the onset of glaciation on a worldwide scale in the Late Miocene, and these forests do not seem to have developed directly into those existing now.

At any event, South America was more accessible to interchange with Africa than with North America until the Eocene, and appears intially to have been populated with a flora and fauna that were shared in common with Africa. From the Eocene onward, interchange with North America has become the predominant theme, culminating in intensity in Late Pliocene and Quaternary time with glaciation and mountain building on a global scale.

Eurasia–Africa

Africa and South America, joined together, rotated away from North America–Europe starting about 180 m.y. BP. During the interval

148–80 m.y. BP, Africa, with at least Italy and Sardinia and possibly also Spain and other areas attached (Vandenberg, Klootwijk, and Wonders, 1978), rotated counter-clockwise relative to Europe and Africa converged some 2800 km on Eurasia (Dewey *et al.*, 1973). The subsequent motion of Africa was north-westward relative to Europe, with continuing counter-clockwise rotation and compression between these continents, resulting in the origins of the Alpine system, until the Paleocene (63 m.y. BP) when Africa and Europe seem to have been connected through Spain, and were also in close proximity in the region of Italy (Hsü, 1977). In other words, although the Tethys Sea seems to have separated Europe from Africa during much of the Cretaceous and Early Tertiary (Hallam, 1973), there were apparently intermittent land connections at various times, interrupted by shallow shelf seas at others. From the Early Paleocene into the Later Eocene (53 m.y. BP), Africa and Europe moved farther apart, with Italy now independent of Africa (Vandenberg *et al.*, 1978). Subsequently, Africa moved about 10° northward to its present position, this convergence resulting in direct connection in the Miocene, 17 m.y. BP, probably in the region of the Balkans. This contact was accompanied by an intensive exchange of both plants and animals between Africa and Europe (Hsü *et al.*, 1977).

The collision of the Arabian Plate, following its rifting from Africa in the Eocene (Garson and Krs, 1976), with Asia, although poorly understood, has resulted in the complex and rugged topography of Turkey and Central Asia (Neev, 1975; Hall, 1976; Hallam, 1976; Trifonov, 1978). There may have been a broad connection between Africa and Asia via the Arabian Peninsula prior to the southward movement of Africa that commenced in the Paleocene. Iran and part of Afghanistan may have occupied a position between Africa–Madagascar and Arabia, definitely forming a portion of Gondwanaland. This block was probably at about 18·5°S 125 m.y. BP and seems to have moved northward with India (Gealey, 1977).

Pathways for Migration within the Southern Hemisphere

The earliest records of angiosperms from Australia are from the Albian (≈115 m.y. BP; Dettmann, 1973; Gould, 1975), some 12 m.y. after their appearance in Eurasia and West Gondwanaland (Africa + South America). Whether this apparent gap is real or not, the pathways by which angiosperms reached Australia and its bordering lands in the Cretaceous are of great biogeographical interest. There are two such pathways, not wholly distinct: (i) the well known cool-temperate one involving migration between southern South America and Australia via Antarctica; and (ii) a more problematical warm-temperate or even subtropical one (at least during the period of the angiosperms) across the area of the present Indian Ocean when the lands now surrounding it were closer together (Raven and Axelrod, 1975). Each will now be reviewed in turn.

South America to Australia

More or less direct overland migration between Australia and East Antarctica, via the South Tasman Rise, seems to have been possible until approximately the close of the Eocene, 38 m.y. BP. At this time, forests of *Northofagus* and austral gymnosperms still occurred in Antarctica, to at least 70°S latitude (Thomson and Burn, 1977). The climates of both areas, and of New Zealand, were warmer and more equable in Late Creataceous and Paleogene time than at present (e.g., Petriella and Archangelsky, 1975; McQueen, 1977)—perhaps comparable to mid-elevations in New Guinea and New Caledonia. They probably persisted until the initiation of glaciation on a continental scale in East Antarctica at about 14 m.y. BP (Kennett, 1978). The possibility of direct overland migration between South America and Australia depends on the nature of West Antarctica prior to the initiation of continental glaciation, and on the possibility of direct connections between the Antarctic Peninsula and Tierra del Fuego along the Scotia Arc (review, Raven and Axelrod, 1975; Rich, 1975a,b). Judged from geophysical evidence and the general geological similarity of southern South America, the Antarctic Peninsula, South Georgia, and the other islands and ridges of the Scotia arc (except the young volcanic South Sandwich Islands), all of these areas are probably fragments of a more or less continuous Mesozoic orogenic belt (Adie, 1977; Winn, 1978). Continental connections between South America and West Antarctica existed during most of the time from at least 125 m.y. BP until the close of the Eocene ($\approx$40 m.y. BP; Herron and Tucholke, 1976; Barker and Griffiths, 1977; Winn, 1978, Kennett, 1978; Katz, pers. comm., is not opposed to this interpretation). West Antarctica, geologically related to the Andes, may have separated from East Antarctica along a tear fault (Craddock, 1973; Adie, 1977; Barker and Griffiths, 1977). The extent and nature of land elevated above the sea at various times is problematical (Craddock, 1973; Grikurov and Lopatin, 1973; Rich, 1975a,b), but evidence is accumulating that continental connections may have existed between East and West Antarctica prior to the Oligocene (Barker and Griffiths, 1977, p. 154).

The balanced nature of the flora and fauna that link southern South America with Australia–Tasmania (for a recent example, Romero and Hickey, 1976), and to a lesser extent, New Zealand, suggest strongly that direct overland migration between the lands has been possible for at least some of the time from the appearance of *Northofagus* in the fossil record, more than 70 m.y. BP (review, Raven and Axelrod, 1972, 1975). These biotas include many parasites and other forms associated with the plants that scarcely could have been spread by long-distance dispersal; some rusts even link two unrelated higher plant hosts in them (Leppik, 1973, pers. comm.). *Northofagus* first appears simultaneously in Australia and New Zealand, and apparently a few million years later in South America (Archangelsky and Romero, 1974). Contrary to the arguments of Darlington (1965, p. 147), *Nothofagus* and the austral gymnosperms disperse very poorly across water barriers (Preest, 1963; Raven and Axelrod, 1972, 1975). In South America, birds regularly eat the nuts of *Nothofagus*; in Tierra del Fuego they include the austral parakeet (*Micropsittace ferruginea* P. L. S. Miller), as well as finches [*Phrygilus patagonicus* (Lowe)] and austral blackbirds (Goodall, pers. comm.; Pisano V., pers. comm.). These birds eat the nuts promptly, however, and are not likely to play a role in dispersing them. Pisano has likewise reported that although there are no direct observations, rodents apparently feed on the nuts of *Nothofagus* in southern Chile, judged from the patterns of reproduction of the seeds. Despite extensive correspondence, I have not been able to find any observations of birds feeding on the nuts of *Nothofagus* in the Old World. The deciduous habit in *Nothofagus* apparently evolved independently in Tasmania and in South America (Hanks and Fairbrothers, 1976), or might have originated in Antarctica in relation to seasonal darkness.

In summary, although the geological evidence is not conclusive for land connections in West Antarctica (cf. Cox, 1978), more or less direct overland migration between South America–Antarctica–Australia and New Zealand seems to have been possible until 80 m.y. BP, with the flora and fauna of New Zealand limited both by its relatively early time of separation and by its consistently high latitude. More or less direct overland migration between Australia and South America via Antarctica seems to have been possible from before the time of origin of the angiosperms until approximately the close of the Eocene ($\approx$40 m.y. BP), but would have involved migration at approximately the latitude of the Antarctic Circle during this entire period of time. Such a relationship explains the balanced and closely related floras and faunas found in each of these regions, and the greater distinctiveness of the biota of New Zealand as compared with the other areas. This appears to have

                    P. H. Raven

been the most likely path for marsupials between South America and Australia (Keast, 1977; see also Raven and Axelrod, 1975), judged in part from their absence in New Caledonia and New Zealand (Martin, 1977). If they had been in South America–Africa early enough to have dispersed across the present area of the Indian Ocean (prior to 110 m.y. BP), they would have been expected on these islands, which separated from Australia–Antarctica 80 m.y. BP. An age of about 70 m.y. for the disjunctions in both marsupials and hyline tree frogs of South America and Australia has been obtained using albumin differences as an evolutionary clock (Maxson *et al.*, 1975). Bees exhibit a similar pattern (Michener, 1979).

Africa to Australia

"Greater India", which may well have included lands now absorbed into the Himalayan system, appears to have separated from south-west Australia and adjacent Antarctica at about 125 m.y. BP (Veevers *et al.*, 1975; Veevers and McElhinny, 1976; Larson, 1977; Norton and Molnar, 1977; Veevers and Cotterill, 1978) and to have moved north-ward at a rate of 14·9 ± 4·5 cm/yr from 70–40 m.y. BP, trailing the developing Ninetyeast Ridge, which was forming over the Kerguelen Hotspot (Sclater *et al.*, 1976; Peirce, 1978). Prior to at least 45 m.y. BP India began to collide with Eurasia, which was then 22° (2500 km) south of its present position, this degree of crustal shortening evidently having taken place during the formation of the Himalayas (Powell and Conaghan, 1975; Molnar and Chen, 1978). India is still moving north-ward at a rate of 5·2 ± 0·8 cm/yr, as it has been for the past 40 m.y. (Veevers and Cotterill, 1978). This movement has resulted in a com-plex mixture of geological and biological elements in the Indian sub-continent, and in some of the most striking land-forms found anywhere on earth (Molnar and Tapponier, 1975). It is not clear whether any genera of plants or animals have been rafted north with India and survived to the present (cf. Schuster, 1972, 1976; Raven and Axelrod, 1975; Ferris *et al.*, 1976; Johnson and Briggs, 1978). Van Valen and Sloan (1977) have hypothesized that dinosaurs may have survived in India until its collision with Asia in the Eocene. At the start of its northward movement, 70 m.y. BP, all of India lay south of 40°S lat.; when it collided with Asia initially, at about the present latitude of Ceylon, its southern end would have been at approximately 20°S lat. At this time, 45 m.y. BP, India would perhaps have been separated from the Seychelles Bank and Madagascar by water gaps not much greater than those that separate Madagascar and Africa at present, and probably about the same distance from the Malay Peninsula–Sunda Islands. Even if its original cool-temperate flora and fauna had been decimated by that time, active evolution and interchange with both east and west would have kept it populated with plants and animals, largely unknown. By the time of its Eocene collision with Asia, India would have been separated from Australia by a water gap of at least 6000 km, greater than the present distance between Madagascar and Australia. There are no families of angiosperms but some 164 genera endemic to the Indian Floristic Region (Rao, 1972; Raven and Axelrod, 1975). When India was in the far south, 125–70 m.y. BP, migration between Australia and Africa was certainly facilitated. The vegetation of Ninetyeast Ridge was similar to that of Australia in Paleogene times (Kemp and Harris, 1975), whereas that of India, initially cool-temperate, rapidly changed as India moved northward from about 60°S lat. to its collision with an extended Asia at about 8°N—a distance of some 4500 km.

In summary, opportunities for migration between Africa and Australia were direct prior to 125 m.y. BP, interrupted but much more direct than at present until India began its rapid northward movement at about 70 m.y. BP, and then increasingly indirect. Judging from the relationships between relatively modern groups that occur both in Africa/Madagascar and in Australia, however, some long-distance dispersal between these two areas has continued to the present day (cf. Raven and Axelrod, 1975). Links such as those reported in *Acacia* (Bell and Evans, 1978) are certainly not ancient but doubtless represent recent dispersal like the other genera we discussed previously. It is after all only about two-thirds of the distance from Madagascar to Australia ($\approx$ 5400 km) that it is from Asia to Hawaii ($\approx$ 8000 km), and yet *Acacia* has dispersed from Asia to Hawaii, probably within no more than a few million years at the very most. During the time of initial appearance and diversification of ancient families such as Proteaceae and perhaps Restionaceae and Myrtaceae, migration between Africa and Australia was probably no more difficult than that between Southeast Asia and Australia at the present day. The Rutales-Sapindineae (Thorne, 1977; Carlquist, 1978), now shown probably to include the anemophilous Australian families Gyrostemonaceae, Stylobasiaceae, and probably Emblingiaceae, as well as the widespread and Australasian Bataceae, also seem to indicate ancient dispersal between Africa and Australia.

Madagascar separated from the coast of Somalia–Kenya prior to 90 m.y. BP (Axelrod and Raven, 1978; McElhinny *et al.*, 1976) and moved southward to its present position. It was probably linked with Africa by a basaltic plateau during the middle Cretaceous (Axelrod and Raven, 1978, p. 87), and this facilitated the passage of some groups of organisms between the two land areas more than others. The Comores might be fragments of such a former connection, considering the paleoposition of Madagascar. The Seychelles are continental fragments that include Precambrian granites and which appear to have broken away from north-west India about 70 m.y. BP (Scrutton, 1976), just as India was initiating rapid northward movement. This appears to account for the presence of such groups as caecilians and of Dipterocarpaceae, subfamily Dipterocarpoideae in the Seychelles and India-Asia and their absence in Madagascar. Mauritius, on the other hand, is a volcanic island no more than 7·8 m.y. old (Baxter, 1975, 1976), and the other Mascarene islands are of comparably recent origin. What is known now about the geology of the Seychelles and Mascarenes provides a ready explanation for Darlington's (1957, p. 524) conclusion that these two groups of islands had very different histories.

Prior to 70 m.y. BP, Africa–Madagascar–Seychelles–India–Antarctica–Australia were much closer together than they are at present, and in some cases actually linked. Early in the differentiation of most groups of angiosperms, there were much better opportunities for migration by this route between Africa and Australia than exist now. For overland migration by a warm-temperate or subtropical route, it is probably necessary to go back to approximately 125 m.y. BP, very early in the history of the angiosperms and almost certainly before the appearance of any existing group at the family level. The probabilities of dispersal between Africa and Australia therefore, seem to have decreased gradually from 125 to 70 m.y. BP, and then more rapidly as India pulled out of the southern complex of lands and began its rapid movement northward. Some elements in the flora of Madagascar may reflect such ancient connections, but the overwhelming majority seem to have been derived from Africa more recently by dispersal over water (Leroy, 1978), as postulated by Darlington (1957), Walker (1972), and others; the distance between Africa and Madagascar at present is only about 420 km, sufficient to allow survival of unusual relict forms, such as lemurs and several endemic families of vascular plants in a climate more equable than that of continental Africa and in relative isolation, but close enough to allow continuous colonization from Africa throughout the entire history of the angiosperms by sweepstakes dispersal (Simpson, 1940, 1943).

Summary

During the Cretaceous and into the Eocene (135–38 m.y. BP), a gap of some 3000 km of largely open water separated Australia from Asia. In the same period, the gap between South America and North America decreased from about 3000 km to approximately half of that distance. Meanwhile, Africa, during the interval 148–80 m.y. BP, converged some 2800 km toward Europe, with continuing compression until the Paleocene (63 m.y. BP), resulting in the origins of the Alpine system. From that time until approximately 53 m.y. BP, Africa moved southward. Therefore, by far the most feasible route for terrestrial organisms that involved overland dispersal or narrow water gaps during the Late Cretaceous (110–65 m.y. BP) was the route between Europe and Africa (Cracraft, 1976). Dispersal poleward was facilitated by the prevailing temperature gradients, which were much less steep than those that prevail at present.

Africa–Europe then was presumably the route taken by such groups as marsupials (Raven and Axelrod, 1972, 1975); dinosaurs (e.g., Galton, 1977; he also considers a Central American link); many gymnosperms; Hamamelidales–Fagales (Endress, 1977), and especially Fagaceae and *Nothofagus* (contra Hanks and Fairbrothers, 1976); Quillajeae sens. str., although they are extinct in Africa if this assumption is correct (cf. Basinger, 1976; Goldblatt, 1976); and all primitive groups of flowering plants such as the suborders Magnoliineae and Laurineae, which link Asia and Australia (Thorne, 1974; Endress, 1977). If they are a natural group, the Pittosporales, which link South-East Asia, Africa, and Australasia (Thorne, 1975), probably originated in and radiated from Africa. The occurrence of marsupials and that of some groups of possibly related dinosaurs (Donnelly, 1975; but see Raven and Axelrod, 1975) in the Late Cretaceous of North and South America seems similarly to indicate dispersal via Europe and Africa, even though fossils documenting this are unknown. The well-marked Upper Cretaceous pollen genus *Aquilapollenites*, which closely resembles some living Santalaceae (Jarzen, 1977), occurred across the northern hemisphere but in the southern hemisphere only in West Africa, this again suggesting the route between northern and southern hemispheres at this time.

At the start of the Eocene (54 m.y. BP), Africa, South America, and Australia were all well separated from the main northern hemisphere land mass, with dispersal between the northern and southern

                    P. H. Raven

hemispheres presumably very difficult for terrestrial organisms. Subsequently, each of the three southern continents moved northward, and direct overland or archipelagic connections were established in each area by the Late Miocene (10–15 m.y. BP), with dispersal increasingly feasible during the preceding 40 m.y. period.

Direct overland connections between North America and Europe (49 m.y. BP) and between Australia and South America ($\approx$40 m.y. BP) persisted into the Eocene. Overland connections between New Zealand–New Caledonia and Australia–Antarctica were severed in the Late Cretaceous (80 m.y. BP). A warm-temperate pathway from Africa to Australia via Madagascar, India, and Antarctica was increasingly indirect from 125 m.y. BP until the Late Eocene (40 m.y. BP) collision of India with Eurasia at about 2500 km south of its present position.

Angiosperms seem to have originated as relatively xeromorphic shrubs (Stebbins, 1965; Doyle, 1977; Doyle *et al.*, 1977) in the interior of the vast continent of West Gondwanaland, formed by the union of Africa and South America in or before the Early Cretaceous (127 m.y. BP). Bees, which have played such a key role in the diversification of angiosperms, also may have originated here (Michener, 1979). From there, they could spread easily into Eurasia–North America, where they appeared almost immediately, and perhaps with more difficulty, or across water barriers, to Australia–Antarctica, where they appeared no later than 115 m.y. BP. When they encountered mesic, warm, equable conditions in Southern Eurasia and elsewhere, and as these conditions spread with the increasing separation of Africa and South America, they apparently gave rise to mesophytic trees and shrubs comparable with those so well represented among unspecialized groups of flowering plants at present (Doyle, 1977). These early derivatives have survived best in cool tropical mountains, as in South-East Asia, New Guinea, and Queensland, and in islands, such as Madagascar and New Caledonia, that have been isolated from their adjacent continents and which have had an equable climate since the middle Late Cretaceous, when modern groups of angiosperms were just beginning to differentiate. Viewed in the context of the world as it was when flowering plants were first spreading, South-East Asia is seen as a distant outpost and a most unlikely center of origin—a conclusion that is still being offered (e.g., Thorne, 1976) on the basis of distributions of plants and continents as we see them more than 125 m.y. after the origin of the group.

Acknowledgments

Grant support provided by the U.S. National Science Foundation has helped to make possible these investigations. I am especially grateful to

I. W. D. Dalziel, M. Galore, N. P. Goodall, A. Graham, A. Hallam, W. Hamilton, K. S. Hsü, H. R. Katz, E. M. Kemp, E. Pisano V., P. V. Rich, and the late E. E. Leppik for useful discussions that have aided materially the formation of these ideas. D. I. Axelrod has kindly reviewed the entire manuscript.

References

Adie, R. J. (1977). The geology of Antarctica: A review. *Phil. Trans. R. Soc. Lond.* Ser. B **279**, 123–130.

Archangelsky, S. and E. Romero (1974). Los registros más antiguos del polen de *Nothofagus* (Fagaceae) de Patagonia (Argentina y Chile). *Bol. Soc. Bot. Méx.* **33**, 13–30.

Armstrong, J. E. and T. K. Wilson (1978). Floral morphology of *Horsfieldia* (Myristicaceae). *Am. J. Bot.* **65**, 441–449.

Austin, P. M. (1975). Paleogeographic and paleotectonic models for the New Zealand geosyncline in eastern Gondwanaland. *Geol. Soc. Am. Bull.* **86**, 1230–1234.

Austin, P. M. (1977). Reply to Griffith (1977). *Geol. Soc. Am. Bull.* **86**, 1206–1210.

Axelrod, D. I. (1972). Edaphic aridity as a factor in angiosperm evolution. *Am. Nat.* **106**, 311–320.

Axelrod, D. I. and P. H. Raven (1978). Late Cretaceous and Tertiary vegetation history of Africa. *In* "Biogeography and Ecology of Southern Africa" (M. J. A. Werger, ed.). Monogr. Biol. 31, 79–130. Junk, The Hague.

Barker, P. F. and D. H. Griffiths (1977). Toward a more certain reconstruction of Gondwanaland. *Phil. Trans, R. Soc. Lond.* Ser. B **279**, 143–159.

Barr, K. W. (1974). The Caribbean and plate tectonics—some aspects of the problem. *Verhandl. Naturf. Ges. Basel* **84**, 45–67.

Barrett, D. M. (1977). Agulhao Plateau off southern Africa. A geophysical study. *Geol. Soc. Am. Bull.* **88**, 749–763.

Basinger, J. F. (1976). *Paleorosa similkameenensis*, gen. et sp. nov., permineralized flowers (Rosaceae) from the Eocene of British Columbia. *Can. J. Bot.* **54**, 2293–2305.

Baxter, A. N. (1975). Petrology of the Older Series lavas from Mauritius, Indian Ocean. *Geol. Soc. Am. Bull.* **86**, 1449–1458.

Baxter, A, N. (1976). Geochemistry and petrogenesis of primitive alkali basalt from Mauritius, Indian Ocean. *Geol. Soc. Am. Bull.* **97**, 1028–1034.

Bell, E. A. and C. S. Evans (1978). Biochemical evidence of a former link between Australia and the Mascarene Islands. *Nature, Lond.* **273**, 295–296.

Berg, C. C. (1978). Cecropiaceae, a new family of the Urticales. *Taxon* **27**, 39–44.

Bowin, C. (1976). Caribbean gravity field and plate tectonics. *Geol. Soc. Am. Spec. Paper* **169**, 1–79, maps.

Broin, C. E. de, F. Aubertin, and C. Ravenne (1977). Structure and history of the Solomon–New Ireland region. Int. Symp. Geodynamics S.-W. Pacific, p. 37–50. Éditions Technip, Paris.

Carlquist, S. (1978). Wood anatomy and relationships of Bataceae, Gyrostemonaceae, and Stylobasiaceae. *Allertonia* **1**, 297–330.

Cox, K. G. (1978). Flood basalts, subduction and the break-up of Gondwanaland. *Nature, Lond.* **274**, 47–49.

Cracraft, J. (1975). Mesozoic dispersal of terrestrial faunas around the southern end of the world. *Mém Mus. Nat. Hist. Nat.* Sér. A **88**, 29–54.

Cracraft, J. (1976). Avian evolution on southern continents: Influences of palaeogeography and palaeoclimatology. Proc. 16th Int. Ornith. Congr., p. 40–52, Canberra.

Craddock, C. (1973). Tectonic evolution of the Pacific margin of Gondwanaland. *In* "Gondwana Geology" (K. S. W. Campbell, ed.) p. 609–618. Australian National University Press, Canberra.

Darlington, P. J. (1957). "Zoogeography: The Geographical Distribution of Animals." Wiley, New York.

Darlington, P. J., Jr. (1965). "Biogeography of the Southern End of the World." Harvard University Press, Cambridge.

Dettmann, M. E. (1973). Angiospermous pollen from Albian to Turonian sediments of eastern Australia. *Geol. Soc. Austral. Spec. Publ.* **4**, 3–34.

Dewey, J. F., W. C. Pitman III, W. B. F. Ryan and J. Bonin (1973). Plate tectonics and the evolution of the Alpine system. *Geol. Soc. Am. Bull.* **84**, 3137–3180.

Donnelly, T. W. (1975). The geological evolution of the Caribbean and Gulf of Mexico—some critical problems and areas. *In* "The Ocean Basins and Margins Vol. 3. The Gulf of Mexico and the Caribbean" (A. E. M. Nairn and F. G. Stehli, eds) p. 663–689. Plenum, New York.

Doyle, J. A. (1977). Patterns of evolution in early angiosperms. *In* "Patterns of Evolution as Illustrated by the Fossil Record" (A. Hallam, ed) p. 501–546. Elsevier, Amsterdam.

Doyle, J. A., P. Biens, A. Doerenkamp and S. Jardiné (1977). Angiosperm pollen from the pre-Albian Lower Cretaceous of Equatorial Africa. *Bull. Cent. Rech. Explor.-Prod. Elf-Aquitaine* **1**, 451–473.

Endress, P. K. (1975). Evolutionary trends in the Hamamelidales-Fagales group. *Plant Syst. Evol.* Suppl. 1, 321–347.

Endress, P.K. (1977). Über Blütenbau und Verwandtschaft der Eupomatiaceae und Himantandraceae (Magnoliales). *Ber. Dt. Bot. Ges.* **90**, 83–103.

Ferris, V. R., C. G. Goseco and J. M. Ferris (1976). Biogeography of free-living soil nematodes from the perspective of plate tectonics. *Science* **193**, 508–510.

Forster, R. (1978). Evidence for an open seaway between northern and southern proto-Atlantic in Albian times. *Nature, Lond.* **272**, 158–159.

Galton, P. M. (1977). The ornithopid dinosaur *Dryosaurus* and a Laurasia-Gondwanaland connection in the Upper Jurassic. *Nature, Lond.* **268**, 230–232.

Garson, M. S. and M. Krs (1976). Geophysical and geological evidence of the relationship of Red Sea transverse tectonics to ancient fractures. *Geol. Soc. Am. Bull.* **87**, 169–181.

Gealey, W. K. (1977). Ophiolite obduction and geologic evolution of the Oman Mountains and adjacent areas. *Geol. Soc. Am. Bull.* **88**, 1183–1191.

Goldblatt, P. (1976). Cytotaxonomic studies in the tribe Quillajeae (Rosaceae). *Ann. Mo. Bot. Gard.* **63**, 200–206.

Gould, R. E. (1975). The succession of Australian pre-Tertiary megafossil floras. *Bot. Rev.* **41**, 453–483.

Graham. A. (1976). Late Cenozoic evolution of tropical lowland vegetation in Veracruz, Mexico. *Evolution* **29**, 723–735.

Graham, A. (1977a). The tropical rain forest near its northern limits in Veracruz, Mexico: Recent and ephemeral. *Bol. Soc. Bot. Méx.* **36**, 13–19.

Graham, A. (1977b). Studies in Neotropical paleobotany. II. The Miocene communities of Veracruz, Mexico. *Ann. Mo. Bot. Gard.* **63**, 787–842.

Griffiths, J. R. (1975). New Zealand and the southwest Pacific margin of Gondwanaland. *In* "Gondwana Geology" (K. S. W. Campbell, ed.) p. 619–637. Australian National University Press, Canberra.

Griffiths, J. R. (1977). Discussion of Austin (1977). *Geol. Soc. Am. Bull.* **86**, 1203–1206.

Grikurov, G. E. and B. G. Lopatin (1973). Structure and evolution of the West Antarctic part of the circum-Pacific mobile belt. *In* "Gondwana Geology" (K. S. W. Campbell, ed.) p. 638–650. Australian National University Press, Canberra.

Haile, N. S., M. W. McElhinny and I. McDougall (1977). Palaeomagnetic data and radiometric ages from the Cretaceous of West Kalimantan (Borneo), and their significance in interpreting regional structure. *J. Geol. Soc. Lond.* **133**, 133–144.

Hall, R. (1976). Ophiolite emplacement and the evolution of the Taurus suture zone, southeastern Turkey. *Geol. Soc. Am. Bull.* **87**, 1078–1088.

Hallam. A. (1973). Distribution patterns in contemporary terrestrial and marine animals. *Spec. Pap. Palaeontol.* **12**, 93–105.

Hallam, A. (1976). Geology and plate tectonics interpretation of the sediments of the Mesozoic radiolarite-ophiolite complex in the Neyriz region, southern Iran. *Geol. Soc. Am. Bull.* **87**, 47–52.

Hamilton, W. D. (1973). Tectonics of the Indonesian Region. *Geol. Soc. Malaysia Bull.* **6**, 3–10.

Hanks, S. L. and D. E. Fairbrothers (1976). Palynotaxonomic investigation of *Fagus* L. and *Nothofagus* Bl.: Light microscopy, scanning electron microscopy, and computer analyses. *In* "Botanical Systematics" (V. H. Heywood, ed.) Vol. 1, p. 1–141. Academic Press, London and New York.

Hergreen, G. F. W. (1974). Middle Cretaceous polynomorphs from northeastern Brazil. Results of a polynological study of some boreholes, and comparison with Africa and the Middle East. *Bull. Serv. Carte Géol. Als. Lorr.* **27**, 101–116.

Herron, E. M. and B. E. Tucholke (1976). Sea-floor magnetic patterns and basement structure in the southeastern Pacific. *In* "Initial Report Deep Sea Drilling Project" (E. D. Hallister, C. Craddock *et al.*, eds) Vol. 35. p. 263. Government Printing Office, Washington.

Hickey, L. J. and J. A. Doyle (1977). Early Cretaceous fossil evidence for angiosperm evolution. *Bot. Rev.* **43**, 3–104.

Hickey, L. J. and R. K. Peterson (1978). *Zingiberopsis,* a fossil genus of the ginger family from Late Cretaceous to Early Eocene sediments of western interior North America. *Can. J. Bot.* **56**, 1136–1152.

Holden, J. C. (1976). Permian-Triassic continental configurations and the origin of the Gulf of Mexico. Comment. *Geology* **4**, 324.

Hsü, K. J. (1977). Tectonic evolution of the Mediterranean basins. *In* "The Ocean Basins and Margins" (A. E. M. Nairn, W. H. Kanes and F. G. Stehli, eds) Vol. 4A, p. 29–75. Plenum, New York.

Hsü, K. J. *et al.* (1977). History of the Mediterranean salinity crisis. *Nature, Lond.* **267**, 399–403.

Irving, E. (1977). Drift of the major continental blocks since the Devonian. *Nature, Lond.* **270**, 304–309.

Jarzen, D. M. (1977). *Aquilapollenites* and some santalalean genera. A botanical comparison. *Grana* **16**, 29–39.

Jenkins, D. A. L. (1974). Detachment tectonics in western Papua New Guinea. *Geol. Soc. Am. Bull.* **85**, 533–548.

Johnson, L. A. S. and B. G. Briggs (1978). Three old southern families—Restionaceae, Myrtaceae and Proteaceae. *In* "Ecological Biogeography in Australia" (A. Keast, ed.). Junk, The Hague.

Jurdy, D. M. and R. Van der Voo (1975). True polar wander since the Early Cretaceous. *Science* **187**, 1193–1196.

Keast, A. (1977). Historical biogeography of the marsupials. *In* "The Biology of Marsupials" (B. Stonehouse and D. Gilmore, eds) p. 69–95. University Park Press, Baltimore.

Kemp, E. M. (1978). Tertiary climatic evolution and vegetation history in the southeast Indian Ocean region. *Palaeogeogr. Palaeoclimat. Palaeoecol.* **24**, 169–208.

Kemp, E. M. and W. K. Harris. (1975). The vegetation of Tertiary islands on the Ninetyeast Ridge. *Nature, Lond.* **258**, 303–307.

Kennett, J. P. (1978). The development of planktonic biogeography in the Southern Ocean during the Cenozoic. *Marine Micropaleont.* **3**, 301–345.

Kumar, N. and R. W. Embley (1977). Evolution and origin of Ceará Rise: An aseismic rise in the western equatorial Atlantic. *Geol. Soc. Am. Bull.* **88**, 686–694.

Ladd, J. W. (1976). Relative motion of South America with respect to North America and Caribbean tectonics. *Geol. Soc. Am. Bull.* **87**, 969–976.

Langenheim, J. H. (1964). Present status of botanical studies of amber. *Bot. Mus. Leafl. Harvard* **20**, 225–287.

Larson, R. L. (1977). Early Cretaceous breakup of Gondwanaland off western Australia. *Geology* **5**, 57–60.

Leppik, E. E. (1973). Origin and evolution of conifer rusts in the light of continental drift. *Mycopath. Mycolog. Applic.* **49**, 121–136.

Leroy, J.-F. (1978). Composition, origins and affinities of the Madagascan vascular flora. *Ann. Mo. Bot. Gard.* **65**, 535–589.

Lowrie, W. and D. E. Hayes (1975). Magnetic properties of oceanic basalt samples. *In* "Initial Reports of the Deep Sea Drilling Project" (D. E. Hayes and L. A. Frakes *et al.*, eds) Vol. 28, p. 869. Government Printing Office, Washington.

Luyendyk, B. F., W. B. Bryan and P. A. Jezek (1974). Shallow structure of the New Hebrides Island Arc *Geol. Soc. Am. Bull.* **85**, 1287–1300.

Maguire, B. *et al.* (1977). Pakaraimoideae, Dipterocarpaceae of the western hemisphere. *Taxon* **26**, 341–385.

Manchester, S. R. (1977). Wood of *Tapirira* (Anacardiaceae) from the Paleogene Clarno Formation of Oregon. *Rev. Palaeobot. Palynol.* **23**, 119–127.

Martin, P. G. (1977). Marsupial biogeography and plate tectonics. *In* "The Biology of Marsupials" (B. Stonehouse and D. Gilmore, eds) p. 97–115. University Press, Baltimore.

Maxson, L. R., V. M. Sarich and A. C. Wilson (1975). Continental drift and the use of albumin as an evolutionary clock. *Nature, Lond.* **225**, 397–400.

McElhinny, M. W. (1973). "Palaeomagnetism and Plate Tectonics." Cambridge University Press, Cambridge.

McElhinny, M. W., N. S. Haile and A. R. Crawford (1974). Palaeomagnetic evidence shows Malay Peninsula was not a part of Gondwanaland. *Nature, Lond.* **252**, 641–645.

McElhinny, M. W., B. J. J. Embleton, L. Daly and J.-P. Pozzi (1976). Paleomagnetic evidence for the location of Madagascar in Gondwanaland. *Geology* **4**, 455–47.

McElhinny, M. W., S. R. Taylor and D. J. Stevenson (1978). Limits to the expansion of earth, moon, mars and mercury and to changes in the gravitational constant. *Nature, Lond.* **271**, 316–321.

McQueen, D. R. (1977). The ecology of *Nothofagus* and associated vegetation in South America. *Tuatara* **22**, 233–236.

Michener, C. D. (1979). The biogeography of the bees. *Ann. Mo. Bot. Gard.* **66**, (3).

Milsom, J. (1977). Preliminary gravity map of Seram, eastern Indonesia. *Geology* **5**, 641–643.

Molnar, P. and W.-P. Chen (1978). Evidence of large Cainozoic crustal shortening of Asia. *Nature, Lond.* **273**, 218–220.

Molnar, P. and P. Tapponnier (1975). Cenozoic tectonics of Asia: Effects of a continental collision. *Science* **189**, 419–426.

Mullins, H. T. and G. W. Lynts (1977). Origin of the northwestern Bahama Platform: Review and interpretation. *Geol. Soc. Am. Bull.* **88**, 1447–1461.

Neev, D. (1975). Tectonic evolution of the Middle East and the Levantine basin (easternmost Mediterranean). *Geology* **3**, 683–686.

Norton, I. and P. Molnar (1977). Implications of a revised fit between Australia and Antarctica for the evolution of the Eastern Indian Ocean. *Nature, Lond.* **267**, 338–340.

Owen, H. G. (1976). Continental displacement and expansion of the earth during the Mesozoic and Cenozoic. *Phil. Trans. R. Soc. London* Ser. A. **281**, 228–291.

Paris, J. P. and J. D. Bradshaw (1977). Paleogeography and geotectonics of New Caledonia and New Zealand in the Triassic and Jurassic. Int. Symp. Geodynamics S.-W. Pacific, p. 209–216. Éditions Technip, Paris.

Paris, J. P. and R. Lille (1977). New Caledonia: Evolution from Permian to Miocene. Mapping data and hypotheses about geotectonics. Int. Symp. Geodynamics S.-W. Pacific, p. 195–208. Éditions Technip, Paris.

Peirce, J. W. (1978). The northward movement of India since the Late Cretaceous. *Geophys. J. R. Astron. Soc.* **52**, 277–31.

Petriella, B. and S. Archangelsky (1975). Vegetación y ambiente en el Paleoceno de Chubut. Acta I Congr. Argent. Paleontol. Bioestratigr. **2**, 257–270.

Petters, S. W. (1978). Mid-Cretaceous paleoenvironments and biostratigraphy of the Benue Trough, Nigeria. *Geol. Soc. Am. Bull.* **89**, 151–154.

Powell, C. McA. and P. J. Conaghan (1975). Tectonic models of the Tibetan Plateau. *Geology* **3**, 727–731.

Preest, D. S. (1963). A note on the dispersal characteristics of the seed of the New Zealand podocarps and beeches and their biogeographical significance. *In* "Pacific Basin Biogeography" (J. L. Gressitt, ed.) p. 415–423. University of Hawaii Press, Honolulu.

Rao, C. K, (1972). Angiosperm genera endemic to the Indian Floristic Region and its neighbouring areas. *Ind. Forest.* **98**, 560–566.

Raven, P. H. and D. I. Axelrod (1972). Plate tectonics and Australasian paleobiogeography. *Science* **176**, 1379–1386.

Raven, P. H. and D. I. Axelrod (1975). Angiosperm biogeography and past continental movements. *Ann. Mo. Bot. Gard.* **61**, 539–673.

Raven, P. H. and D. I. Axelrod (1978). Origin and relationships of the California flora. *Univ. Calif. Publ. Bot.* **72**, 1–134.

Ravenne, C., G. Pascal, J. Dubois, F. Dugas and L. Montadert (1977). Model of a young intra-oceanic arc: The New Hebrides Island Arc. Int. Symp. Geodynamics S.-W. Pacific, p. 63–78. Éditions Technip, Paris.

Rich, P. V. (1975a). Changing continental arrangements and the origin of Australia's non-passeriform continental avifauna. *Emu* **75**, 97–112.

Rich, P. V. (1975b). Antarctic dispersal routes, wandering continents, and the origin of Australia's non-passeriform avifauna. *Mem. Nat. Mus. Victoria* **36**, 63–126.

Romero, E. J. and L. J. Hickey (1976). A fossil leaf of Akaniaceae from Paleocene beds in Argentina. *Bull. Torrey Bot. Club* **103**, 126–131.

Rosen, D. W. (1975). A vicariance model of Caribbean biogeography. *Syst. Zool.* **24**, 431–464.

Schuster, R. M. (1972). Continental movements, "Wallace's line" and Indomalayan-Australasian dispersal of land plants: Some eclectic concepts. *Bot. Rev.* **38**, 38–86.

Schuster, R. M. (1976). Plate tectonics and its bearing on the geographical origin and dispersal of angiosperms. *In* "Origin and Early Evolution of Angiosperms" (C. B. Beck, ed.) p. 48–138. Columbia University Press, New York.

Sclater, J. G., B. P. Luyendyk and L. Meinke (1976). Magnetic lineations in the southern part of the Central Indian Basin. *Geol. Soc. Am. Bull.* **87**, 371–378.

Scrutton, R. A. (1976). Fragments of the earth's continental lithosphere. *Endeavour* **35**, 99–103.

Simpson, G. G. (1940). Mammals and land bridges. *J. Wash. Acad. Sci.* **30**, 137–163.

Simpson, G. G. (1943). Mammals and the nature of continents. *Am. J. Sci.* **241**, 1–31.

Smith, A. G. and J. C. Briden (1977). "Mesozoic and Cenozoic Palaeocontinental Maps." Cambridge University Press, Cambridge.

Srivastava, S. K. and G. E. Rouse (1970). Systematic revision of *Aquilapollenites* Rouse 1957. *Can. J. Bot.* **48**, 1591–1601.

Stebbins, G. L. (1965). Probable growth habit of the earliest flowering plants. *Ann. Mo. Bot. Gard.* **52**, 457–468.

Stevens, G. R. (1977). Mesozoic biogeography of the south-west Pacific and its relationship to plate tectonics. Int. Symp. Geodynamics S.-W. Pacific, p. 309–326. Éditions Technip, Paris.

Sultan, I. Y. Z. (1978). Mid-Cretaceous plant microfossils from the northern part of the Western Desert of Egypt. *Rev. Palaeobot. Palynol.* **25**, 259–267.

Tedford, R. H. (1975). Marsupials and the new paleogeography. *Soc. Econ. Paleontol. Minearol. Spec. Publ.* **21**, 109–126.

Thomson, M. R. A. and R. W. Burn (1977). Angiosperm fossils from latitude 70°S. *Nature, Lond.* **269**, 139–141.

Thorne, R. F. (1974). A phylogenetic classification of the Annoniflorae. *Aliso* **8**, 147–209.

Thorne, R. F. (1975). Angiosperm phylogeny and geography. *Ann. Mo. Bot. Gard,* **62**, 362–367.

Thorne, R. F. (1976). Where and when might the tropical angiospermous flora have originated? *Gard. Bull. Singapore* **29**, 183–189.

Thorne, R. F. (1977). Some realignments in the Angiospermae. *Plant Syst. Evol.* Suppl. **1**, 299–319.

Trifonov, G. G. (1978). Late Quaternary tectonic movements of western and central Asia. *Geol. Soc. Am. Bull.* **89**, 1059–1072.

Valen, L. Van and R. E. Sloan (1977). Ecology and the extinction of the dinosaurs. *Evol. Theory* **2**, 37–64.

Vandenberg, J., C. T. Klootwijk and A. A. H. Wonders (1978). Late Mesozoic and Cenozoic movements of the Italian Peninsula: Further paleomagnetic data from the Umbrian sequence. *Geol. Soc. Am. Bull.* **89**, 133–150.

Veevers, J. J. and D. Cotterill (1978). Western margin of Australia: Evolution of a rifted arch system. *Geol. Soc. Am. Bull.* **89**, 337–355.

Veevers, J. J. and M. W. McElhinny (1976). The separation of Australia from other continents. *Earth Sci. Rev.* **12**, 139–159.

Veevers, J. J., C. McA. Powell and B. D. Johnson (1975). Greater India's place in Gondwanaland and in Asia. *Earth Planet. Sci. Lett.* **27**, 383–387.

Voo, R. Van der, F. J. Mauk and R. B. French (1976). Permian-Triassic continental configurations and the origin of the Gulf of Mexico. *Geology* **4**, 177–180.

Walker, A. (1972). The dissemination and segregation of early primates in relation to continental configuration. *In* "Calibration of Hominoid Evolution" (W. W. Bishop and J. A. Miller, eds) p. 195–218. Scottish Academic Press, Edinburgh.

Winn, R. D., Jr. (1978). Upper Mesozoic flysch of Tierra del Fuego and South Georgia Island: A sedimentologic approach to lithosphere plate restoration. *Geol. Soc. Am. Bull.* **89**, 533–547.

Zanten, B. O. van (1973). A taxonomic revision of the genus *Dawsonia* R. Brown. *Lindenbergia* **2**, 1–48.

Ziegler, A. C. (1977). Evolution of New Guinea's marsupial fauna in response to a forested environment. *In* "The Biology of Marsupials" (B. Stonehouse and D. Gilmore, eds) p. 117–138. University Park Press, Baltimore.

History of Flora, Vegetation and Climate in the Colombian Cordillera Oriental During the last Five Million Years

T. van der HAMMEN

University of Amsterdam, Netherlands

Introduction

Tropical high mountains are extremely interesting areas from both ecological and phytogeographical points of view. They may be compared with natural laboratories where many fundamental problems of evolution, migration, adaptation, etc. can be studied.

Although observations on the actual distribution of organisms and their relation with environmental factors provides us with an increasingly broader knowledge, we have become increasingly aware of the necessity to assemble historical data to arrive at a proper understanding of the actual situation; it has become quite clear that vegetation has not only dimensions in space, but also in time.

One of the most extensive areas of high mountain vegetation in the tropics lies in the northern Andes, and from an important section of this area, namely from the Colombian Cordillera Oriental, we have at our disposal a relatively large quantity of relevant historical data. We will try to give, after a short account of the present situation, an overall view of our present knowledge of the history of this area. This knowledge is based mainly on palynological–paleoecological and geological studies of Quaternary sediments found at present at elevations between 2000 and 4800 m.

Present-day Flora and Vegetation-belts

Above the zone of lower tropical vegetation (rain forest, savannahs, or more xerophyte vegetation) reaching to an elevation of about 1000 m, follows the Subandean belt (about 1000–2300 m). The composition of the Subandean forest is already markedly different from that of the tropical forest, a number of tropical genera (such as *Byrsonima, Mauritia, Bombax,* etc.) being almost completely absent. The abundance of a number of good pollen-producers such as *Acalypha, Alchornea* and *Cecropia* is important from a pollen analytical point of view. These genera mostly do not extend into the next zone, the Andean belt (about 2300–3300/3500 m). In this, belt forests of *Weinmannia* and *Quercus* dominate, but such genera as *Alnus, Myrica, Podocarpus, Clusia, Rapanea, Juglans, Hedyosmum,* etc., are also frequently represented. While many elements of the Andean forest are of neotropical origin, other ones (such as *Alnus Myrica* and *Juglans*) are of northern or (*Weinmannia*) southern origin. On the wetter slopes, the Andean forests may be developed as cloud forests, i.e. the forest floor and the tree trunks are completely covered with bryophytes.

The tropical high mountain flora *par excellence* is found in the high Andean Páramo belt (3300/3500 m to about 4800 m). It is subdivided into three zones: the Subpáramo, the Páramo proper and the Superpáramo. In the (lower) Subpáramo, shrubs (and small trees) mostly belonging to Compositae, Ericaceae, *Polylepis, Aragoa, Hypericum, Miconia,* and other taxa, are abundant. In the proper (grass) Páramo, open grasslands dominate, *Calamagrostis* and *Swallenochloa* being the most important representatives of the Gramineae. The characteristic Espeletiinae, also present in the Subparamo and sometimes in the lower Superpáramo, are of very frequent occurrence in the grass-Páramo, and define to a large extent its physiognomy. In the Superpáramo (above 4000/4200 m) the vegetation cover becomes incomplete as the result of soil movements caused by the daily process of alternating freezing and thawing. In the Páramo flora, the diverse phytogeographic origin is even more manifest than in the Andean forest

Cleef (see p. 175) made a careful analysis of the vascular Páramo flora of the Cordillera Oriental and grouped the genera into seven phytogeographic "elements". Approximately half of the 260 genera are of tropical or local (Andean) origin, and 40% of temperate origin. Of this last mentioned group, about 10% is of austral–antarctic and about 10% of certainly holarctic origin. About 8% of the genera are endemic in the Páramo. The relation between certain vegetation types of the Páramo and some subantarctic types, such as cushion plant communities of higher plants (Cleef, 1978), are also interesting.

Historical Data

The history of the Andean flora and vegetation begins with the appearance of the first hills and ridges in the area of the Andean geosyncline.

Towards the end of the Cretaceous most of the present area of the Eastern Cordillera was already raised above sea level. Upward movements during the Paleocene gave rise to several of the Paleozoic massifs, thus subdividing the area into several more or less separate basins. The separating hills and ridges do not seem to have been very high. During several intervals of the Lower and Middle Tertiary, further tectonic movements must have taken place resulting in the appearance of new, or of more pronounced, hill-ridges. Here again nothing is known of the exact elevations of these hills. The appearance of *Podocarpus* was most probably concomitant with the formation of these hills, and conceivably during several intervals hills a little over 1000 m high may have existed locally (van der Hammen, 1961).

In the Miocene, the area suffered further folding and faulting and local upheavals, but at the beginning of the Pliocene, the area of the present high plain of Bogotá was still within the tropical belt. The presence of *Podocarpus* and *Weinmannia* seems to indicate the local development of a (lower) montane zone in the mountains east of the high plain, formed by the further upheaval of the hills already present in that area (van der Hammen *et al.*, 1973).

During the Pliocene the proper and final upheaval of the Cordillera took place. The area of the actual high plain of Bogotá gradually became raised to about 2500 m, the areas to the east and south up to over 4000 m, and the area of the Sierra Nevada del Cocuy up to more than 5500 m. In the meantime, sedimentation in the Bogotá basin went on and since several intervals of this "Tilatá-formation" contain pollen, we have been able to distinguish several episodes of this upheaval. We know of the presence of a tropical lowland flora in the lower part, of a Subandean forest vegetation, and, finally, of an Andean forest vegetation. Extensive mountain forests with *Weinmannia* as a dominant had already developed by that time. One of the first abundant northern elements, which appeared when the Cordillera Oriental is near its present elevation, is *Myrica*. The Isthmus of Panamá was also formed during the Pliocene, and by connecting North and South America provided an easy passway for this genus, at the present time in Panamá even found as low as a little above sea level (van der Hammen *et. al.*, 1973; van der Hammen, 1974).

The first indications of the existence of open high Andean vegetation are from the Upper Pliocene, namely the higher representation of, e.g. Gramineae, Compositae and *Hypericum*. The first real Páramo flora was found in sediments near the top of the Tilatá formation. There are indications that the upper part of the Andean forest belt had not yet become well-developed and that the forest limit was, comparatively, lower than it is today. The pollen diagram, however, indicates (at 2800 m) stands of completely open grass vegetation, and we have to accept a climate which was cooler than it is at the present time. It seems reasonable to place this cold phase at the very beginning of the Pleistocene, but a late Pliocene age cannot altogether be excluded. In this primitive type of Páramo vegetation we recorded, apart from abundant Gramineae, *Polylepis, Aragoa, Hypericum, Miconia, Plantago, Polygonum*, pollen of the *Ranunculus* type, *Valeriana, Myriophyllum* and *Jamesonia*. These elements are partly of local neotropical, partly of wide tropical, austral-antarctic, and partly of cool temperate origin. The presence of *Aragoa*, a Páramo endemic is especially noteworthy, as is the presence of the high Andean genus *Polylepis* (van der Hammen *et al.*, 1973).

Apparently the basal sediments of a sequence of 200 m of lake sediments below Bogotá are only slightly younger. From here until present time continuous sediments and pollen diagrams are present. These pollen diagrams show a continual alternation of Andean forest vegetation and Páramo vegetation, representing a long sequence of glacials and interglacials.

The earlier Páramo flora of the sequence shows the same species as mentioned above and in addition *Geranium*, Caryophyllaceae and two types of *Lycopodium*. Slightly later *Gunnera, Gentianella* and *Lysipomia* appear. The Andean forest is enriched with *Styloceras* and *Juglans*. Now and then a pollen grain of *Alnus* is found, indicating that a member of this genus was growing not far away. The sequence described here probably belongs to the Lower Pleistocene.

Near Guasca, a vulcanic ash was found in sediments corresponding with the lower part of this Lower Pleistocene sequence. Fission-track dating (by J. Boelstorff) yielded an age of $3 \cdot 62 \pm 0 \cdot 67$ million years. Despite the large statistical error, the sample can hardly be younger than 3 million years. This would back-date clear glacial periods and the Lower Pleistocene to that age. Recent data on the age of the Lower Pleistocene in Europe and of glaciations in Patagonia confirm the relatively ancient age of the Early Pleistocene "glacials" (e.g. Zagwijn, 1974).

At a depth of 150 m in the sequence of lake sediments near Bogotá, or at greater depths further to the centre of the basin of the high plain, the pollen record indicates that *Alnus* migrated into the area and became an important element of the local stands of vegetation. We take this point provisionally as the beginning of the Middle Pleistocene. We do not yet have an absolute date for this beginning, but recent fission-track and potassium argon dates from vulcanic ashes intercalated in the lake sediments of the Middle and Upper Pleistocene, suggest an age by extrapolation, of 1·5 million years or more. The sequence of glacials and interglacials continues in the "Middle" and "Upper" Pleistocene, new floral elements appearing now and then.

The first appearance of *Quercus*, another immigrant from the north, is most striking. We know from the above mentioned dates that this happened about 1 million years ago (and not 250 000 years ago as we once estimated). Soon this genus became an important element in the forest of the western part of the Cordillera Oriental. It seems to have got an opportunity when, at the end of a glacial and the beginning of an interglacial time, large open areas above 2000 m once more became available for invasion by forest (van der Hammen, 1964, 1974; van der Hammen *et al.*, 1973).

We know much more about the last interglacial–glacial cycle (roughly the last 130 000 years), and especially about the Late Glacial and the Holocene (the last 13 000 years) (van Geel and van der Hammen, 1973; Schreve Brinkman, 1978; van der Hammen, 1978; and the general review and earlier literature review in van der Hammen, 1974). During the last interglacial and the early last glacial interstadials, extensive marsh forests of *Alnus* and *Weinmannia* surrounded the lake of the "Sabana de Bogotá". The slopes of the surrounding mountains were covered by Andean forest. After the first major cold phase (Lower Pleniglacial), the lake level rose during a later phase of the Middle Pleniglacial. Weinmannia then disappeared almost completely from the high plain. In the latest part of the Middle Pleniglacial (some 30 000 years ago) and the first part of the Upper Pleniglacial (until some 20 000 years ago) *Polylepis* forest was very abundant in the area (at about 2600 m above sea level), and probably formed a belt between the proper Andean forest and the open Páramo. Some 20 000 years ago the entire area became covered with grass Páramo, and the lake-levels fell considerably. Open Páramo vegetation prevailed even as low as 2000 m on the eastern flank of the Cordillera Oriental. From the available evidence a lowering of average annual temperature of 6–7°C or more can be calculated for the high plains. The annual rainfall must have been less than half of the present amount. It may be noticed here that the decrease of temperature in the tropical lowland most probably was considerably less (3°C?).

At the beginning of the Late Glacial there was a considerable increase in rainfall, and the lake levels rose again. During the warmer

Guantiva-interstadial the area of the high plains and surroundings became reforested, but it took some time before the fresh soils from the glacial time developed into mature forest soils. Therefore, a pioneer of eroded soils, *Dodonaea*, is fairly common during this interstadial. The reforestation is interrupted by a short last cold phase: the El Abra stadial, when Subpáramo vegetation was present on the surrounding slopes.

Now the Holocene began, some 10 000 years ago, and Andean forest established itself more permanently. In the area of Fuquene *Quercus* was common in the beginning, but later the forest became more diversified, and *Weinmannia* increased. This may be related to the gradual build-up of a humus layer. Actual differentiation (*Quercetum, Weinmannia* forest, etc.) may be related to similar and other differences in soil conditions, in annual rainfall and seasonality. In different parts of the Cordillera at different elevations, it was established that the average annual temperature during the Middle Holocene must have been somewhat higher than the present one ($\approx 2°C$). The altitudinal forest limit was several hundred metres higher, and so was the limit between the Subandean and the Andean forest belt. Some 3000 years ago the climate became slightly cooler and corresponded with the present. Short periods of relatively low lake levels, correlated with periods of lower rainfall or higher evaporation (higher temperatures), occured around 5000 and 2000 years ago.

From about 2500 years ago, there has been an increasing influence of man on the stands of vegetation due to the agricultural activities of the Indian population. After the Spanish conquest, this influence became disastrous because of deforestation, burning and subsequent soil erosion.

Conclusions

The geological history seems to be of considerable importance for the understanding of the flora and vegetation of the Colombian Cordillera Oriental, and for tropical mountains in general. Because of the Pliocene upheaval of the area, new climatic zones were created, the higher ones forming islands with a colder climate in a sea of tropical and subtropical climates. These montane belts and "islands" during and after the uplift gradually became populated by new flora elements, forming new types of vegetation.

The history of the flora of the Subandean belt may have started well back in the Tertiary, the history of the Andean flora in the Upper

Miocene or Lower Pliocene, and the history of the Páramo flora in the Middle to Upper Pliocene. Because of the gradual adaptation to the new environments, it may be expected that the limits of certain vegetation belts (from a floristic point of view) were originally lower and gradually rose to their present position. There are clear indications for a lower altitudinal forest limit in the Upper Pliocene. At the same time when the Cordillera rose to its present elevation, in the Pliocene, the Isthmus of Panamá, connecting North and South America, was formed. Species of lower elevations could pass easily, those of higher elevations conceivably used the "stepping stones" of isolated higher areas.

Immigration has taken place since a relatively early stage from the south (austral–antarctic region), and since the formation of the isthmus, from the north (holarctic region). The probability of immigration into the higher zones was much greater during glacial periods when the source areas often lay nearer to the equator than they do today, and when the surface of the receiving montane islands was a mutliple of that of today. Half of the genera of the Páramo, and more of the lower montane zones, are of (neo) tropical or Andean origin, respectively, or endemic. During the Quaternary, the area was subjected to a succession of alternating glacial and interglacial climates. During glacial times more or less isolated Páramo "islands" (or "islands" of the Andean forest) became connected with others to form much larger islands.

The alternating possibility of evolution in isolation and exchange of taxa seems to have been very important for speciation, particularly in the Páramo belts. The current research on long sections in the area is aimed at a better understanding of the relations of geological history and climatic change with immigration, adaptation and evolution of the floral elements and of the changing composition and ecology of the erstwhile vegetation types.

References

Cleef, A. M. (1978). Characteristics of neotropical Páramo vegetation and its Subantarctic relations. *Erdwiss. Forsch.* **13**, 365–390.
Geel, B. van and T. van der Hammen (1973). Upper Quaternary vegetational and climatic sequence of the Fuquene area (Eastern Cordillera, Colombia). *Palaeogeog. Palaeoclimatol. Palaeoecol.* **14**, 9–92.
Hammen, T. van der (1961). Late Cretaceous and Tertiary stratigraphy and tectogenesis of the Colombian Andes. *Geol. Mijnbouw* **40**, 181–188.
Hammen, T. van der (1964). A pollen diagram from the Quaternary of the

Sabana de Bogotá (Colombia) and its significance for the geology of the Northern Andes. *Geol. Mijnbouw* **43**, 113–117.

Hammen, T. van der (1974). The Pleistocene changes of vegetation and climate in tropical South America. *J. Biogeog.* **3**, 3–26.

Hammen, T. van der (1978). Stratigraphy and environments of the Upper Quaternary in the El Abra corridor and rock shelters (Colombia, South America). *Palaeogeog. Palaeoclimatol. Palaeoecol.* **25**, 11–162.

Hammen, T. van der, J. H. Werner and H. van Dommelen (1973). Palynological record of the upheaval of the Northern Andes, a study of the Pliocene and Lower Quaternary of the Colombian Eastern Cordillera and the early evolution of its high-andean biota. *Rev. Palaeobot. Palynol.* **16**, 1–122.

Schreve Brinkman, E. J. (1978). A palynological study of the Upper Quaternary sequence in the El Abra corridor and rock shelters (Colombia, South America). *Palaeogeog. Palaeoclimatol. Palaeoecol.* **25**, 1–109.

Zagwijn, W. H. (1974). The Pliocene–Pleistocene boundary in Western and Southern Europe. *Boreas* **3**, 75–97.

II. General Phytogeography of Tropical Floras

Some Geographic Trends in Morphological Variation in the Asian Tropics and their Possible Significance

P. S. Ashton

Arnold Arboretum, Harvard University, USA

Introduction

The distribution of Dipterocarpaceae subfamily Dipterocarpoideae spans the full extent of the Asian tropics, extending southwards to the Seychelles and south-east to New Guinea and the Louisiades archipelago. With 13 genera and $\approx$470 species they reach exceptional species diversity in the mixed rain forests of Sundaland (Malaya, Sumatra, Borneo and West Java); there they are well represented in the understorey and, as a family, dominate the emergent stratum. These trees dominate the fire-climax deciduous forests of north-eastern India and Indo-Burma too. Dipterocarps are also particularly well known on account of their importance in the timber industry. On this account they provide us with a uniquely complete record of distribution and variability from which trends of potential general significance to other tropical tree families may be drawn.

Geographical Distribution of Taxa

Dipterocarps can be grouped into a small number of distribution patterns. These not only conform with the distribution of climatic and other ecological factors but, as shall be seen, each group manifests a more or less distinctive pattern of variation at around the species level. We are now beginning to perceive some of the underlying biological causes for this.

Wallace's line, which runs east of the Philippines and between Borneo and Celebes and Bali and Lombok, is a major phytogeographic boundary for dipterocarps which cannot be explained in terms of climatic differences: of the 10 genera and 380 species in Malesia as a whole five genera and only 26 species occur east of it. About 280 species occur in Borneo yet only eight in Celebes 125 km to the east (Ashton, 1979a). Though but three species straddle the line, the 23 other eastern species are closely related and often vicarious with others in the west. In the Dipterocarp flora of peninsular India, Sri Lanka and the Seychelles there are eight genera, three of which (and in addition a major section of *Shorea*), are endemic; and there are ≈60 species. Though all but three of the latter are also endemic to the region, once again there are remarkable examples of vicarism not only between Sri Lanka and southern India but between that region and the Far East. These include *Dipterocarpus hispidus* Thw. of Sri Lanka and *D. baudii* Korth. of South-East Asia and Sumatra, *D. glandulosus* Thw. of Sri Lanka and *D. costatus* Gaertn. f. of South-East Asia, *S. stipularis* Thw. of Sri Lanka and *S. hypochra* Hance of South-East Asia, and *Hopea parviflora* Bedd. of South India and *H. odorata* Roxb. of South-East Asia.

Of the 13 Asiatic genera, four of those with species east of Wallace's line (*Dipterocarpus, Vatica, Hopea,* and *Shorea*) are among the five which occur in both peninsular India and the Far East. The fifth, *Anisoptera*, occurs in the Tertiary fossil record of South-East India, along with *Dryobalanops* which is presently endemic to aseasonal Sundaland. *Cotylelobium* is also confined to aseasonal climates but occurs in both Sundaland and South-West Sri Lanka (Ashton, 1977a), the only example of such disjunct distribution at generic level. Among these contemporary wides, all but *Cotylelobium* occur in both aseasonal and seasonal evergreen forests. *Dipterocarpus* and *Shorea* even have a few representatives in fire-climax savanna woodland though these are without exception closely allied to rain forest species.

The same trends occur in the sections of the two large genera *Hopea* and *Shorea*. In *Hopea* both sections contain one widespread subsection with species in both seasonal South-East Asia and New Guinea, and another which is endemic to aseasonal western Malesia including the Philippines; only one subsection occurs in India and Sri Lanka. Of the 11 sections of *Shorea*, section Doona with 10 species is endemic to aseasonal South-West Sri Lanka and six are confined to aseasonal western Malesia; among the latter section Pachycarpae, of 10 species, is endemic to Borneo. Among the remaining four sections Anthoshoreae is found in both India and Sri Lanka and east of Wallace's line,

Brachypterae is concentrated in western Malesia but extends into weakly seasonal parts of the Philippines and Moluccas, Pentacme occurs throughout the Philippines and in Indo-Burma but shuns Sundaland, and Shorea—though centred in South-East Asia—is well represented in Sri Lanka and contains the two fire-resistant species of the genus inhabiting the "dry dipterocarp forests" of Indo-Burma. These two, which are vicariants east and west of the Burmese Arakan ranges, are closely allied to the widespread South-East Asian and West Malesian evergreen species *S. guiso* (Blco) Bl.

From these facts I deduce the following:

(i) The high endemicity of supraspecific taxa in peninsular India, Sri Lanka and the Seychelles bespeaks a long dipterocarp history; I have argued elsewhere (Maguire and Ashton, 1977; Ashton, 1979a) that this region may be the cradle of the subfamily.

(ii) The frequency of vicarism in widespread genera between West Malesia and both India with Sri Lanka and New Guinea suggests that intermittent migration has continued until recently; in the absence of endemism above species level a recent arrival of dipterocarps into New Guinea is suggested.

(iii) The absence of supraspecific taxa confined to fire-climax dry dipterocarp forests, the close affinity of the nevertheless distinctive dipterocarp flora there with evergreen forest species, and the restriction of this component to North-East India and Indo-Burma suggests a recent origin.

(iv) The distribution of *Cotylelobium* suggests that a belt of aseasonal climate has occurred between peninsular India, and Malesia at some time since the Oligocene when dipterocarps first appear in the Asian forest record; nevertheless the other widespread genera are all well represented in seasonal evergreen forests and it appears that migration through a corridor of such forests has been more frequent, probably until recently.

It is the better explored region west of Wallace's line that now merits detailed attention. Within the aseasonal region of western Malesia, which encompasses most of the Malay Peninsula, Sumatra except the north-west and extreme south, Borneo except the south-east, and the eastern Philippines several families of distribution patterns are detectable (Ashton, 1964, 1972, 1979b):

(i) Wides of the lowland forest on the prevailing yellow/red soils; occurring at least in Malaya, Sumatra and Borneo, and sometimes the Philippines. These are of two types, those that do, and those that do not, transgress the seasonality boundary and extend into Indo-Burma, or the Philippines, Java or Bali. Satellites of these are the wide exclusively montane elements, and a group, of some 25 throughout Malesia, which occur in the slightly seasonal margins of the aseasonal zone and which may or may not extend out into the true seasonal evergreen forests.

(ii) Wides of the peat swamps; occurring throughout the range of their habitat, especially in West Malaya, East Sumatra and South and North-West Borneo.

(iii) Wides in coastal and freely draining, infertile leached yellow soils in mixed rain forests; their ranges vary in extent, but most occur in the subcoastal hills of North-West Borneo between the Kapuas drainage and Kota Kinabalu, and some in the Anambas, Riouw archipelago, the east coast hills of Sumatra near Kuantan, and down the east coast of Malaya; occasionally also in North-West Sumatra and Perak state, North-West Malaya. These form part of the Riouw Pocket Flora of Corner (1978). This distribution pattern is obscured both by high levels of endemism and vicarism, and by the extension of some members up the dry ridge crests of the Malayan inland hills.

(iv) Riparian wides.

Endemicity occurs in all these habitats. It is least in the peat swamps, probably because of their intermittent continuity in the Pleistocene. It is low on the prevailing yellow-red lowland soils where it is generally expressed as island-wide vicarism: rarely between Sumatra and Malaya, more commonly between them and Borneo, and generally (i.e. with a higher representation than the wides) between the Philippines and Borneo. Such vicarism prevails in the mountains. This would appear to reflect the history of hill land continuity. Other habitats have a higher rate of endemicity; on the relatively xeric leached sites and the riparian fringes there is some vicarism, but both here and in two habitats concentrated in Borneo—skeletal clay soils on steep hillsides and podsols—the majority of endemics are either most closely allied to wides of the yellow-red soils nearby or have no obvious affinities. A very few of the podsol, or heath forest, element, occur also in eastern Malaya.

Within the immensely diverse Bornean dipterocarp flora, with ≈ 280 species, there is a remarkable degree of local endemism on the xeric yellow and white coastal soils, and to a lesser extent the river banks. This may be attributable to the fragmentation of these habitats, the hills by river valleys and the rivers by hills, over a prolonged period (Ashton, 1972). Further minor endemism occurs on the North-East Borneo ultrabasics.

Geographical Patterns in Biological Characteristics

The only sharp discontinuity in the ecological and geographical ranges of the subfamily exists between the evergreen forests and the fire-climax dry dipterocarp savanna woodlands. Species of the latter, as might be expected, are noted for their bark, which superficially markedly differs from that of their evergreen forest congeners in being thick and ruggedly fissured—though these characters seem to be influenced by burning frequency and habitat aridity. Those that have been tested have some seed dormancy, exceptional in the family, while cryptocotylar germination occurs in those species belonging to groups (e.g. *Shorea* sects. Shorea and Anthoshoreae) in which it is not the norm (Maury, 1978). These species also coppice unusually readily, and their frequently burned saplings develop prominent taproots.

There are overall geographical trends throughout the subfamily too, though each is studded with exceptions. As might be expected, species of the seasonal regions have a greater tendency to deciduousness, though no dipterocarp is more than shortly so; the widespread *Dipterocarpus gracilis* Bl. is shortly deciduous in Thailand for instance, though fully evergreen in the aseasonal part of its range. Curiously, species of mesic sites are more frequently deciduous than those of xeric, a general phenomenon among tropical Asian tree species: thus, in the Cardamone chain of southern Cambodia *D. dyeri* Pierre in the moist valleys is briefly deciduous whereas *D. costatus* Gaertn. f. on the sandy hills nearby remains staunchly evergreen.

More intriguing is the degree of hairiness. There is an overall trend towards glabrousness from seasonal to aseasonal regions, reaching an extreme in such understorey taxa as *Vatica* and many *Hopea* but also *Shorea* sect. Richetioides where both emergent and main canopy species show the same trend. The tomentum seems to disappear from leaves first, then twigs, young shoots and finally inflorescences and floral bracts. The fimbriate unicellular hairs and distinctive multicellular tufts are typical Malvalian characters in the family; their loss would therefore appear to be a derived condition. Does this imply that the Dipterocarpaceae originated in the seasonal tropics, a view supported by the distribution of the more generalised African and South American subfamilies?

Yet more evocative are the trends—and here, it must be admitted, exceptions abound—in flower and fruit characters. Flower size is remarkably constant within genera, and the sections of *Shorea*, with the limited exception of *Dipterocarpus*. Larger flowered taxa are exclusively species which present them above the forest canopy. It is interesting that the larger flowered genera *Vateria, Vateriopsis, Dipterocarpus, Anisoptera* and *Parashorea* as well as the larger flowered sections of *Shorea* (Shorea, Anthoshoreae and Pentacme) comprise the full complement of those supraspecific taxa exclusively confined to the canopy which extend into the Asian seasonal tropics. A few emergents in the predominantly understorey and small-flowered genus *Hopea* also do so however, and the large-flowered genera *Dryobalanops* and *Upuna* are restricted to aseasonal climates.

Anther size and stamen number broadly follow the same trend (Ashton, 1979a). *Vateria, Vateriopsis, Dipterocarpus, Parashorea, Shorea* sect. Pentacme—but also *Dryobalanops, Cotylelobium, Neobalanocarpus* and *Stemonoporus* and *Shorea* sects. Rubellae and Richetioides subsect. Polyandrae which are confined to the aseasonal zone—have large elongate and bright yellow anthers. Anthers are smaller and white and subglobose to ellipsoid in other taxa, but among these they are largest in sects. Shorea and Anthoshoreae, and also sect. Ovales with one species in aseasonal West Malesia; *Hopea plagata* Vidal, with the largest anthers in its genus, is a canopy species of Philippine seasonal evergreen forests. Among the smallest are those of *Shorea* sect. Richetioides, in which the number of loculi is also halved. Stamen numbers vary between 5–110, 15 being most general. There are more than 15 in *Vateria, Vateriopsis,* most *Dipterocarpus, Anisoptera* sect. Anisoptera, *Dryobalanops, Shorea* sects. Shorea, Rubellae, and six spp. of Anthoshoreae all of which have representatives in seasonal localities—but also *Shorea* sect. Ovalis, one species in Brachypterae, and three species in *Hopea* two of which occur in seasonal sites. According to Muller (in Ashton, 1979) there is a relationship between dipterocarp flower size and pollen size, even within the genus *Dipterocarpus*; on present evidence, therefore, stamen number may give a better indication of pollen production than anther size.

In a very few species staminal number is reduced below 15; these include six species of *Hopea* and three of the 13 wingless-fruited species of *Shorea* sect. Richetioides to be discussed, where there are 10; also *Stemonoporus elegans* Thw., *S. cordifolius* Thw., and *Vatica pentandra* Ashton which is known from a single collection, where there are but five. Of these, two are local endemics in Sri Lanka, one in Malaya and seven in Borneo; all present their flowers beneath the forest canopy. *Hopea sangal* Korth., also with 10 stamens, flowers above the canopy and is widespread from Java and Borneo to southern Thailand, while *H. acuminata* Merr, is vicarious with it and does likewise in the Philippines.

A striking difference in dipterocarp flowering phenology exists between the seasonal and aseasonal sectors. In the former the flowering of populations is annual, though flowering and fruiting intensity varies from year to year as in other trees and individuals do not necessarily flower each year. Though sporadic flowering occurs each year here and there in the aseasonal tropics, and some—especially riparian—species and provenances flower more frequently than others, most is confined to spectacular gregarious flowerings at intervals of up to 10 years when the great majority of species and nearly all mature individuals flower more or less intensely. It is not always realised that this phenomenon is by no means unique to dipterocarps and occurs, for instance, in several leguminous genera. Such flowerings are synchronous over a short period within populations, and sequential among related species, the whole process extending over several months (Ng, 1977; Chan, 1978; Wood, 1956; Ashton, 1969). Only certain understorey *Stemonoporus* of Sri Lanka appear to flower sporadically over long periods and at frequent intervals without close synchronisation. It is of particular interest that the climatic boundaries between annual and non-annual flowering seem to coincide closely with the boundaries of regions of exceptional dipterocarp species diversity.

The abundance of dipterocarps in aseasonal western Malesia and their extraordinary flowering phenology poses an obvious problem for successful pollination. *Stemonoporus, Shorea* sect. Doona, *Neobalanocarpus,* and *Dryobalanops* are visited by the ubiquitous meliponid bees of the genus *Trigona*; the latter three, though gregarious, should not pose an impossible problem for pollination by these insects, and all flower more frequently than other genera. *Dipterocarpus* is said to be pollinated by honey bees (*Apis*) in Dry Dipterocarp forest (T. Smitinand, pers. comm.) though we have not found them to do so in the aseasonal tropics where a variety of insects appear to visit; the latter is so also in *Shorea* sect. Pachycarpae (S. Appanah, pers. comm.). Thrips (*Thysanoptera*) have been confirmed as the pollinators of *Shorea* sect. Mutica (Chan, 1978); they have also been found in the flowers, and there is evidence that they are pollinators, of *Hopea* and *Shorea* sect. Richetioides. Their short life-cycle and apparent fecundity (S. Appanah, in prep.) could provide the means to build up the numbers required. Our knowledge of dipterocarp pollinators is recent and still fragmentary, but it can be seen that bees appear to visit the species with large yellow elongate anthers, thrips the small white anthers; there may be a correlation between pollinator and pollen size. Bee pollination seems to prevail in the seasonal tropics and Sri Lanka.

There is no obvious trend in fruit size with geographical or ecological distribution, though in several riparian species with water borne fruit they are large and have thick often corky pericarps. There are, though, trends related to the reduction in the usually aliform fruit sepals; this has occurred many times and in several genera. It too is frequently associated with water dispersal, but also occurs particularly among species that present their flowers beneath the canopy (Ashton, 1979b) and there mostly occurs in local endemics. A striking contrast, for instance, is provided in the wet zone of Sri Lanka between the 10 species of *Shorea* sect. Doona, all of which have aliform fruit sepals, and the 15 *Stemonoporus* species which do not. Both have submontane and lowland elements, yet not a single species of the former is local within the aseasonal south-west while every one of the latter is, often being confined to a single range of hills or forest and exhibiting remarkable vicarism over short distances. *S. cordifolius* and *S. elegans*, both of which happen to be endemic to the southern slopes of Adam's Peak in one of the wettest forests in Asia, possess two-celled ovaries; throughout the rest of the subfamily normal flowers possess three-celled ovaries. In all species each cell contains two ovules. It would appear, then, that staminal number also gives a guide to the pollen: ovule ratio, and that this ratio tends to decline in the humid tropics, especially among the local endemics.

Geographical Trends in Intraspecific Variation

Here, too, among several identifiable patterns the most distinct contrast is that between seasonal and aseasonal regions (Symington, 1943; Ashton, 1979a, c), and particularly between dry dipterocarp forest species and others. Two features stand out in seasonal regions: great variability, particularly in tomentum within species, and the existence of "hybrid swarms" by which I mean groups of trees with characters intermediate between two widespread and distinct species growing in the vicinity. The intraspecific variability is sufficiently striking to have been given taxonomic status in such species as *Shorea roxburghii* G. Don, *S. siamensis* Miq., *A. costata* Korth. and *Dipterocarpus obtusifolius* Teysm. ex Miq. but I believe it is too continuous for these varieties to be maintained. The more glabrous forms prevail in the least seasonal parts of the ranges of all but *A. costata*; much local variability occurs in the more seasonal areas where it is similarly said to be related to soil aridity (e.g. Kerr, 1911). Hybridization between species occupying the same habitat can occur, as between *D. obtusifolius* and *D. tuberculatus* Roxb. in dry dipterocarp forest; but most seems to be between species occupying differing but adjacent habitats, as for example between the evergreen forest species *S. guiso* (Blco.) Bl. and *S. obtusa* Wall. of dry dipterocarp forest, and between *D. intricatus* Dyer of dry dipterocarp forest and the riparian *D. alatus* Roxb. Hybridization is not rare, either, among species prevailing on different soils within seasonal evergreen forest, especially in *Dipterocarpus* (e.g. Parker, 1931).

The pattern of morphological variation prevailing in the aseasonal is the converse of that in the seasonal tropics: taxa as a rule exhibit remarkable local uniformity, but there is profuse allopatric geomorphological and ecotypic differentiation. The degree of this differentiation varies considerably; this is well exemplified within the large sections of *Shorea*. Most species of sects. Shorea and Brachypterae are so distinct that few obvious groupings can be recognized on morphological comparisons. Nevertheless, in sect. Shorea several pairs of vicariants exist in which the main (though not the only) diagnostic characters are small apparently non-adaptive differences in the number of stamens. Examples are *Shorea collina* Ridl. of Malaya and *S. lunduensis* Ashton of Borneo; *S. ciliata* King of Malaya and *S. brunnescens* Ashton of Borneo, *S. sumatrana* (Sloot. ex Thor.) Sym. ex Desch of Sumatra and Malaya and *S. seminis* (De Vr.) Sloot. of Borneo, *S. lumutensis* of Malaya and *S. crassa* Ashton of Borneo, *S. astylosa* Foxw of the Philippines and *S. domatiosa* Ashton of Borneo. Among the six species in sect. Anthoshoreae with more than 15 stamens the same allopatric variation in staminal number occurs, but in addition there is more distinctive and equally constant differentiation of leaf characters which allow, for instance, the vicariants of slightly seasonal regions *S. polita* Vidal (Philippines), *S. montigena* Sloot. (Moluccas) and *S. gratissima* Dyer (Malaya, Borneo) to be recognized even when sterile. The remaining three species are yet more distinctive, and their characters can be equated with the limiting factors of the individual habitats they occupy. That same pattern is repeated in section Rubellae. In section Mutica, which is almost confined to aseasonal Sundaland where the species are often common and even gregarious, all species have 15 stamens; a core group of 13 species occurs in Sumatra, the Malay peninsula, and in all but two in Borneo as well. Two occur on xeric sites, four in peat swamps and seven in lowland mixed dipterocarp forest. Among the last-mentioned the very distinctive *S. acuminata* Dyer of Sumatra and Malaya is replaced by *S. quadrinervis* Sloot. of Borneo. *S. parvifolia* Dyer and *S. macroptera* Dyer, which grow together with the former, are differentiated into geographical subspecies; these exist too in the sole member of sect. Ovalis, and this pattern of specific and infraspecific vicarism occurs also in the 15 species of sect. Anthoshoreae with 15 stamens. The remaining 12 species in sect. Mutica are endemic to Borneo; some are very local, some widespread, and each occupies a distinct though overlapping geographical range; five clearly share a common origin with *S. macroptera,* seven with *S. parvifolia.* The pattern in this section is repeated in other dipterocarp taxa notably parts of *Dipterocarpus, Hopea* and *Vatica.* Without experimental evidence differentiation between a species and a subspecies must rest on morphological criteria and is essentially arbitrary even if, as here, consistent. It is interesting in this context that two geographical subspecies of *S. macroptera* coexist in some Sarawak forests yet no hybridization was recorded during enumerations of their populations (Ashton, 1969). Extremes of local differentiation, as a rule in several correlated but morphologically independent characters and therefore regarded by me to indicate full speciation, occur in *Shorea* sect. Richetioides and among certain groups of *Hopea.* In both this endemism prevails among species that flower within the canopy. Those several species in which the characteristic alate fruit sepals of the family are vestigial are local endemics with few exceptions; it will be recalled that staminal number is reduced in several of these, and the number of anther loculi halved in sect. Richetioides.

In all examples described so far from the aseasonal tropics taxa are characterized by allopatric differentiation, internal uniformity and constancy of differentiating characters (Ashton, 1969). In summary, it appears that the great species diversity within the aseasonal zones might have resulted from fragmentation of the variation of a few widespread ancestral forms; these might originally have been comparable to the few widespread and variable species that still constitute the majority in the flora of the Asian seasonal zone.

By implication, hybridization is rare in the aseasonal zone. A few examples of individuals or small groups with characters intermediate between two species do exist, but the ease by which they are recognized emphasises their rarity. They too appear generally to occur between parents which inhabit differing soils. *Shorea* sect. Pachycarpae, with 10 species and endemic to Borneo—which is surprising in view of its intermittent connection with Sumatra and Asia during the Pleistocene—is none the less an exception. All species but *S. mecistopteryx* Ridl. are unusually variable; the variation is often highly localized and allopatric but more or less continuous; local morphological hybrid populations exist between several. In this instance it may be supposed that allopatric differentiation has occurred but reimmigration has been too frequent in the shifting Bornean landscape for breeding barriers in some instances to be completed. There are two other analogous examples: the group of river bank species *Vatica rassak* (Korth.) Bl., *V. umbonata* (Hook. f.) Burck and *V. stapfiana* King also appear to hybridize rather frequently; all are semigregarious and inhabit the shifting alluvial banks of rivers; they tend to flower more frequently than forest species. Again, species which transgress climatic boundaries tend not only to be more variable in the seasonal part of their range, but to hybridize at or near climatic boundaries. Examples are *Dipterocarpus hasseltii* Bl. which seems to hybridize with *D. retusus* Bl. in Bali and *D. gracilis* in the Philippines, and *Anisoptera costata* whose variability in the north of Malaya suggests local hybridization, possibly with *A. curtisii* Dyer ex King (Ashton, 1979c).

It is clear that diversification has been more favoured in some parts of aseasonal western Malesia than others. Some allopatric differentiation has occurred between the east coast populations and those of the rest of Malaya within *S. macrantha* Brandis, *S. singkawang* Miq., *D. lowii* Hook. f., *Parashorea densiflora* Sloot. ex Sym. and others, and these segregates are broadly associated with centres of exceptional species diversity, in this case in the eastern coastal hills and the coastal hills of Perak in the north-west. By far the richest centre of both local endemism and species diversity is in North-West Borneo. By contrast there are 45 species in the edaphically relatively uniform and fertile Philippines, 45% of which are endemic, but local endemicity is low in spite of the climatic diversity. Instead, many species are exceptionally variable; this variation can be regional and correlated with climate (e.g. *S. falciferoides* Foxw., *S. contorta* Vidal, *A. thurifera* (Blco.) Bl.) or local and at least in some cases correlated with drainage (e.g. *S. polysperma* (Blco.) Bl., *S. almon* Foxw.). It would appear that a combination of periodic isolation and high soil diversity enhances both the process of speciation and overall diversity in the aseasonal tropics.

What, then is a Dipterocarp Species?

The morphological variability of the dipterocarps of the seasonal tropics implies that frequent change in the distribution of habitats and geographical boundaries in evolutionary time has favoured variability but rarely provided sufficiently prolonged isolation barriers for fertility barriers to evolve. The frequent occurrence of hybridization albeit local might suggest the same; nevertheless hybridization in this zone often occurs between species which are morphologically considerably more distinct than species within groups that rarely or never hybridize in the aseasonal tropics such as, for instance, the seven Shoreas of sect. Mutica which cohabit the Malayan mixed dipterocarp forests. We may surmize that hybrid populations are generally eliminated by competi-

                    P. S. Ashton

tion as they will lack, unlike their parents, a niche to defend. In spite of
their variability, and the occurrence of frequent hybridization, the
dipterocarps of the seasonal tropics are easy to identify within that zone
and we have no reason to question their nature as outbreeding biological species. A similar explanation could be suggested for the variability
in *Shorea* sect. Pachycarpae, though here it is the constantly changing
distribution of Bornean mountains and soils at the margin of the Sunda
shelf that could be the cause of the instability. Clearly these hypotheses
are ripe for testing.

But what of the multitude of other species in the aseasonal tropics? At
first sight their allopatric differentiation and the clear discontinuities in
variation by which they are distinguished suggest them to be true
biological species. The seedlings of species in *Shorea* sect. Mutica are
distinguishable and well known to foresters, yet hybrids seem to be
exceptionally rare. It would seem that the species are incompatible
with one another, and Chan (1978) has indirect evidence that this is so
though he has shown some of them to be highly self-incompatible. The
recent demonstration that some dipterocarps are at least facultatively
apomictic through adventive embryony (Kaur *et al.*, 1978), and the
frequency of polyembyony and triploidy suggest that this is by no
means uncommon, provides a further unexpected mechanism for dipterocarp diversification. It could be significant that of the seven species
in which apomixis is inferred or demonstrated three possess geographical subspecies, three (*Hopea subalata* Sym. with *H. mesuoides* Ashton
(unexamined), and *Shorea resinosa* Foxw. with *S. agami* Wood ex Ashton)
are vicariants and one (*H. latifolia* Sym.) is widespread but very closely
related to another widespread species occupying a different habitat.
The remaining species (*H. odorata*) interestingly enough is a gregarious
river-bank species of seasonal Indo-Burma, overlapping with but
essentially replaced by *H. sangal* in West Malesia. It is difficult to see
why this differentiation should be allopatric if apomixis were obligate.
Facultative apomixis could proliferate fit genotypes and accelerate
ecotypic differentiation and short-term evolution, but what part can it
play in the continuing evolution of a family of angiosperms in an
environment which presumably experiences gradual yet inexorable
and continuous, rather than sudden or oscillatory, change (Ashton,
1977b)? How widespread is apomixis in the rain forest? Is the loss of fruit
wings, reduction of staminal number and anther size, reduction in
ovule number and the reduction in pollen: ovule ratios which is so
clearly related with local endemicity part of a syndrome of characters
associated with self-pollination as Cruden (1977) suggests? So far we
have no evidence of high levels of self-compatibility among rain forest

species, and the existence of the same trend in large dioecious genera, e.g. *Garcinia* (Clusiaceae), cannot be explained thus; yet at least some *Garcinia*, and *Syzygium* (Myrtaceae) in which the same trends exist are agamospermous—could agamospermy be associated instead?

At present we are merely scratching the surface, but there seems no doubt that previous suppositions (e.g. Stebbins, 1958; Bawa, 1974; Bawa and Opler, 1975) that xenogamy prevails in climax woodland, and tropical forest in particular, must be brought into question.

References

Ashton, P. S. (1964). Ecological studies in the mixed dipterocarp forests of Brunei State. Oxford For. Mem. 25.

Ashton, P. S. (1969). Speciation among tropical forest trees: Some deductions in the light of recent evidence. *Biol. J. Linn. Soc.* **1**, 155–196.

Ashton, P. S. (1972). The quaternary geomorphological history of western Malesia and lowland forest phytogeography. Trans. 2nd Aberdeen–Hull Symp. Mal. Ecol. p. 35–42.

Ashton, P. S. (1977a). Dipterocarpaceae. *In* "A revised Handbook to the Flora of Ceylon 1" (M. D. Dassanayake, ed.) Vol. 2 p. 166–196. University of Ceylon, Peradeniya.

Ashton, P. S. (1977b). A contribution of rain forest ecology to evolutionary theory. *Ann. Mo. Bot. Gard.* **64**, 694–705.

Ashton, P. S. (1978). Flora Malesiana precursores: Dipterocarpaceae. *Gard. Bull. Sing.* **31**, 5–48.

Ashton, P. S. (1979a). Dipterocarpaceae. *In* "Flora Malesiana" (C. G. G. J. van Steenis, ed.) (in press).

Ashton, P. S. (1979b). Phylogenetic speculations on Dipterocarpaceae. *Mém. Mus. Hist. Nat. Paris.* (in press).

Bawa, K. S. (1974). Breeding systems of tree species in a lowland tropical community. *Evolution* **28**, 85–92.

Bawa, K. S. and P. A. Opler (1975). Dioecism in tropical forest trees. *Evolution* **29**, 167–179.

Chan, H. T. (1978). The reproductive biology of some Malaysian dipterocarps. Ph.D Thesis, University of Aberdeen.

Corner, E. J. H. (1978). The freshwater swamp forest of South Johore and Singapore. *Gard. Bull. Sing.* Suppl. 1.

Cruden, R. W. (1977). Pollen: ovule ratios: A conservative indicator of breeding systems in flowering plants. *Evolution* **31**, 32–46.

Gan, Y. Y., F. W. Robertson, P. S. Ashton, E. Soepadmo and D. W. Lee (1977). Genetic variation in wild populations of rain-forest trees. *Nature, Lond.* **269**, 323–325.

Kaur, A., C. O. Ha, K. Jong, V. E. Sands, H. T. Chan, E. Soepadmo and P. S. Ashton (1978). Apomixis may be widespread among trees of the climax rain forest. *Nature, Lond.* **271**, 440–442.

Kerr, A. F. G. (1911). Dipterocarpaceae of northern Siam. *J. Siam Soc.* **8**, 1–24.

Maguire, B. and P. S. Ashton (1977). Pakaraimoideae, Dipterocarpaceae of the wester Hemisphere. II. Systematic, geographic and phyletic considerations. *Taxon* **26**, 341–385.

Maury, G. (1978). Diptérocarpacées de fruit à la plantule. Thesis, Paul Sabatier University, Toulouse.

Ng, F. S. P. (1977). Gregarious flowering of dipterocarps in Kepong, 1976. *Malay. For.* **40**, 126–137.

Parker, R. N. (1931). Illustrations of Indian forest plants. Part II: Five species of *Dipterocarpus*. *Ind. For. Rec. (Bot.)* **16**, 1–16.

Stebbins, G. L. (1958). Longevity, habitat and release of genetic variability in higher plants. *Cold Spring Harb. Symp. Quant. Biol.* **23**, 365–378.

Symington, C. F. (1943). Foresters manual of dipterocarps. *Mal. For. Rec.* **16**.

Wood, G. H. S. (1956). The dipterocarp flowering season in North Borneo, 1955. *Malay. For.* **19**, 193–201.

The Flora and Vegetation of Tropical Africa Today and Tomorrow

J. P. M. BRENAN

Royal Botanic Gardens, Kew, England

In October 1977 I had the privilege and pleasure of being asked to present a contribution on some aspects of the phytogeography of tropical Africa at a symposium on the biogeography of Africa, held at the Missouri Botanical Garden. This is not by way of self-advertisement but rather as an excuse. When I read the statement in the first circular about the present Symposium that "papers will be planned to concentrate on the following subjects: history of tropical floras, present distribution of vegetation types," etc., I realized that I had already covered in some detail the five topics enumerated for tropical Africa, and that there was thus a distinct danger of a not very profitable rehash. This apologia, which I hope you will excuse, will thus explain why this paper may seem a little unbalanced, and will serve as an introduction to what I hope to cover.

During the past 50 years, and particularly since the Second World War, tropical Africa has been subjected to an intensity of study in plant ecology and floristics that is certainly without previous parallel for Africa itself, and quite possibly for any other large tropical region. The tasks are far from complete, but the present seems a particularly good time to take stock of the position. I would like to focus on tomorrow more than today, as there are many lessons to be learned from previous experience that might guide our future policy more surely.

Although the tropics of Cancer and Capricorn are clear boundaries on the map, they are not nearly so useful in the field or in assessing scientific data. In particular, plants and animals do not seem aware that there are boundaries. The area to be dealt with here is thus delimited politically from Senegal, Mali, Niger, Chad and the Sudan in the north to Angola, Botswana, Rhodesia and Mozambique in the south. The total area is just over 20 000 000 sq km.

Good (1974) estimated the total vascular plant flora at approximately 30 000 species, and although the basis for his calculation is not clear, it seems a reasonable estimate. In comparison with other large tropical areas, Africa seems to have a relatively poor flora. I have discussed the evidence and reasons for this elsewhere (Brenan, 1978) and do not wish to go into the problem in greater depth now, more than to say that the present tropical African flora appears to be impoverished mainly as a result of past climatic fluctuations that seem to have been more drastic than, say, in South-East Asia or South America.

The degree of uniqueness—endemism—of the tropical African flora is quite high. At the generic level over 40% are endemic to tropical Africa, and, if Africa is extended to include South Africa and Madagascar, the total rises to over 58%. For the more widespread genera there is a much stronger relationship eastwards than westwards: 18·6% reach Asia and/or Australia and only 3·5% also occur in South America. The pantropical element (12·1%) is certainly significant, although to some extent affected by introductions by man.

Although the poverty of the tropical African flora has been emphasized, there are significant fluctuations within Africa. Contrast between the outstanding floristic richness of the south-eastern Cape region and less rich northern and more tropical parts of South Africa is well-known and well-documented. In tropical Africa the differences are less striking and the geographical patterns more subtle. Using endemism as an index (Brenan, 1978), the following countries are considered to have a very high area/endemism index: Cameroun, São Tomé, Principe, Annobon, Gabon, Cabinda, Zaïre, Ethiopia, Tanzania, Zanzibar and Angola. Probably Equatorial Guinea and Congo (Brazzaville) could be added to this list, but evidence is insufficient.

The following are medium/high: Guinea Republic, Liberia, Togo, West and Central Nigeria, Fernando Po, Somali Republic, Uganda, Kenya, Mozambique and Zambia. The remainder are low or very low in endemism.

The general floristic picture can be summed up by stating that the semi-desert to savannah regions north of the equator from the Sudan westwards, are uniformly low or very low in endemism, with poor floras. Somalia and Ethiopia have a much richer flora, particularly Ethiopia, and, with better data, Somalia would probably enter the very high category. The savannah regions south of the equator are considerably richer both floristically and in endemics than north of the equator, with Angola and adjacent southern Zaïre especially so.

The rain-forest areas, especially well-developed in West and Central Africa, but extending in a limited patchy way to East Africa, appear to have a comparatively rich flora near the Gulf of Guinea in the west (Gabon, Cameroun) but to become gradually more impoverished eastwards to eastern Zaïre and Uganda. However, certain small isolated areas in Kenya and Tanzania have, in comparison, a rich flora with strong endemism.

The high African mountains, outstanding literally and metaphorically, deserve at least a brief mention. The well-known giant species of *Senecio* and *Lobelia* alone make their flora extraordinary. Though the endemism is high (81%) the Afro-alpine flora is numerically poor with only 278 species (Hedberg, 1968). Isolation and local endemism have played a part, but the similarities between the high alpine flora of the different African mountains seem more striking than the differences.

Although detailed knowledge of the vegetation of tropical Africa is still very far from complete, the outlines seem to command some general acceptance. The works of White (1965), Chapman and White (1970) as modified and synthesized by Wickens (1976), have resulted in the recognition of five regions:

(i) Sudano-Zambezian—a large region corresponding to tropical savannah and subdivided into five domains.

(ii) Guinea-Congo—corresponding to the evergreen or partly evergreen forest areas, and subdivided into three domains.

(iii) Afro-montane ⎫ altitudinally different zones of the mountain
(iv) Afro-alpine ⎭ areas.

(v) Saharo-Sindian—desert and semi-desert north of the equator.

These regions have been well described and delimited and it does not seem worthwhile discussing them in closer detail, except to make the point that the distribution maps of many widespread African species show some remarkable correspondences with these regions and domains.

A good recent general account of African vegetation has been given by Schnell (1976), and the vegetation has been well mapped (Vegetation Map of Africa South of the Tropic of Cancer, 1959), of which a thorough revision has been recently completed under the auspices of UNESCO and AETFAT (White, 1976). On the whole, then, there is now available a good general picture of the main vegetation types and phytogeographical zones of tropical Africa. That this has been achieved in comparatively so short a time is both a testimony to the intensity of research in tropical Africa, and also to the way in which international cooperation between individual research workers in different countries has been fostered and encouraged by the activities of the Association pour l'Etude Taxonomique de la Flore d'Afrique Tropicale (AETFAT).

Much of tropical Africa is covered by published floras, either for single countries or for regions (geographical aggregations of countries). Without in any way attempting to give a complete survey of all current African floras, it may nevertheless be useful to summarize the content and degree of completeness of some of these floras (Table 1).

Some of the works mentioned are scarcely floras, though they do give a reasonably critical catalogue of the plants of the areas covered. This comment applies, for example, to work on São Tomé (Exell, 1944) and to the enumeration by Cufodontis (1953–1972) covering an important area of North-East Africa. Conversely, Andrews (1950–1956), though a descriptive flora, is uncritical and out-of-date. I later make further comment on both on the incompleteness and on the range of works included under the term "flora".

How well is the flora of tropical Africa known? Léonard (1975) gave revealing statistics about the progress of our knowledge of the phanerogamic flora of tropical Africa during the period 1953–1973. During that time 7478 new species and 391 new genera were described (later unpublished records complete the picture to the end of 1976). Léonard concluded that tropical African flora is still far from being completely known since new discoveries are constantly and frequently being made, but that during the period 1966–1973 there was a clear and significant slowing down of the rates of published discovery of new taxa in comparison with the statistics for the period 1953–1965. The reasons for this diminution are not analysed, but it may be surmized that increased knowledge of the flora is one factor, and that a lessening of the flow of new material, especially from less accessible areas, may be another.

Léonard (1965) also published a map, with a textual commentary, on the extent of floristic exploration in Africa south of the Sahara. This has been recently revised by Hepper (1971). Although the data on which such maps are based are very crude, and subjective assessment has been a frequent and important factor, nevertheless the large tracts of tropical Africa still (and correctly) categorized as poorly known floristically are undoubtedly an expression of objective truth. There is still a long way to go before any sizeable continuous tract of tropical Africa can claim to be as well known floristically as even the most remote parts of Europe. What remains to be accomplished is formidable, and there is a strong element of urgency to get it finished.

Table 1. Some tropical African floras.

Reference	Countries covered	No. of spp. published	Approx. %
"Flora of West Tropical Africa" (Hutchinson and Dalziel 1954–1968)	West Africa from Senegal, Mali and Niger, South to North West Cameroun	7326 (Hepper, 1971)	100
"Flore du Cameroun" (Aubréville and Leroy (1963–1975)	Cameroun	1038 (Letouzéy, 1976)	20–25
"Flore du Gabon" (Aubréville and Leroy 1961–1972)	Gabon	1435 (Floret, 1976)	25
"Catalogue of the Vascular Plants of S. Tomé" (Exell, 1944)	Sâo Tomé, Principe, Annobon	683	100
"Flore du Congo, du Rwanda et du Burundi Flore d'Afrique Centrale" (Robyns et al., 1948–1977)	Zaïre, Rwanda, Burundi	3921 (Brenan, 1978)	39
"The Flowering Plants of the Sudan" (Andrews, 1950–1956)	Sudan Republic	3137 (Brenan, 1978)	100
"Enumeratio Plantarum Aethiopiae Spermatophyta" (Cufodontis 1953–1972)	Ethiopia, Somalia, Djibouti	6323	100
"Flora of Tropical East Africa" (Turrill et al., 1952–1977	Uganda, Kenya, Tanzania, Zanzibar	4412	40
"Flora Zambesiaca" (Exell et al., 1960–1971)	Botswana, Zambia, Rhodesia, Malawi, Mozambique	2228	40
"Conspectus Florae Angolensis" (Exell et al., 1937–1970)	Angola, Cabinda	2137 (Exell and Gonçalves, 1973, supplemented by personal counting)	40

The sixth plenary meeting of AETFAT in Uppsala (Hedberg, 1968) focused on the needs for conservation of vegetation in Africa south of the Sahara. The follow-up to this meeting (Hedberg, 1976), though showing some progress, was far from encouraging. Change in the environment, often drastic in its scale, usually reflects short-term political expediency rather than any sense of long-term stewardship. In the investigation and publication of floras, especially where fragile tropical ecosystems are concerned, time may be vitally important.

Other factors reinforce this:

(i) Developing countries need to exploit their plant resources carefully and knowledgeably, often against a background of increasing population, a shortage of employment and an economy in need of financial help. Short-term financial gains argue very cogently against measures for conservation. If the evidence on which more rational policies can be based is made available as quickly as possible, then not only will developing countries benefit but also humanity in general. In this way vegetation stands a better chance of being husbanded rather than devastated, to the benefit of future generations.

(ii) Political change, especially in developing countries, may be rapid and damaging. Sound policies may be annulled by political change alone.

(iii) The cost of research and its publication are increasing dramatically. In Britain we have been faced by the situation in which technical work for the benefit of Third World countries can be priced so high on commercial grounds that it is too expensive to be purchased by most of those for whose benefit it has been produced. Increases in the cost of printing show no signs of stopping, and not infrequently outpace inflation.

What are the taxonomic and floristic needs of developing countries such as many of those in tropical Africa, and to what extent are our policies in Europe meeting those needs? In the first place the point needs to be made strongly that floras are still greatly needed. It is easy to assume that the position of western Europe, well, sometimes lavishly, endowed with descriptive floras, is typical of the rest of the world. For the majority of countries in tropical Africa there is no complete modern flora and indeed no complete flora wholly written in the last 100 years. For some countries the last general coverage was the "Flora of Tropical Africa", of which the first volume was published in 1868, and which is still not quite complete!

For a country to exploit fully and rationally its plant species and vegetation a flora is a necessity, yet for many countries there is not even a modern inventory.

Basic alpha taxonomy is sometimes considered old-fashioned and insufficiently innovatory by European universities, though biosystematics are still "with-it". However, basic taxonomy may often be more in tune with the immediate needs of developing countries than other esoteric but more popular disciplines of botany. Although there are a few honourable exceptions, tropical taxonomy, and even tropical botany, are sadly neglected.

Taxonomic training in Europe is often sought by students from tropical Africa. Too often we fail to provide the training that is going to be most relevant to the functions and tasks they will need to perform when they return home. Taxonomy cannot be carried out satisfactorily on any significant scale without herbaria. Training is needed both at the research and technical level to ensure that these valuable, unique and vulnerable collections may be preserved for posterity. A few years of neglect can spell ruin. Flourishing and efficient national regional herbaria in the tropics can perform important advisory and research functions for the countries in which they are situated, and can also help to accelerate the production of floras.

Mention was made earlier of the diversity of works bearing the title "flora". As much information is transmitted to developing countries under this title, its vagueness and imprecision are noteworthy. Floras of countries or regions within tropical Africa range from unannotated catalogues or lists to treatments that are not clearly distinguishable from full-scale descriptive monographs. There are numerous intermediate stages. The main point to be made is that a number of large and important tropical African floras are still woefully incomplete; the tasks of completing these must be of high priority. Manpower and finance are all too often deficient and affected drastically and capriciously by political developments in Africa unrelated to the need for the work. European countries still hold reference collections of such importance, size and uniqueness that they must be considered indispensable to any major regional taxonomic research in tropical Africa. In addition there is a wealth of accumulated expertise in taxonomy. European countries, through their appropriate ministries and research councils, should be prepared to support the contribution these countries can make to fulfil a moral and scientific obligation.

Plants can only be properly understood and fully used if they bear names which are in general use and with which relevant published information is linked. Usually these will be the internationally accepted scientific Latin names. The need, often loudly voiced in Europe, for stabilization of generic and specific names, is even stronger in tropical countries without proper floras or inventories, where instability or multiplicity of names for one species may be a positive obstacle to the retrieval of relevant information.

It is inevitable that, in a live and developing discipline, differences of taxonomic opinion will occur, and these will certainly lead to instability of names. Nevertheless changes are too often made on the basis of incomplete research, and there is sometimes a failure by taxonomists to appreciate the responsibility they have towards users. Again, floras have shown themselves to be powerful agents in promoting general acceptance and hence stability of names.

Referring to Table 1 it is difficult to avoid the impression that research is taking too long. Some of the factors causing this have already been mentioned. Nevertheless is there anything that taxonomists can do to help the situation? Again I would emphasize both the need for overseas countries to have this information quickly and the element of urgency imposed on taxonomists by the speed of alteration and destruction of natural vegetation.

Let us start with an instructive example: the first edition of the "Flora of West Tropical Africa" was published between March 1927 and February 1936. The number of species covered was 5639. The second edition (Hutchinson and Dalziel, 1954–1968) was completed between August 1954 and May 1972, with 7326 species. This is a complete flora that has passed through two editions.

Comparison of these figures with those given for other tropical African floras in Table 1 will show that other major floras are often relatively slower in their time:species ratio. An important reason for this difference must be that the flora of West Tropical Africa relies on conciseness and keys, while others are often much more elaborate in their treatment, with full descriptions and often lavish specimen citation. Some of these works, for example the "Flora of Tropical East Africa" (Turrill *et al.*, 1952–1977) are of the highest scientific quality and value.

However, in the future and for new floras, might some compromise with these high standards be forced upon us? Should not we seriously consider the merits of favouring floras that are more concise but less time-consuming to produce and less costly to buy? Might the monographic revision be better as a separate work in its own right rather than an integral part of a flora?

With the recent growth of nationalism, a flora of one country is often a better candidate for national finance and support than a flora of a region of which the contributing country may be only a portion. In other words, the scientific case for a regional flora may be strong, but the political appeal of single-country floras may be stronger. Enough money to pay for preparation and publication, and people available to write the work within a reasonable time are ultimately the controlling factors. In Africa this tendency seems especially likely, and taxonomic research must come to terms with it. Cooperation and coordination may become increasingly important to ensure uniformity of treatment. These questions deserve serious consideration.

Perfection will always be out of reach but none of the developments forecast or recommended here need be incompatible with a very high standard of taxonomic research. I make a strong plea that we should plan our future floras with care and coordination, and a clear sense of the priorities both from the producer's and user's points of view.

References

Andrews, F. W. (1950–1956). "The Flowering Plants of the [Anglo-Egyptian] Sudan" Vols 1–3. T. Buncle, Arbroath, Scotland.

Aubréville, A. and J. F. Leroy (1961–1972). "Flore du Gabon." Muséum National d'Histoire Naturelle, Paris.

Aubréville, A. and J. F. Leroy (1963–1975). "Flore du Cameroun." Muséum National d'Histoire Naturelle, Paris.

Brenan, J. P. M. (1978). Some aspects of the phytogeography of tropical Africa. *Ann. Mo. Bot. Gard.* **65**, 437–478.

Chapman, J. D. and F. White (1970). "The Evergreen Forests of Malawi." Commonwealth Forestry Institute, University of Oxford, England.

Cufodontis, (1953–1972). "Enumeratio Plantarum Aethopiae Spermatophyta."

Exell, A. W. (1944). "Catalogue of the Vascular Plants of S. Tomé (with Principe and Annobon)." British Museum (Natural History), London.

Exell, A. W. and M. L. Gonçalves (1973). A statistical analysis of a sample of the Flora of Angola. *Garc. Orta, Sér. Bot.* **1**, 105–128.

Exell, A. W., F. A. Mendonça, A. Fernandes and E. J. Mendes (eds) (1937–1970). "Conspectus Florae Angolensis, 1–4." Junta de Investigaçôes do Ultramar, Lisboa, Portugal.

Exell, A. W. *et al.* (1960–1971). "Flora Zambesiaca." Crown Agents, London.

Floret, J. J. (1976). Flore du Gabon. *Boissiera* **24**, 575–580.

Good, R. D'O. (1974). "The Geography of the Flowering Plants" 4th edn. Longman, London.

Hedberg, I. and Hedberg, O. (1968). Conservation of vegetation in Africa south of the Sahara. *Acta Phytogeog. Suecica* **54**.

Hedberg, I. (1976). Follow-up of the AETFAT meeting at Uppsala in 1966 on conservation of vegetation in Africa south of the Sahara. *Boissiera* **24b**, 437–441.

Hepper, F. N. (1971). Progress of the flora of West Tropical Africa. *Mitt. Bot. Staatssamml. München* **10**, 25–26.

Hutchinson, J. and J. M. Dalziel (1954–1968). "Flora of West Tropical Africa" (2nd edn) Vols 1–3 (revised by R. W. J. Keay and F. N. Hepper). Crown Agents, London.

Léonard, J. (1965). Carte du degré d'exploration floristique de l'Afrique au sud du Sahara. *Webbia* **19**, 906–914.

Léonard, J. (1975). Statistiques des progès accomplis en 21 ans dans la connaissance de la flore phanérogamique africaine et malgache (1953–1973). *Boissiera* **24a**, 15–19.

Letouzey, R. (1976). Flore du Cameroun. *Boissiera* **24**, 571–3.

Robyns, W. *et al.* (1948–1977). "Flore du Congo du Rwanda et du Burundi" (1948–1970), continued as "Flore d'Afrique Central (Zaire–Rwanda–Burandi)". Jardin Botanique, Brussels.

Schnell, R. (1976). "Introduction à la Phytogéographie des Pays Tropicaux 3–4. La Flore et la Végétation de l'Afrique Tropicale, 1–2." Gaulthier-Villars, Paris.

Turrill, W. B., E. Milne-Redhead and R. Polhill (eds) (1952–1977). "Flora of Tropical East Africa." Crown Agents, London.

Vegetation Map of Africa South of the Tropic of Cancer (1959). Published on behalf of l'Association pour l'Etude Taxonomique de la Flore d'Afrique Tropicale with the assistance of UNESCO. Oxford University Press, Oxford.

White, F. (1965). The savanna woodlands of the Zambezian and Sudanian domains. *Webbia* **19**, 651–681.

White, F. (1976). The vegetation map of Africa. The history of a completed project. *Boissiera* **24b**, 659–666.

Wickens, G. E. (1976). The flora of Jebel Marra (Sudan Republic) and its geographical affinities. Kew Bull. Add. Ser. 5. Her Majesty's Stationery Office, London.

Distribution Patterns of Lowland Neotropical Species with Relation to History, Dispersal and Ecology, with Special Reference to Chrysobalanaceae, Caryocaraceae and Lecythidaceae

G. T. PRANCE

New York Botanical Garden, USA

Introduction

Many of the phytogeographic conclusions which we make are based on present day distribution maps of the species which we study. We are all familiar with taxonomic monographs which have distribution maps accompanied by various comments on history, dispersal and ecology of the species, and with broader phytogeographic papers based largely on species distributions.

The paper by Professor Brenan in this symposium demonstrated that the African flora is now extremely well-known in comparison with the neotropics. Many interesting conclusions can be reached from such a well-known flora. In this paper I want to examine the status and validity of distribution patterns in the lowland neotropics. This region remains one of the least explored parts of the world, and with the relative lack of other evidences (such as pollen and fossil) our conclusions for the lowlands, must be drawn from distribution maps of an under-explored area. For example, I have pin-pointed Pleistocene forest refugia (Prance, 1974c), and in this process one of the greatest dangers is locating centres of good collection, rather than centres of endemism or diversification. However, if we consider all these limitations carefully, it is quite possible to draw significant conclusions, especially if adequate personal field work has been carried out in the area. After giving some warnings about the use of distribution maps in the neotropics, I intend to use maps to reach a few conclusions.

Collecting in the lowland neotropics has reached a level at which we can draw significant and interesting conclusions about the vegetational history of the region. However, we must be aware of the pitfalls, and consider such factors as the degree of collecting in an area; the ecology of the species under study and the distribution of suitable habitats; the geological history of the area, and the dispersal mechanism of the species. Dr Kalkman, in his paper made a plea for collectors to gather more data about dispersal. I strongly endorse this because the interpretation of distribution patterns often does not make sense without dispersal data.

The Status of Collecting in the Lowland Neotropics

I will give a few examples from my studies of the Chrysobalanaceae to illustrate the inadequate state of the inventory of the neotropics. A more detailed account is given in Prance (1978b).

In 1972, I monographed the Chrysobalanaceae for Flora Neotropica (Prance, 1972) treating 323 species known at that date. In the six years since then, much new material has been collected throughout the range of the family. This has resulted in the description (Prance, 1973, 1974a, b, 1976a, 1977, 1978a, 1979) of 25 more new species and one new subspecies, and two new species by Lundell (1974), see Table 1. This addition of 27 new species in a six-year-period represents an 8·36% increase. Details of these new species and their habitats are given in Table 1. This shows that many of the new species came from the lowland forests of Amazonia, and Panama, two under-explored areas. Other monographers are finding a similar growth rate to their species lists which only happens in poorly collected areas.

Figure 1 shows the collecting history of *Couepia longipendula* Pilg., which I termed a local endemic to the Manaus area in the 1972 monograph. In subsequent collecting I have shown that it is a widespread species in Amazonia, as is shown on the map.

Figure 2 shows the occurrence of the genus *Hirtella* by degree square in South America. From this generic representation *Hirtella* would appear to be reasonably well collected throughout the region. There are only a few species of *Hirtella* in the cerrado of Central Brazil, and I would therefore expect the dots to be more scattered in that area as they are on Fig. 2. Figure 3, however, shows the number of species that have been collected from each degree square. The number of species which occur in the squares including the Amazonian cities of Manaus and Belém where I have concentrated my field work is most striking. These

Table I. New Species and Subspecies of Chrysobalanaceae described since the "Flora Neotropica" monograph in 1972 (Prance, 1972a).

Species	Locality	Habitat	Date of Collection of the type
Couepia dolichopoda Prance (1974b)	Peru–Loreto	Forest on terra firme	1972
Couepis edulis Prance (1973)	Brazil–Amazonas	Forest on terra firme	1971
Couepia glabra Prance (1973)	Brazil–Amazonas	Forest on terra firme	1971
Couepia marlenei Prance (1974a)	Brazil–Amazonas	Forest on terra firme	1972
Hirtella arenosa Prance (1976a)	Brazil–Amazonas	White sand campina	1968
Hirtella conduplicata Prance (1976a)	Brazil–Amazonas	Forest on terra firme	1973
Hirtella magnifolia Prance (1979)	Brazil–Amazonas	Forest on terra firme	1976
Hirtella revillae Prance (1979)	Peru–Loreto	White sand swamp forest	1976
Licania aracaensis Prance (1976a)	Brazil–Amazonas	Montane forest, 1000 m altitude	1975
Licania cabrerae Prance (1976a)	Colombia–Antioquia	Montane forest, 2000 m altitude	1957
Licania cecidiophora Prance (1978a)	Peru–Loreto	Forest on terra firme	1974
Licania cuatrecasasii Prance (1979)	Colombia–Valle	Upland forest, 1000 m altitude	1972
Licania chiriquiensis Prance (1976a)	Panama	Cloud forest, 1250 m altitude	1975
Licania fasciculata Prance (1979)	Panama	Lowland forest	1972
Licania furfuracea Prance (1976a)	Venezuela–Bolívar	Forest beside stream	1971
Licania jefensis Prance (1976a)	Panama	Cloud forest	1969
Licania jimenezii Prance (1973)	Surinam	Forest on terra firme	1971
Licania joseramosii Prance (1979)	Brazil–Amazonas	Campina forest	1976
Licania kallunkii Prance (1979)	Panama	Lowland forest	1975
Licania marlenei Prance (1976a)	Brazil–Amazonas	Forest on terra firme	1972
Licania montana Prance (1976a)	Venezuela–Lara	Cloud forest, 1500 m altitude	1975
Licania morii Prance (1976a)	Panama	Lowland rain forest	1975
Licania octandra (Hoffmgg. ex R. & S.) Kuntze subsp. grandifolia Prance (1974a)	Brazil–Amazonas	Forest on terra firme	1973
Licania pakaraimensis Prance (1976a)	Venezuela–Bolívar	Montane forest, 1400 m altitude	1973
Licania stewardii Prance (1976a)	Brazil–Amazonas	Campina forest	1974
Licania sp. nov.	Ecuador	Forest on terra firme	1969
Parinari sp. nov.	Brazil–Bahia	Coastal forest	1969
Licania guatemalensis Lundell (1974)	Guatemala	Rain forest	1971
Licania mexicana Lundell (1974)	Mexico	Forest beside stream	1943

                    G. T. Prance

two areas coincide with the concentration of collecting around the cities, and it is probable that many other grid squares in Amazonia contain at least the 16 species which have been collected in the Manaus square. Can we conclude that the area around Manaus is particularly rich or that it is well collected? Actually both factors are important. In

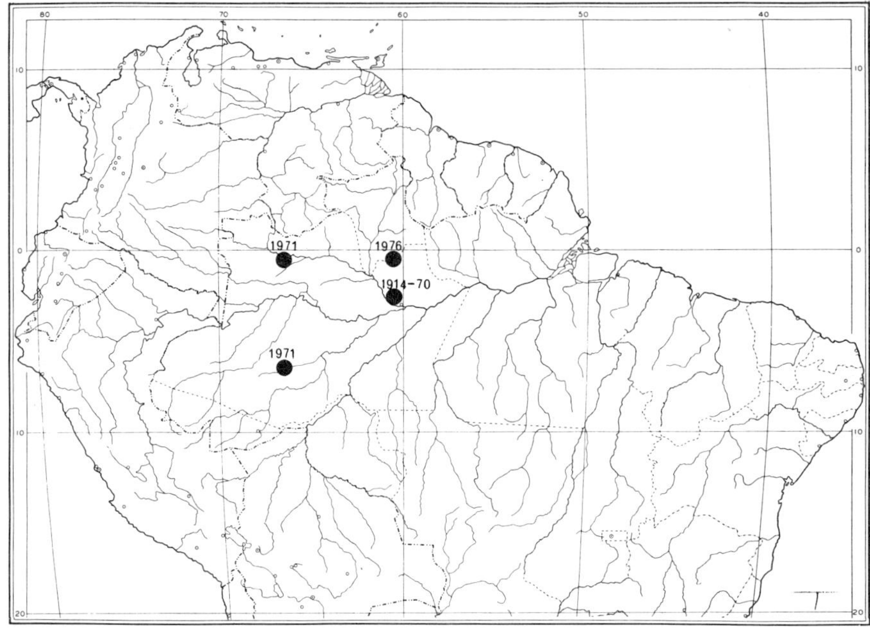

Fig. 1. The collecting history of *Couepia longipendula* Pilg. (Chrysobalanaceae), showing how recent collections have considerably extended the range of a species that from its discovery in 1914 until 1970 was thought to be a local endemic of the Manaus region.

addition to being well collected, the area around Manaus has much habitat diversity, with two large rivers with different water types, many white sand campinas and a large area of forest to the north that has apparently been rather stable throughout the recent climatic changes of the post-Pleistocene (Irion and Absy, 1978). Figure 4 gives a similar grid square distribution of the collection of the genus *Couepia* to show that *Hirtella* is not an isolated case. An exactly similar pattern of collection is shown for *Couepia* where 18 of the 55 species have been collected in the Manaus square. These maps serve not only as a warning to drawing conclusions about distributions from present day maps, but also for the pin-pointing of undercollected areas. It is to be hoped that the Brazilian Projeto Flora Amazônica and other collecting programmes with which we are associated can fill in some of the gaps on these maps.

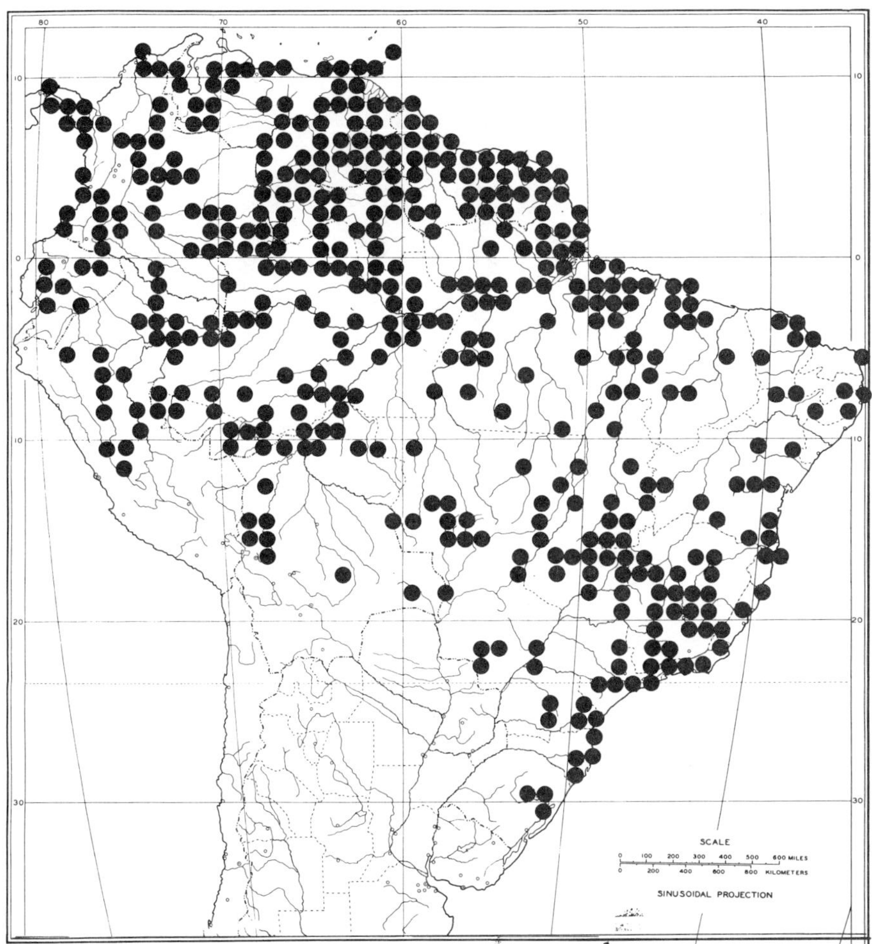

Fig. 2. Collecting of the genus *Hirtella* (Chrysobalanaceae) in South America showing presence in a one degree grid square. (The maps in Figs 2–14 represent presence of a species in a one degree grid square and not the exact locality of any specific collection.)

Disjunct Distribution and History, Ecology and Dispersal

When we are faced with a disjunction like that of *Mouriri oligantha* Pilg.
given in Fig. 5, we must ask whether this represents poor collecting or a
genuine disjunction. Specialists collecting in the intervening areas

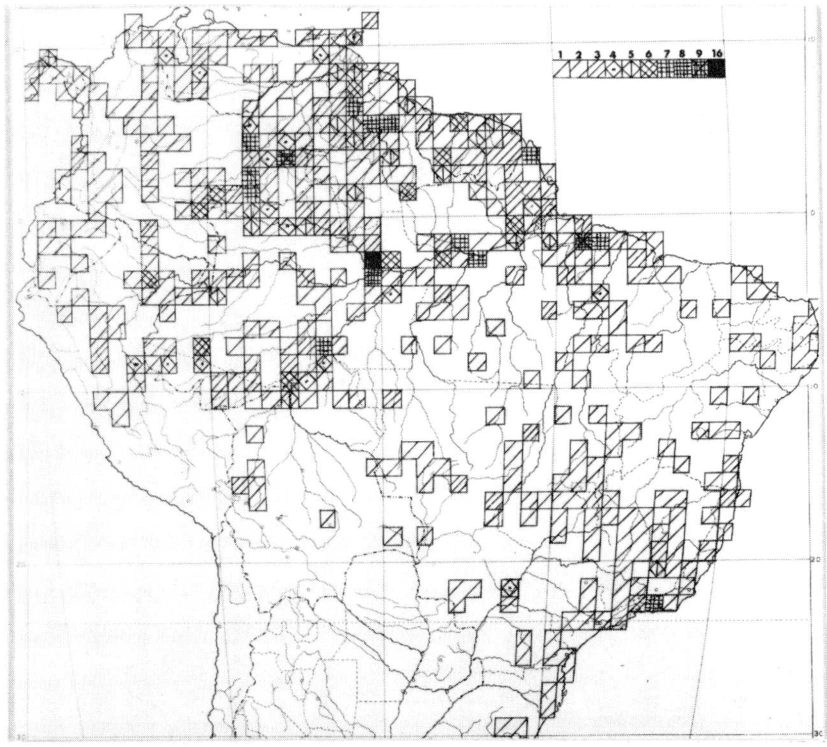

Fig. 3. Collection of the genus *Hirtella* by degree grid square showing the
number of species collected in each square. Note the number of species
collected in the vicinity of Manaus (16) and Belém (9) where I have concentrated my field work. This map helps to determine undercollected areas both
for future collecting and to avoid misinterpretation of distribution map data.

between apparent disjunctions are the best equipped to prove or dis-
prove the disjunction. In the case of *Mouriri oligantha* intensive field
work in the intervening area, and a visit by the monographer of the
genus, Thomas Morley, have failed to fill in the gap on the map. This
disjunction is repeated in several other cases such as *Couepia parillo* DC.
(Chrysobalanaceae), and is probably a genuine one. We can then begin

to question why these species are divided into two disjunct populations. The answer probably lies in the climate changes of the Pleistocene.

Figure 6 shows the distribution of *Licania affinis* Fritsch (Chrysobalanaceae) which has long been known as a common species in the Guianas, but has recently been collected in Panama (Prance,

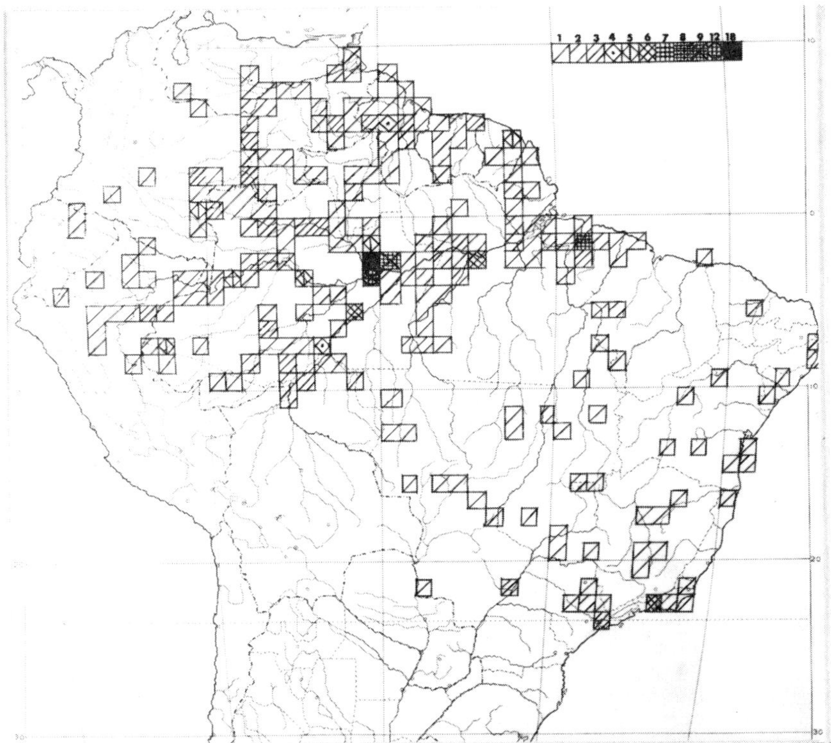

Fig. 4. Collecting of the genus *Couepia* (Chrysobalanaceae) by degree grid square, showing the number of species collected in each square. The same areas as shown in Fig. 3 for *Hirtella* have been the focii of collecting.

1974a). This disjunction lies between the relatively well collected area of northern Venezuela and has also been found in many other families, e.g. Myristicaceae (Gentry, 1975), and *Caryocar nuciferum* L. (Prance, 1976b). This is a well established genuine disjunction. There is an area of much drier forest between these two areas of tropical moist forest in Panama and the Guianas. *Licania affinis* has a large drupaceous fruit which is unlikely to be dispersed over a long distance. This disjunction is accounted for by climate changes with the intervention of a drier

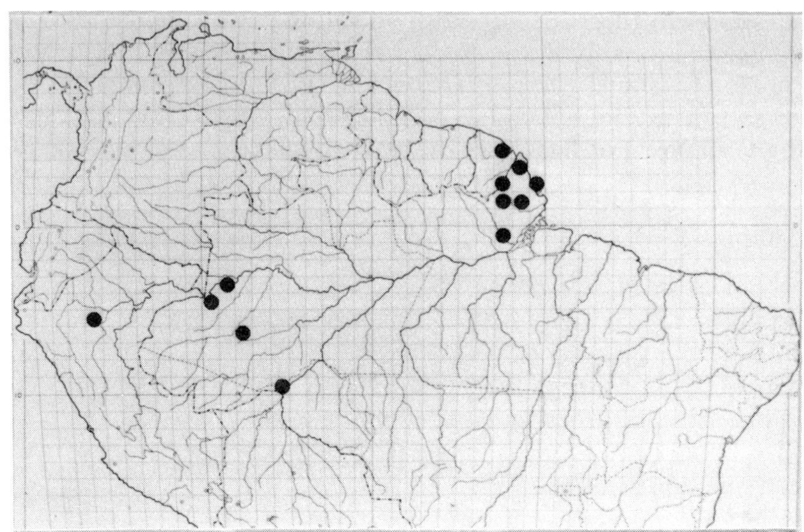

Fig. 5. The disjunct distribution of *Moriri oligantha* Pilg. (Melastomataceae). Since much collecting has been carried out in the area between the two populations, and since this disjunction is repeated in many other cases, it is probably a genuine disjunction.

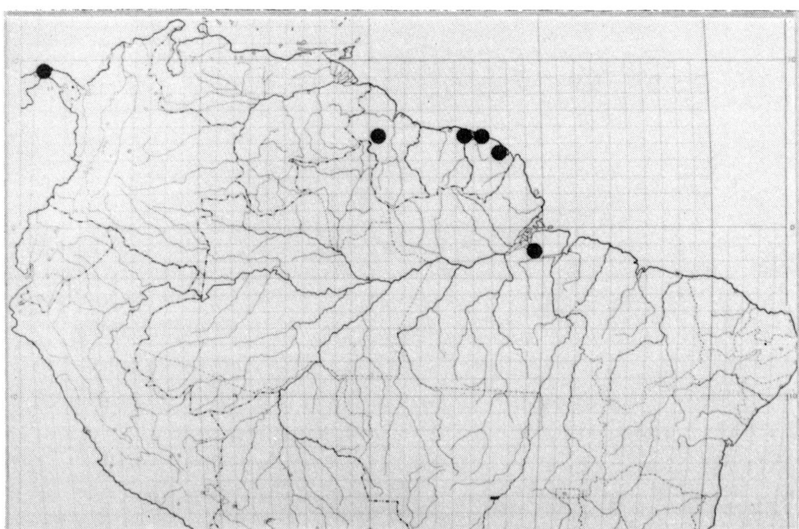

Fig. 6. The disjunct distribution of *Licania affinis* Fritsch (Chrysobalanaceae) by degree square presence. This disjunction has been found in several different groups (see text).

region between the two areas where *L. affinis* and other species still occur.

Figure 7 is a distribution map of *Couratari macrosperma* A. C. Smith which is one of the tallest emergent trees of the southern part of the Amazonian forest, where it is abundant. The map shows that it also

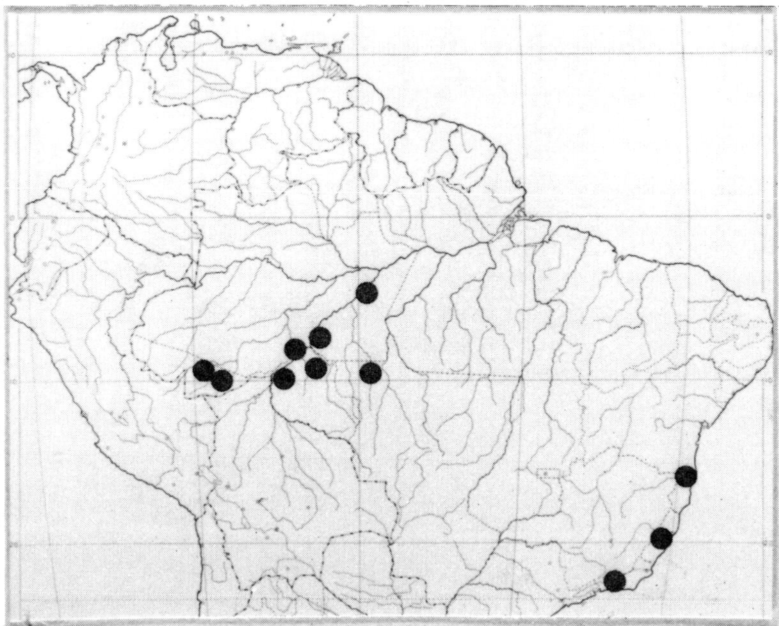

Fig. 7. The disjunct distribution of *Couratari macrosperma* A. C. Smith (Lecythidaceae). This distribution is also known in other species. Since *Couratari* has a winged fruit it is questionable whether this disjunction is caused by a former continuous distribution when there was a greater forest cover, or whether it is an example of long distance dispersal.

occurs in the coastal forests of Rio de Janeiro and Espírito Santo, a distribution pattern which is common in many other species. The two forest areas where this species grows are separated by the cerrado of the Planalto of Central Brazil. There is a strong link between the forests of eastern Brazil and those of Amazonia which has occurred through the gallery forests and through a greater forest connection in the past in times of more humid climate. However, *C. macrosperma* has a light winged seed and is a wind-dispersed species. The problem of interpretation is whether this distribution represents long distance dispersal or whether it is a disjunction relict from a more continuous forest which is known to have occurred. The long distance dispersal of *C. macrosperma* seems quite possible because it is a very tall tree with a winged seed.

Figure 8 is the distribution of *Hirtella glandulosa* Spreng. (Chrysobalanaceae) which is a savanna species abundant in the margins of the cerrado of Central Brazil. It also extends into the Guianas

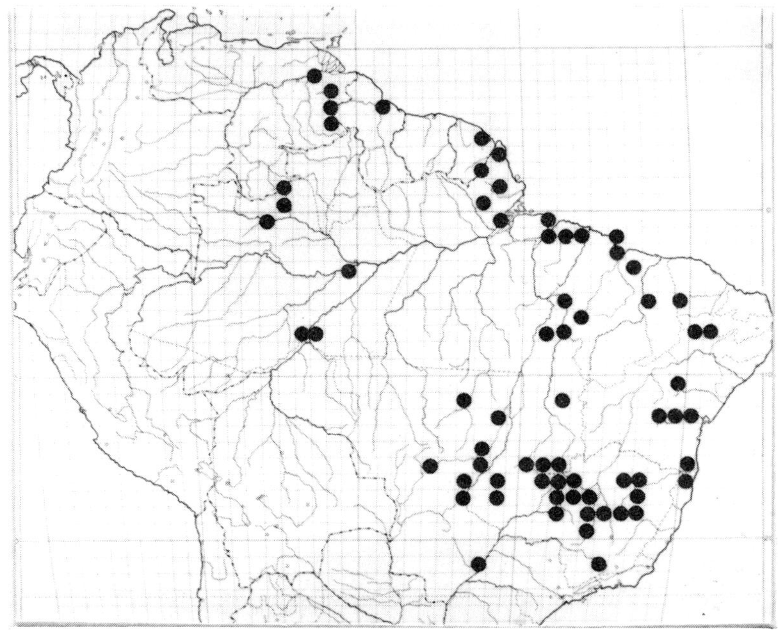

Fig. 8. Distribution of *Hirtella glandulosa* Spreng. (Chrysobalanaceae) by degree square presence. This common species of the cerrado of Central Brazil extends into the Guianas and Venezuela in savanna habitats. The distribution reflects well the distribution of major savanna areas in Amazonia.

and Amazonian savannas. The distribution map of the species represents quite well the availability of suitable open habitats for this species. I would not expect to fill in very many of the blank grid squares on the map by further collecting. *H. glandulosa*, like many other savanna species, has a rather small drupaceous fruit with a soft edible mesocarp which is eaten by birds. The question whether this species is a relict of more continuous open areas or the result of long distance dispersal is open. It is interesting to note the number of savanna and white sand campina species which have long distance dispersal capability, in marked contrast to the rain forest species, which generally have large heavy diaspores. The long distance dispersal capability of many savanna and campina species means that some extrapolations which have been made about their species distribution patterns as a relict of more continuous vegetation are not always true. Macedo and Prance (1978) have shown that 75·67% of the woody species studied in an Amazonian white sand campina had an obvious capacity for long distance dispersal. This enables species to distribute between the individual campinas which are small isolated islands of white sand within the rain forest area. Significant conclusions can only be drawn from the heavy fruited savanna and campina species.

Caryocar brasiliense Camb. is a characteristic cerrado species of Central Brazil. This species has a very large heavy drupaceous fruit which is unlikely to be dispersed over long distances. Recently I have collected this species well within the Amazonian rain forest area at Serra Cachimbo in Pará. This area of savanna and campina is related phytogeographically to both the cerrado and the Amazonian campinas. It is isolated from the main body of the cerrado to the south by tall forest. The disjunct population of the heavy fruited *Caryocar brasiliense* almost certainly is a relict from the time when the Cachimbo area was linked continuously to the cerrado, and not separated by a wide band of rain forest as it is today (arrowed in Fig. 15). Several other heavy fruited cerrado plants also occur in the Cachimbo campinas such as *Parinari obtusifolia* Hook.f. (Chrysobalanaceae).

Figure 16 shows the distribution of *Caryocar gracile* Wittm. a species which is confined to white sand campina forest (caatinga) of the upper Amazon. This type of forest only occurs in the rather disjunct areas where *C. gracile* is marked on the map in the upper Rio Negro and in two places near to the Amazon river. Thus *C. gracile* would not occur in the intervening area because of its specific habitat requirement. Distribution maps are often quite meaningless without discussion and adequate mapping of habitat occurrence.

These few examples given above and taken from three plant families serve to show how important it is to consider distribution, dispersal, ecology and history together since there have been many historic factors affecting the present day distribution patterns in the neotropics. One of the tendencies and dangers among biogeographers is the attempt to explain distribution patterns in light of only one of the important historic occurrences rather than a consideration of as many as possible. There is also too often an over emphasis on one theory, for example, vicariance. The tendency is to jump on the current bandwagon. For example, there has been an overemphasis on the effects of Pleistocene forest refugia in lowland America to the exclusion of habitat data. Refugia were certainly important, but, other factors were equally

                      G. T. Prance

important on distribution, evolution and species diversification. For example, tectonic history, which is covered by Dr Raven, the uplift of the Andes (in the Miocene), and the union of the north and south American sub-continents with their subsequent biological influence on one another (only 5·7 million years ago). The present day distribution

Fig. 9. The distribution of *Parinari excelsa* Sabine in South America.

patterns especially in large genera, are a complicated reticulum often influenced by several of the above factors. Even the Pleistocene and Holocene climate changes have occurred several times and in differing degrees each time making the refuge situation very complicated indeed.

The Genus *Parinari*

I will review briefly the distribution of all the species of two small genera in order to discuss some of the problems outlined above: *Parinari* (Chrysobalanaceae) and *Caryocar* (Caryocaraceae). These genera are typical of many other predominantly lowland genera which I have studied in the Chrysobalanaceae and Lecythidaceae, and they are not too large to treat in the limited space available.

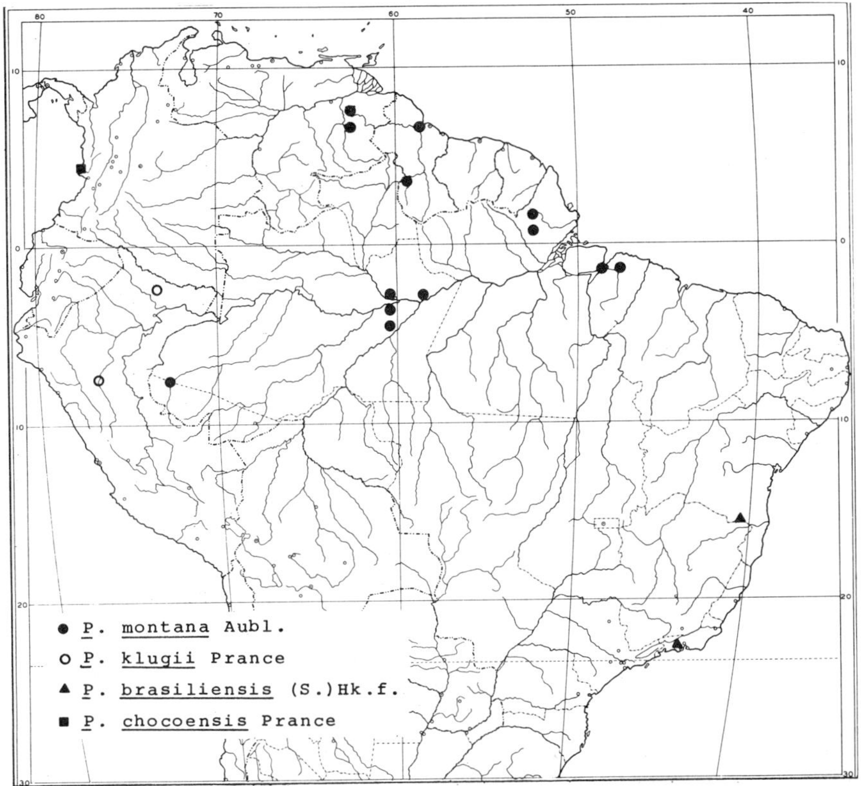

Fig. 10. Distribution of species of *Parinari*.

Parinari, a pantropical genus, has 18 species distributed from northern Colombia to eastern Central Brazil around Rio de Janeiro. Table 2 gives a summary of the geographic distribution and habitat of each species, and Figs 9–14 give an idea of the overlap of species ranges. *Parinari* is an interesting genus because the species are very closely related and little differentiated on a worldwide basis. The species now recognized have very small morphological differences with a few exceptions on each continent. Many of the species occupy habitats very distinct from each other. The distribution of the genus would indicate

Table 2. Phytogeographic and habitat breakdown of *Parinari* (Chrysobalanaceae). Top = distribution and habitat of species; Middle = breakdown by habitat; Bottom = breakdown by geographic regions.

Species	Geographic distribution	Habitat
1. *P. campestris*	Venezuela, Guianas, Brazil North of Amazon	River and savanna margins
2. *P. montana*	Venezuela, Guianas, Central and East Amazon, Brazil	Forest on terra firme
3. *P. rodolphii*	Venezuela, Guianas, Brazil–Amapá	Forest on terra firme, várzea
4. *P. excelsa*	Colombia, Venezuela, Guianas, Amazonia, East Brazil	Forest on terra firme
5. *P. sprucei*	Western Amazonia	Várzea and igapó
6. *P. occidentalis*	Bolivia, Brazil–Acre	Forest on terra firme
7. *P. pachyphylla*	Colombia, North Venezuela	Upland dry forest
8. *P. brasiliensis*	Brazil–Rio de Janeiro	Forest
9. *P. klugii*	Peru–San Martín, Loreto	Forest
10. *P. maguirei*	Guyana local endemic	Savanna
11. *P. littoralis*	Bahia, endemic	Coastal wet forest
12. *P. parvifolia*	Guyana, East Amazonia	Forest on terra firme
13. *P. cardiophylla*	Central Amazonian Brazil	Forest on terra firme
14. *P. parilis*	Peru–Loreto	Várzea forest
15. *P. chocoensis*	Colombia–Chocó	River banks
16. *P. obtusifolia*	Central Brazil	Cerrado
17. *P. romeroi*	Pacific Ecuador, Colombia	Forest
18. *P. sp. nov.*	Bahia	Coastal wet forest

Habitats	Total no. species	Species (from top section)	
		Local	Widespread
Forest on terra firme Guianas and Amazon	6	6, 9, 12, 13	2,4
Open river and savanna margins	1		1
Terra firme and Várzea	1		3
Várzea and igapó	2	5, 4	
Upland drier forest	1	7	
South-east coastal forest Brazil	3	8, 11, 18	
Savanna (Guiana)	1	10	
Pacific Coast forest	2	15, 17	
Cerrado	1	16	

Geographic region	Total no. species	Species (from top section)
Venezuela, Guiana, North-East Amazon	3	1, 3, 12
Venezuela, Guiana, Central to East Amazonia	1	2
Widespread Colombia–South Brazil	1	4
Central Amazonia	1	13
Western Amazonia	4	5, 6, 9, 14
North Venezuela and Colombia	1	7
South-east Brazillian coast	3	8, 11, 18
Guyana endemic	1	10
West of Andes Colombia, Ecuador	2	15, 17
Central Brazil	1	16

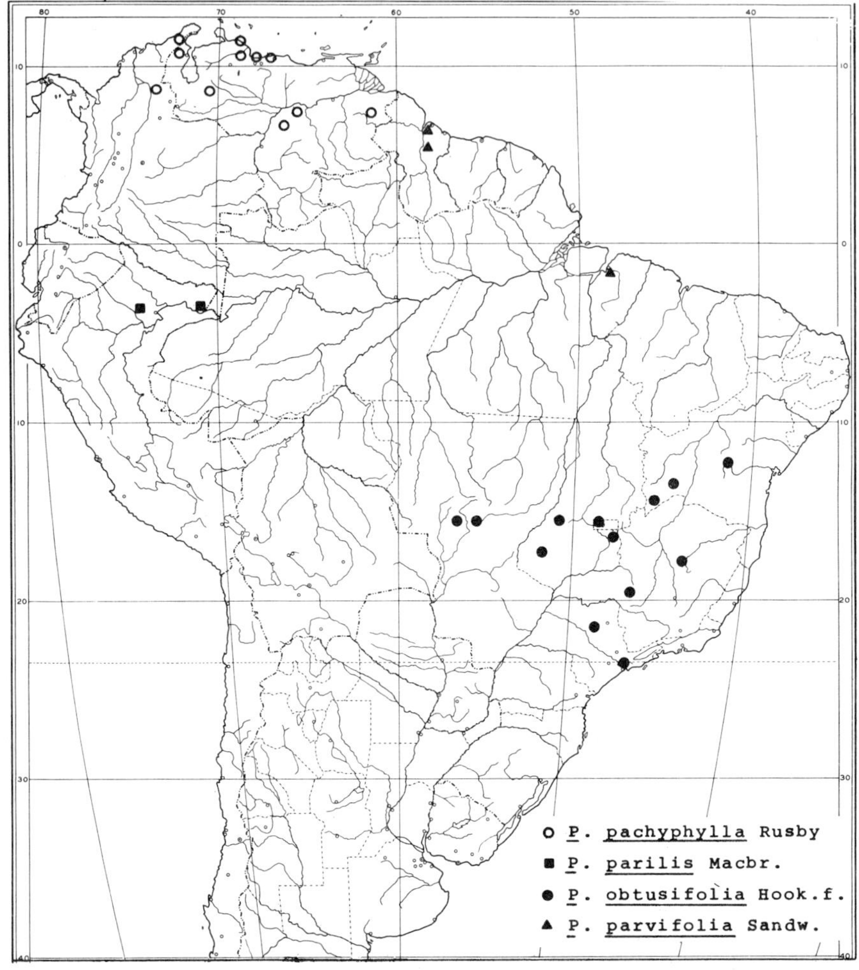

Fig. 11. Distribution of species of *Parinari*.

that it is an old one, dating back to the time when there was greater connection or at least proximity between Africa, America and Asia. Table 2 shows that the 18 neotropical species of *Parinari* occur in nine different habitats, with the greatest number of species (six) in the rain forest on non-flooded ground of Amazonia and the Guianas. In this habitat and region there is considerable sympatry between the species, as is common in most medium to large sized genera which occur in Amazonia. In *Parinari* there is one widespread morphologically rather variable species (or ochlospecies), *P. excelsa*. The occurence of one or

Fig. 12. Distribution of species of *Parinari*.

more such species is a common feature of most large genera which I have studied. *P. excelsa* also occurs in Africa where it is widespread and variable.

Parinari has adapted well to the various habitats available in the region. In addition to the several species which are large trees of the forest, there is a low suffrutex adapted to the cerrado (*P. obtusifolia* Hook.f.), a low savanna species endemic to the Kaietur savannas of Guyana (*P. maguirei* Prance), a species of the upland forests of the Venezuelan–Colombian border (*P. pachyphyllum* Rusby) and a species of the open lighter area of savanna and river margins (*P. campestris* Aubl.).

There are also species of *Parinari* in the areas of rain forest which are today isolated from that of Amazonia. *Parinari* has two species, in the forest of the Pacific coast of Colombia and Ecuador, west of the Andes, and three species in the wet coastal forest of eastern Brazil in Bahia to

Fig. 13. Distribution of *Parinari campestris* Aubl.

Rio de Janeiro. These species are isolated today, but, the one wide-spread species, *P. excelsa*, also extends its range into the eastern coastal forest of Brazil. The isolation of both the Pacific coastal forests and those of eastern Brazil is most frequently at the species level. These regions share most genera in common with Amazonia, but have many different endemic species. The comparatively recent uplift of the Andes and the intervening drier climate of north-eastern Brazil has given time for specific rather than generic differentation in woody groups. This is true in the other genera of Chrysobalanaceae and Lecythidaceae which I have studied, for example *Hirtella* and *Licania* (Chrysobalanaceae) and *Gustavia* and *Eschweilera* (Lecythidaceae).

The largely woody lowland Amazonian genera tend to have only a few species in the cerrado of central Brazil, but they do usually have at least a few such as *Parinari obtusifolia* and *Caryocar brasiliensis*. This is in

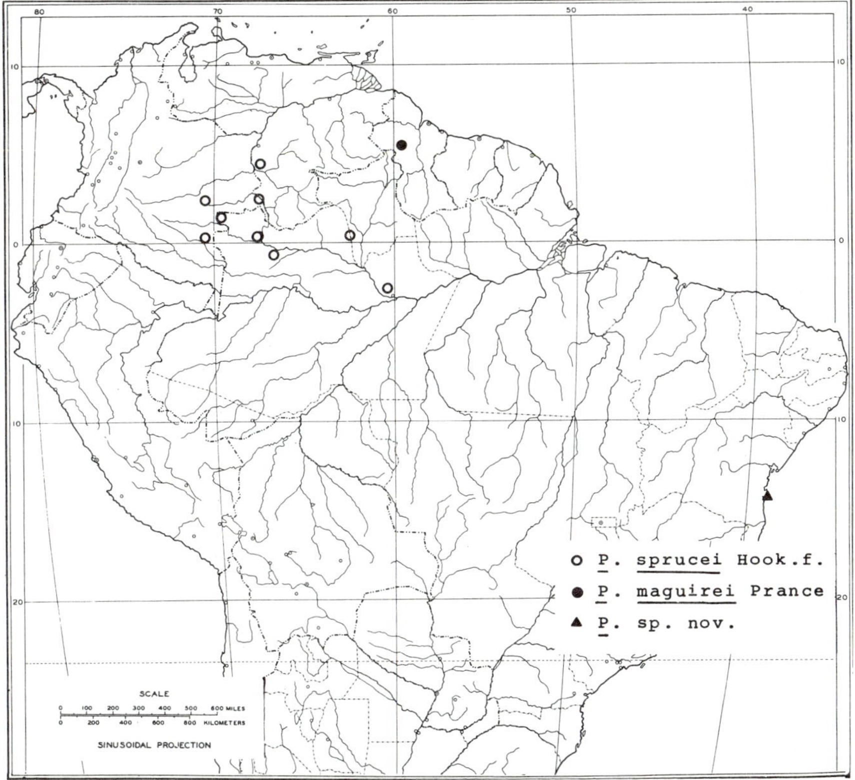

Fig. 14. Distribution of species of *Parinari*.

marked contrast to various shrubby genera whose centre of diversification is in the cerrado, for example, many genera of Leguminosae as *Cassia*, *Mimosa* and *Bauhinia*.

The Genus *Caryocar*

Caryocar has an equally wide range as *Parinari* and extends into Central America as far North as Costa Rica (Figs 15–18). It is absent from the

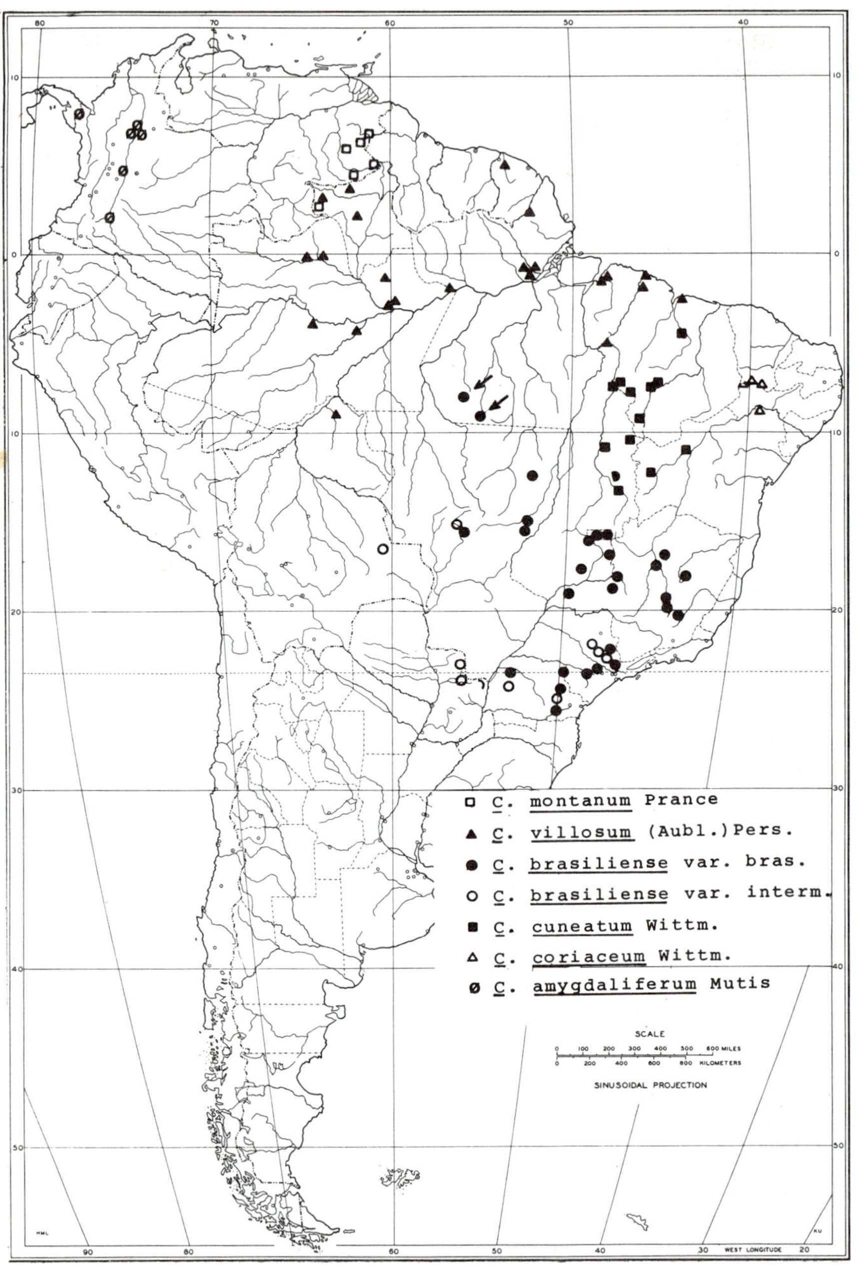

Fig. 15. Distribution of species of *Caryocar*. The arrows show *C. brasiliense*, a cerrado species, growing well within the Amazon Forest at Serra do Cachimbo.

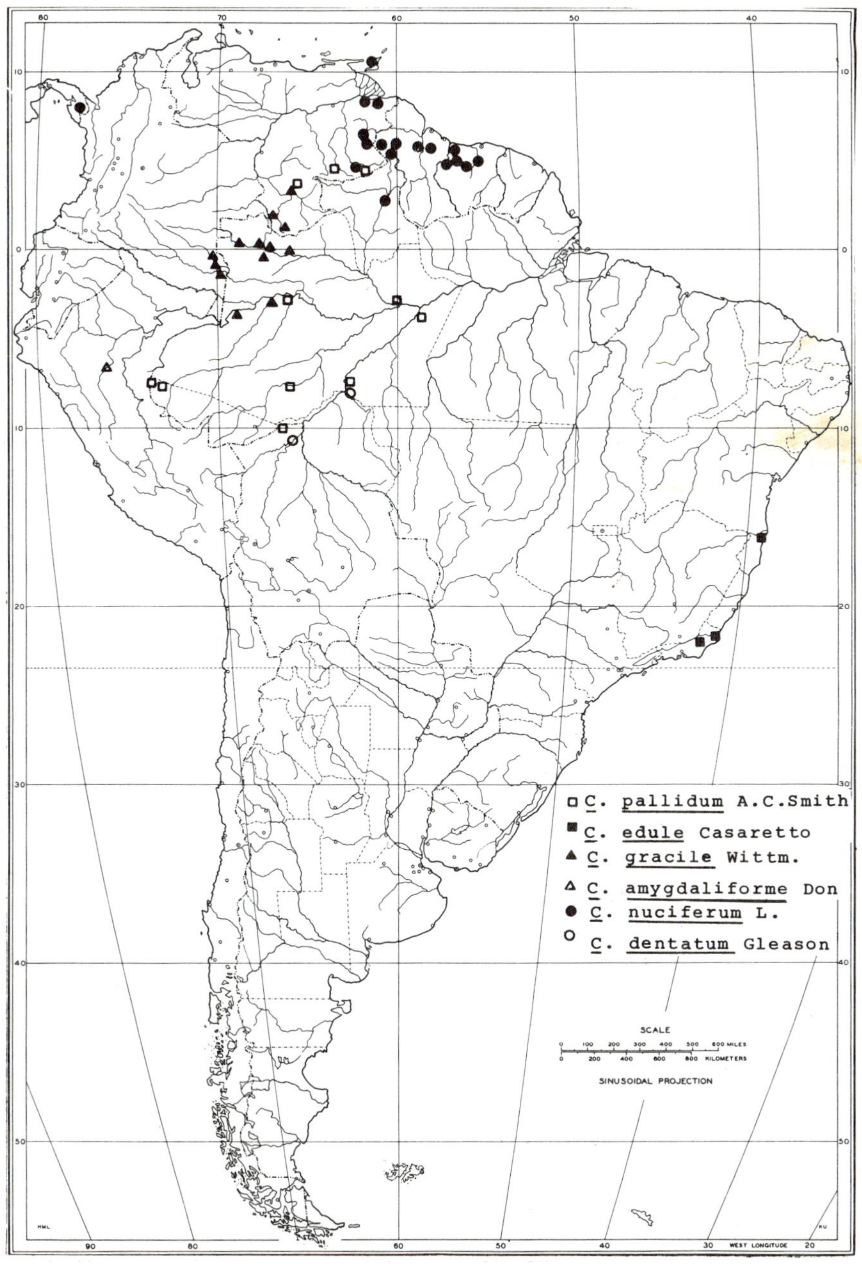

Fig. 16. Distribution of species of *Caryocar*, note the Panama/Guiana disjunct distribution of *C. nuciferum* L. which is similar to that of *Licania affinis* (Fig. 6).

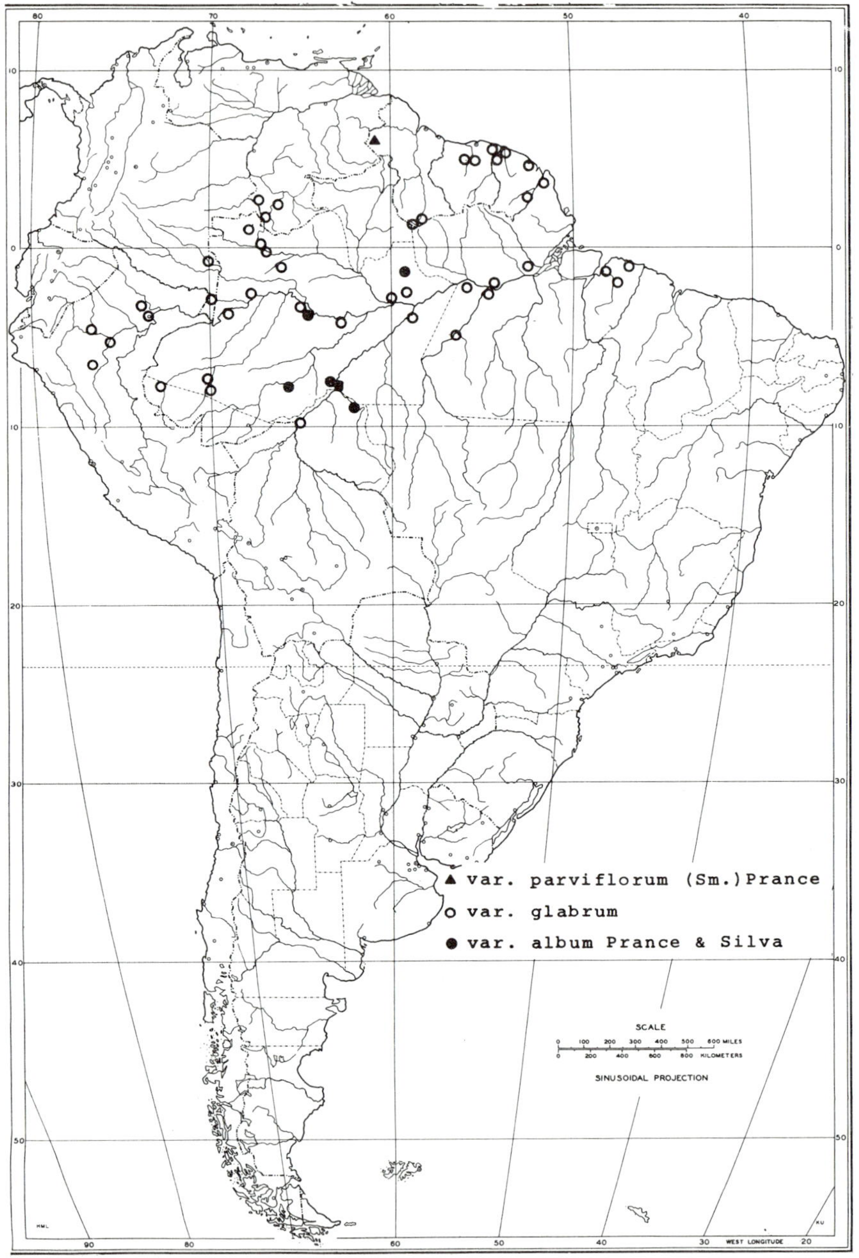

Fig. 17. Distribution of *Caryocar glabrum* (Aubl.) Pers.

Pacific coastal forest of South America. *Caryocar* is an exclusively neo-tropical genus and family, and therefore, probably of much more recent origin than the pantropical *Parinari*. Table 3 shows that the 15 species of *Caryocar* occupy seven different habitats. Like *Parinari*, the greater number of species (eight) occur in forest on non-flooded ground of Amazonia and the Guianas where there is considerable sympatry between species. There is also a widespread ochlospecies, *C. glabrum* (Aubl.) Pers. with a slightly more restricted distribution than *Parinari excelsa*.

Caryocar has also adapted well to the different habitats with *C. brasiliensis* a suffrutex of the cerrado, *C. gracile* Wittm. in the white sand campina forest of upper Amazonia, *C. montana* Prance in montane forest generally over 1000 m altitude, and *C. coriaceum* Wittm. and *C. cuneatum* Wittm. occuring in the much drier deciduous forest of north-eastern Brazil. There is also one species, *C. edule* Casaretto in the Atlantic coastal forest of Brazil, but none in the Colombian Pacific*. There is one species endemic to the Magdalena valley of Colombia (*C. amygdaliferum* Mutis) and one in Central America (*C. costaricense* Donnell Smith).

The general tendencies mentioned above for *Parinari* and *Caryocar* occur in all other woody Amazonian genera of over ten species which I have studied. For example, similar patterns are found in *Licania* with 168 species and *Eschweilera* with over 100 species. In these larger genera there is much more sympatry especially in the rain forest areas of the Pacific and Atlantic coasts and in Amazonia. What has enabled a large number of sympatric species in these genera to occur together?

The long evolutionary history of *Parinari* in the neotropics began with the movement of the continents when the African, Asian and American elements of the genus became isolated from one another. The minimal amount of morphological differentiation (floral, fruit and vegetative) since then is striking in the case of *Parinari*. Another significant event for the evolution of *Parinari* was millions of years later, the uplift of the Andes. This isolated populations on either side of the mountains lead-ing to the differentiation of *P. chocoensis* and *P. romeroi* in the Pacific coastal forest. There were considerable climate changes, both through the uplift of the Andes and through the worldwide fluctuations which led to the Pleistocene glaciations. The uplift of the Andes led to the development of a drier area in Northern Colombia and Venezuela and consequent isolation of the forests of Panama from those of the Guianas. Many species in this area were divided into two populations accounting for the disjunction of *Licania affinis* (Fig. 6), and *Caryocar nuciferum* (Fig. 16). The complete isolation is a more recent event than the uplift of the Andes judging by the number of species which have not

* While this paper was in press, *C. nuciferum* has been collected in Pacific Colombia (Chocó) by A. Gentry.

Table 3. Phytogeographic and habitat breakdown of *Caryocar* (Caryocaraceae): Top = distribution and habitat of each species; Middle = breakdown by habitat; Bottom = breakdown by geographic regions.

Species	Geographic distribution	Habitat
1. *C. brasiliensis*	Central Brazil, Paraguay	Cerrado
2. *C. coriaceum*	North-East Brazil	Deciduous woodland
3. *C. villosum*	Fr. Guiana, Central, East and South Amazonia	Forest on terra firme
4. *C. nuciferum*	Guiana/Panama disjunct	Forest on terra firme
5. *C. cuneatum*	North-East Brazil	Deciduous woodland
6. *C. gracile*	Colombia, Venezuela, Brazil, North-West Amazonia	White sand campina forest
7. *C. amygdaliforme*	Amazonian Peru	Forest
8. *C. glabrum*		
v. glabrum	Guianas, West to East Amazonia	Forest on terra firme
v. parviflorum	Guianas, Central Amazonia	Forest on terra firme
v. album	Guiana local endemic	Forest on terra firme
9. *C. montanum*	Venezuela–Bolívar, Brazil–Roraima, local	Upland forest, ± 1000 m altitude
10. *C. pallidum*	South Venezuela, Brazil–Amazonia	Forest on terra firme
11. *C. edule*	Central East Brazil coast	Forest beside stream
12. *C. costaricense*	Costa Rica, Panama	Forest
13. *C. dentatum*	Bolivia, Brazil–Rondônia	Forest on terra firme
14. *C. amygdaliferum*	Panama, Colombia–Magdalena Valley	Forest
15. *C. microcarpum*	Guianas, West to East Amazonia	Várzea, igapó

Habitats	Total no. species	Species (from top section)
Forest on terra firma (Guiana, Amazon)	8	3, 4, 7, 8, 10, 12, 13, 14
Várzea, Igapó	1	15
Montane forest Guayana	1	9
North-East Brazil deciduous forest	2	2, 5
East coast forest, Brazil	1	11
White sand campina forest	1	6
Cerrado	1	1

Geographic region	Total no. species	Species (from top section)
Venezuela, Guiana, Amazonia	4	3, 8, 10, 15
Guiana/Panama	1	4
Panama, North Colombia	1	14
Guayana Highlands	1	9
Western Amazonia	2	6, 7
Central Amazonia	1	12
South-East Amazonia	1	13
Central South-East Brazil	1	11
North-East Brazil	2	2, 5
Central Brazil, Paraguay	1	1

                    G. T. Prance

differentiated in marked contrast to the Pacific coastal forests compared to Amazonia.

The change to a much drier climate in north-eastern Brazil also isolated the forests of Atlantic coastal Brazil from those of Amazonia. This also led to the differentiation of species. However, there are still

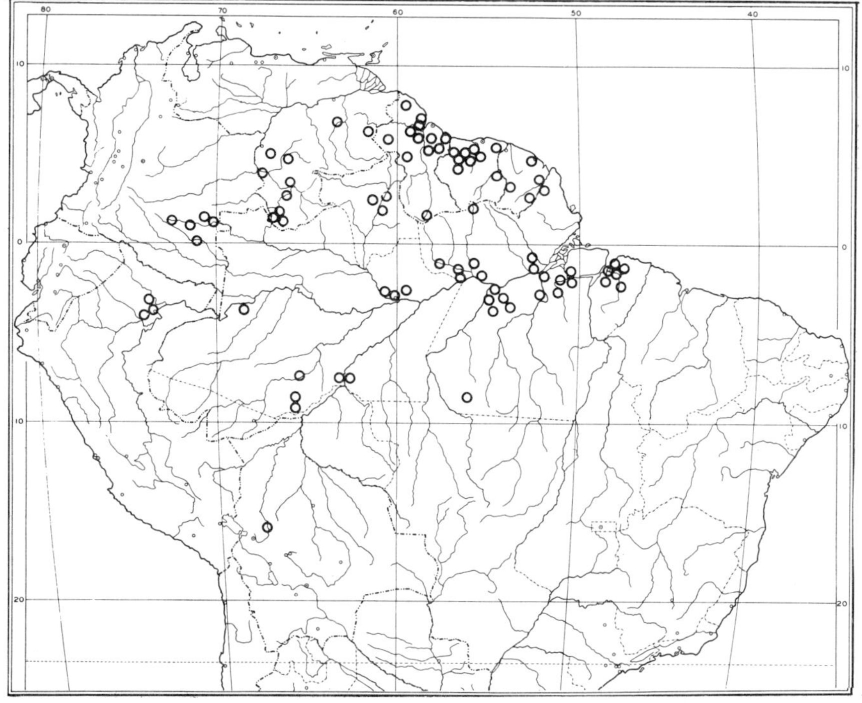

Fig. 18. Distribution of *Caryocar microcarpum* Ducke.

many species distributed across the barrier which have not differentiated such as *Parinari excelsa*, *Hirtella bicornis* Mart. and Zucc. (Chrysobalanaceae) and the 130 other species listed in Lima (1953) as common to Amazonia and the forests of eastern Brazil. Other climate changes, such as those in the Pleistocene and later, are less obvious today since the vegetation has returned to a continuous rain forest in the areas affected. The Amazon forest was much impoverished and reduced in area by periods of drier climate and was even replaced by savanna in certain areas. For further details see Haffer (1969), Brown (1977), Prance (1974c).

The isolation of rain forest species into several disjunct allopatric populations in the Pleistocene and later has certainly contributed greatly to the species diversification of the lowland forest, but it is not the only explanation for the large number of sympatric species in genera of the Amazon rain forest. Other factors are equally important, such as predator pressure (Janzen, 1970), the variety of niches

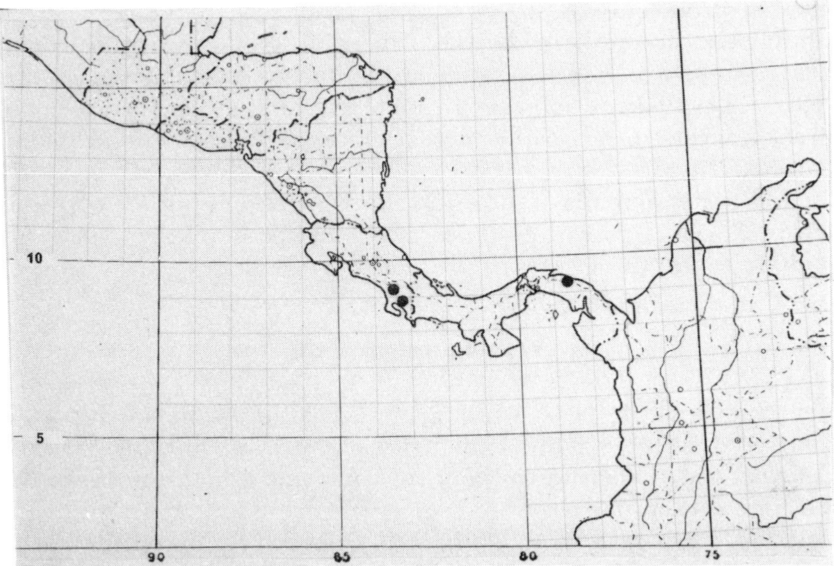

Fig. 19. Distribution of *Caryocar costaricense* Donnell Smith.

(Richards, 1969; Ashton, 1969), phenological isolation, adaptation in coevolution with specific pollinators etc. It was shown above that species of both *Parinari* and *Caryocar* have adapted to various different niches such as savanna, cerrado, white sand, inundated forest. In this case niche adaptation has certainly been a major part of speciation. However, it cannot alone explain the large number of rain forest species which grow together in such a niche. Here the refuge theory is a much more likely explanation.

In conclusion, much data can now be inferred from our present knowledge of plant distributions in the lowland neotropics, but the pitfalls of a patchy data base must be recognized and avoided. For the interpretation of maps we must consider the level of collecting in the area, the habitat occupied by the species under study and the distribution of the region. This should also stimulate us to more effective and intensive collection for the next few years while we still have forest to study. We must not forget that unless we act quickly, and also encourage natives of the tropical countries to act quickly, that we will never be able to fill in the gaps in our knowledge of plant distribution patterns. These gaps must be filled for us to really understand the evolution and dynamics of the tropical rain forest.

Because of the importance of distribution maps, the Organisation for Flora Neotropica plans to give some emphasis to them. They are planning the production of a series of distribution maps for the neotropical area similar to the Distributiones Plantarum Africanarum (Bamps, 1969). The distribution maps of Chrysobalanceae (but not of Caryocaraceae) in this paper have been prepared in a similar manner to those of Africa, using the grid square method. These will be used to initiate a series of maps of the grid square occurence of the neotropical species. The availability of such a series of maps should be a great help to phytogeographic studies of the neotropical flora.

Acknowledgements

This paper was prepared while I held a visiting professorship at the Botanical Institute of the University of Aarhus which is gratefully acknowledged. Recent field work in Amazonia which produced data used here was supported by National Science Foundation grant INT 75–19282. I am grateful to William C. Steward for preparing most of the maps.

References

Ashton, P. S. (1969). Speciation among tropical forest trees: Some deductions in the light of recent evidence. *Biol. J. Linn. Soc.* **1**, 155–196.

Bamps, P. (1969). "Distributiones Plantarum Africanarum 1. Introduction" by F. Dermaret, later volumes by different authors. Jardin Botanique, Brussels.

Brown, K. (1977). Geographical patterns of evolution in neotropical forest Lepidoptera (Nymphalidae: Ithomiinae and Nymphalinae—Heliconiini). *In* "Biogéographie et Evolution en Amérique Tropicale" (H. Descimon, ed.) Publication, No. 9, p. 118–160. Laboratoire de l'Ecole Normale Supérieure, Paris.

Gentry, A. (1975). Additional Panamanian Myristicaceae. *Ann. Mo. Bot. Gard.* **62**, 474–479.

Haffer, J. (1969). Speciation in Amazonian forest birds. *Science* **165**, 131–137.

Irion, G. and M. L. Absy (1978). Paleoclimate in Central Amazonia as reflected in Quaternary sediments. Abstracts of the Tenth International Congress on Sedimentology, Jerusalem.

Janzen, D. H. (1970). Herbivores and the number of tree species in tropical forests. *Am. Nat.* **104**, 501–528.

Lima, D. de A. (1953). Notas sôbre a dispersão de algumas espécies vegetais do Brasil. *An. Soc. Biol. Pernambuco* **11**, 25–49.

Lundell, C. L. (1974). Studies of American plants VI: Rosaceae. *Wrightia* **5**, 39–40.

Macedo, M. and G. T. Prance (1978). Notes on the vegetation of Amazonia II. The dispersal of plants in Amazonian white sand campinas: The campinas as functional islands. *Brittonia* **30**, 203–215.

Prance, G. T. (1972). Chrysobalanaceae. *Flora Neotrop.* **9**, 1–410.

Prance, G. T. (1973). New and interesting Chrysobalanaceae from Amazonia. *Acta Amazon.* **2**(1), 7–16.

Prance, G. T. (1974a). Supplementary studies of American Chryso-Balanacea. *Acta Amazon.* **4**(1), 17–23.

Prance, G. T. (1974b). A new Peruvian species of chiropterophilous *Couepia* (Chrysobalanaceae). *Brittonia* **26**, 302–304.

Prance, G. T. (1974c). Phytogeographic support for the theory of Pleistocene forest refuges in the Amazon basin, based on evidence from distribution patterns in Caryocaraceae, Chrysobalanaceae, Dichapetalaceae and Lecythidaceae. *Acta Amazon.* **3**(3), 5–28.

Prance, G. T. (1976a). Additions to neotropical Chrysobalanaceae. *Brittonia* **28**, 209–230.

Prance, G. T. (1976b). Caryocaraceae. Flora of Panama. *Ann. Mo. bot. Gard.* **63**, 541–546.

Prance, G. T. (1977). Two new species for the Flora of Panama. *Brittonia* **29**, 154–158.

Prance, G. T. (1978a). Insect galls and human ornamentation: The ethnobotanical significance of a new species of *Licania* from Amazonas, Peru. *Biotropica* **10**, 81–86.

Prance, G. T. (1978b). Floristic inventory of the tropics: Where do we stand? *Ann. Mo. Bot. Gard.* **64**, 659–684.

Prance, G. T. (1979). New and interesting species of Chrysobalanaceae. *Acta Amazon.* **9**(1).

Richards, P. W. (1969). Speciation in the tropical rain forest and the concept of niche. *Biol. J. Linn. Soc.* **1**, 149–153.

Tropical Floristic Botany—Concepts and Status—with Special Attention to Tropical Islands

F. R. FOSBERG

Smithsonian Institution, Washington, USA

Introduction

By botany, for the purposes of this paper, I will, arbitrarily, say I mean the organized study of plants as organisms, as contrasted with, on the one hand, folk-knowledge especially of medicinal and agricultural plants, on the other, cellular and molecular biology. It may be argued, and freely admitted that this is a limited, and perhaps artificial concept of botany, but it is one that can be discussed in the context of my title.

By tropical, again for my present purpose, I will arbitrarily say we mean the regions of the earth where there is not even a limited annual season of dormancy due to continued low temperatures. This definition becomes fuzzy on the boundaries of the tropics, in areas of strongly oceanic climates such as the coasts of western Europe and western America, in very high tropical mountains, and in certain extra-tropical islands such as Bermuda, the Bonins, some south temperate Atlantic islands, and northern New Zealand. Again, this definition serves my purposes and tells you what I mean. I could, and would, argue for other definitions in other contexts.

Floristic botany is that branch of taxonomic botany that deals with the plants of limited or defined areas or habitats, as contrasted with systematic or monographic botany, in its narrow sense, which deals, or should deal, with entire groups of plants, at whatever systematic rank or level. It will be seen that this comprehends more than mere floristic lists. It forms the basis of plant geography and provides many of the data for the study of plant evolution. This idea will appear more clearly as the paper continues.

It was probably inevitable that taxonomic botany arose and had its major development in the temperate, even the cold temperate, areas of the earth (an idea I must credit to my friend William T. Stearn, renowned Linnean scholar and botanical historian as well as taxonomist); it has flowered as a part of the western European cultural pattern. One must, of course, go even farther, and point out that this is true of the whole of modern science, as contrasted with philosophy. Whether this is due, as has been argued by some, to the alternation of seasons when natural phenomena can be readily and comfortably observed in nature with seasons when it is more suitable to sit in warm quarters and think and write about these observations, or whether it may be due to an environment where knowledge was essential for survival, or whether it may even be merely a historical accident, is perhaps a matter for the philosophers to ponder. Suffice it to say that systematic botany did originate and develop in the north temperate zone, and that this fact has had important consequences.

The floras of temperate regions are meagre ones compared with those of the tropics. I have often made the perhaps inexact generality that though only 10% of the kinds of plants grow in the temperate zones, 90% of the botanical work has been done there. The relative simplicity of temperate floras has made it possible for botanists to perceive patterns, to develop schemes for the organization and identification of the kinds of plants. The incredible profusion and diversity of tropical plant life would doubtless have defeated the very genesis of such attempts. Linnaeus knew, in 1753, about 5900 species of plants, and was able to create his "sexual system" which served botanists for the first few decades of "systematic botany" until more "natural" systems emerged. These came, very likely, as a result of the earlier artifical organization of the taxa by Linnaeus. The association of like kinds reduced the chaos and made it possible to perceive alternative arrangements.

Viewed from a tropical perspective, the resulting temperate zone picture of the plant world has certain peculiarities. The herbaceous habit is often thought of as the norm, with a minority of tree and shrub species. These are divided into deciduous (broad-leafed) versus evergreen (needle-leafed). Seasonality is obvious and basic. Angiosperms were seen as evolving from Ranunculaceae and Alismataceae. Many families were thought of as predominantly herbaceous, even many of the dicotyledonous ones. Techniques were influenced—specimens were collected in vascula and pressed in books. Herbaria were assembled—with an incalculably great influence on the development of botany. They unfortunately acquired a "stamp-collecting" flavour, which has not yet been entirely outgrown. Botany acquired a reputation as a suitable amusement for ladies of leisure. Even professional botany has not entirely outgrown its historically temperate characteristics. Plant physiology is still essentially the physiology of herbaceous plants. Plant anatomy as it is taught is essentially the anatomy of herbs. Plant genetics, admittedly as much from practical as historical reasons, is largely the genetics of herbaceous plants, mostly annuals (*Zea, Oenothera, Madia, Galeopsis, Clarkia*, cereals, etc.)

The impressions of even a botanically trained person, on first visiting the tropics are those of overwhelmed astonishment. The myriads of unfamiliar kinds of plants, whole new families, woody members of "herbaceous" families: Rubiaceae, Boraginaceae, Compositae, Violaceae, Nyctaginaceae, etc., numerous vines—herbaceous twiners and woody lianas, tree ferns, profusion of epiphytes, all build a picture of strangeness and hopeless complexity.

The reality gradually becomes apparent. Most dicotyledonous families are predominantly woody, as are even some monocots: palms, *Dracaena*, bamboos, and many others. The primitive families are seen to be Annonaceae, Degeneriaceae, Theaceae and Winteraceae or Magnoliaceae sensu lato, all woody. The number of kinds of plants seems endless and it takes a great mental effort to fit what one sees into a framework developed to accommodate the temperate plant world.

One's botany needs rethinking, concepts must grow to accommodate what one sees. In his efforts to bring this about he learns that not too many aeons ago tropical conditions were more widespread. Seasonality becomes as much a matter of wetness and dryness as of warmth and cold. Ecological complexity becomes orders of magnitude greater. Morphology deals with woody and fleshy-herbaceous plants, vines, and epiphytes as much as or more than with the traditional herbaceous annuals. Plant physiology becomes largely tree physiology. The taxonomic system can no longer remain so neat and orderly, the numbers and diversity of plants are too great, specimens are less satisfactory, and knowledge is too incomplete. Our start in the tropics was late, and much information has been lost or has become much harder to get at, due to the widespread destruction of tropical vegetation. Tropical taxonomy is still almost entirely in the alpha stage, except in a few spots where there are long-established institutes and laboratories.

It also becomes evident that the angiosperms originated under tropical conditions from woody ancestors. Theories of where and how this took place are proliferating. All or most of them probably contain some of the truth, but fossil evidence in the tropics to support them is comparatively scarce. This leaves the field open for speculation.

Continental drift (plate tectonics) complicates matters. Godwanaland and the *Glossopteris* flora stimulate much theorizing. The persistence of primitive features, to be expected in the tropics if they evolved under tropical conditions, provides some phenomena from which to develop theories. A few people with long tropical experience and great insight, but with incomplete information, make brilliant deductions, which displease others possessing different parts of the available data. The tropical botanist, even when he goes back to the temperate zone, can never be the same again.

One further generality must be stressed. Although the multistratal, tall rain forest of the tropical lowlands must be regarded as the central core of tropical vegetation, and will continue to serve as the model around which concepts of tropical vegetation are formed, it is by no means the only, or even the predominant, vegetation of the tropics. Tropical landscapes and their vegetation change from this ancient climax along several gradients. These are, of course, mental contructs to aid our understanding, but in places they are also very real and visible.

One of the most obvious is that of available moisture. Ecosystems exist in the tropics along the entire range from a constant large excess of moisture to deserts so extreme that vegetation usually seems completely lacking. This moisture gradient is inextricably intertwined with that of seasonality in the occurrence of the moisture and in the nature of this occurrence—rainfall versus condensation—dew and fog drip. Another obvious gradient is that of slope, and associated with it, drainage. The effect of the moisture gradient is profoundly modified by slope, exposure, and drainage.

Related to all of these is a chemical gradient from acid soils to highly alkaline and saline—resulting from excess or deficiency of water—leaching and loss of basic ions versus evaporation and accumulation of these ions. The effects of this difference on vegetation is profound. Variations in soil texture and chemistry that influence vegetation are too numerous and complex to outline here.

Finally there is the gradient of altitude above sea-level. Vegetational plant geography had its origins in Humboldt's observations in tropical mountains. Elevation, functioning chiefly but not entirely through its effects on the other gradients mentioned above, adds immeasurably to the complexity of the tropical vegetation pattern and to the diversity of tropical ecosystems.

In my context here, the important aspect of the resulting mosaic is that each of these ecosystems, both at the concrete and at the abstract level, has its flora. Lest the above convey an idea that this is a static system, let me point out that climatic and physiographic change, on a long time scale, volcanism, landslides, fire, and hurricanes, on a short, catastrophic scale, and the activities of animals, most of all, man, introduce a continuing and almost unimaginable dynamism to the cinema.

Tropical Floristic Botany

My topic is, specifically, tropical floristic botany. Floristics has almost nowhere attained the status of anything like an exact science. Even in well-known areas with small floras, like the British Isles, the Netherlands, and Scandinavia, uncertainties are introduced by such plant groups as *Hieracium*, *Taraxacum*, and *Rubus*, and by different genus and species concepts. How much more must this be true in the tropics, with their enormous floras, inadequately known.

Among the conditions that determine the inadequacy of our knowledge and understanding of tropical plants are the size of the floras, difficulty and discomfort of exploration, problems of preservation of specimens, health problems, poverty of the people, impermanence of institutions, and the destruction of vast areas of vegetation before many of the plants that make it up have even been discovered. The descriptions of most tropical plant species have been written by botanists who have had little or no field experience with them, and who have had to work with inadequate specimens. Most of the collecting and study of tropical floras has been by botanists from temperate countries, who are at best only visitors or temporary residents in the tropics. Most of the large and important collections are in temperate institutions. This last is actually fortunate, as the problems of maintaining and protecting collections in warm, and especially warm and moist, regions are formidable (Fosberg and Sachet, 1965).

Results of these circumstances are several, and of importance to us here. For much of the tropics we have only the first generation of floras, often only annotated lists, many of them woefully out of date. The second generation are mostly struggling to be born, or to continue. Instead of building on the occasional outstanding first generation floras, much past local floristic work has at best amounted to compiling from these works, and often merely talking about writing floras. Outstanding exceptions to this have in the past been few, but this is being rapidly changed. Many modern floras of tropical areas are being written, such as the Flora Malesiana, the Kew family of African floras, the Flora of Madagascar, the Flora of Thailand, the Flora of Gabon, the

Flora of Nueva Galicia, the Flora of Fiji, the Flora of New Caledonia, the Flora of Taiwan, the Flora of Jamaica, the Flora of Panama, the Flora of the Lesser Antilles, the Flora Hawaiiensis, the Flora of Guatemala, the Flora of the Marquesas, the Flora of Micronesia. the Flora of Ceylon, the Aldabra Flora, and many others. The coverage, however, of vast areas, is sparse and anything but adequate. And the fact that a flora is being written does not ordinarily mean that collecting has been thorough. It is usually patchy at best, and in many tropical countries certain important habitats have been largely or entirely cleared and brought under cultivation. In Ceylon, for example, some species formerly indicated as not rare, have rarely, if ever, been found recently despite intensive collecting during the past decade. Their habitats have been almost entirely converted to tea and rubber plantations and irrigated rice fields.

Phytogeography in the tropics is only beginning to show vague outlines, in spite of the amount that has been written about it. In no tropical region is there adequate information on which to base conclusions, or even convincing theories. In most there probably never will be, as the destruction of vegetation is proceeding at such a pace that botanical collecting cannot hope to be adequate (Raven 1976). If we add to this unreliable taxonomy in many groups, lack of comparability of concepts, obsolete or incomplete floras, and the ready acceptance of faulty determinations, the books written on the subject do not impress me. The approach taken by van Balgooy in his volumes of maps of Pacific plant areas, though they are only as complete as his data will allow, is an exception to the above generalization.

Finally, on the general tropical level, we must consider the phenomenon of the emergence of what may be called the "pan-tropical flora". Under completely natural conditions most species have definitely restricted geographical and ecological ranges. Few of them, especially in the tropics, in pre-human times seem to have had wide distributions except where certain habitats were very uniform and extensive and where there were no effective barriers to dispersal. About the only plants with pantropical ranges were certain strand species, such as *Suriana maritima*, *Ipomoea macrantha*, *Sesuvium portulacastrum*, and *Ipomoea pes-caprae*, and these were probably much fewer than is generally believed. *Dodonaea viscosa*, an extraordinarily adaptable species, may have been another exception. With the advent of man this situation began to change, and is still changing at an accelerating rate. Plants are being carried around by man, both deliberately and accidentally. Pioneer habitats are being created and vastly expanded. Cultivation is almost universal. Three or four general categories of plants have profited, and have extended their distribution, often tropics-wide.

First are cultivated food, drug, and fiber plants such as the coconut, manioc, maize, cotton, sugar cane, *Sambucus mexicana*, and *Cinchona calisaya*. Most, but not all, of these depend on man for their continued occupancy of their now very wide ranges.

Second are a great assortment of cultivated ornamentals. One sees the same *Bougainvillea glabra*, *Allamanda hendersonii*, hybrid *Hibiscus*, and *Codiaeum variegatum* any place he goes in the tropics. There are literally hundreds of such species. Many of these, also, depend on man for their continued existence outside their native ranges. However, many others are able to escape, establish themselves and join one or other of the next two categories, as weeds or adventives.

Third are ecologically pioneer species that are able to occupy, and vastly increase their numbers and ranges, in the open or disturbed habitats created by man by his agricultural, grazing, logging, and other activities. Some of these have been deliberately carried around by man for their useful or ornamental characteristics. Many others, however, are "camp-followers," carried accidentally, or by means of their special dispersal mechanisms. Many of them are aggressive weeds. Some exist in enormous numbers; others would, except that they compete with each other. Some are extremely serious pests in cultivated land. Some are of great value in retarding erosion on denuded land.

The fourth category, much smaller, but very destructive, are plants, deliberately or accidentally carried around by man, that are able to invade closed, rarely even undisturbed, vegetation. Notable among these are *Clidemia hirta*, *Psidium cattleianum*, *P. guajava*, *Paederia foetida*, and *Mikania scandens* sensu lato, which occupy large areas in some tropical lands, especially islands. Certain island endemic species seem doomed to early extinction because of the occupation of their habitats by species of this sort. Many have disappeared already.

These four groups are dominating the tropical landscape more and more. Where the process will stop it is hard to know, but the prospects for any significant areas of original tropical vegetation to persist far into the future are bleak.

Tropical Island Floras

In a broad areal sense, tropical islands range in size from tiny rocks to such enormous mini-continents as Madagascar, New Guinea and Borneo. Some of these large ones have been isolated from continental land for a very long time, e.g. Madagascar, New Caledonia and Socotra.

     F. R. Fosberg

They have weird spectacular floras that make one feel as though he is on another planet. They have all suffered and are suffering devastating changes, and their remarkable plants are disappearing rapidly. Fortunately, modern floras are being written, however belatedly, of the first two. Although collecting is going on, many plants are probably already gone, some before they have even been discovered.

An intermediate category of isolation is represented by such large islands as New Guinea, the eastern Indonesian islands such as Celebes and Halmahara, Ceylon, and the Greater Antilles. Their floras show a considerable degree of endemicity and many plants of great evolutionary interest. The Flora Malesiana is covering some of these islands; floras are being written of Jamaica and Ceylon; Puerto Rico has a detailed, if somewhat old flora, and a modern book on trees.

Large, more strictly continental islands or groups, such as the Greater Sunda Islands, Hainan, Taiwan, the Ryukyu Islands, Zanzibar, the Bahamas and Trinidad, have floras closely related to those of nearby continents, though still with many endemic species. Most of these have floras in preparation, or in the case of Okinawa, recently published.

The islands already mentioned, as well as many others, belong to one of two great categories of islands. They are "continental", as contrasted with the other category, "oceanic" islands. This distinction is based on whether the islands have at one time been joined to, or parts of, continents.

Oceanic islands, in addition to never having been connected with any continental lands, have several other features in common: (i) they are all small, Hawaii and Viti Levu are perhaps the largest, (ii) with very few exceptions they are of volcanic origin, with or without fringes, terraces, or caps of limestone of organic origin, (iii) they have no truly indigenous mammals, except a few bats and marine mammals, (iv) their floras (and faunas) are small, compared with those of like areas on continents at comparable latitudes, (v) they have almost all suffered devastation at the hands of man.

Islands of this kind occur in all three tropical oceans, most in the Pacific, a fair number in the Indian Ocean, and a few in the Atlantic. It is difficult to be certain about some of the West Indies, but it is possible that some of the Lesser Antilles may never have had continental connections. Islands with continental rocks, have in some cases, been completely submerged since being separated from a continental mass. These are, in most respects, effectively oceanic islands.

We will now consider, in general terms, the geographical and biological features of oceanic islands, the origin and nature of their vegetation and floras, and the current destruction and modification of island vegetation. Especial attention will be given to the present state of knowledge of oceanic island floras, to the work that is being done, and what is required to achieve a reliable and adequate information base for understanding of floristic plant geography and ecosystem dynamics of islands.

Brief Geography of Oceanic Islands

Two classes of oceanic islands are commonly distinguished—high and low islands. Low islands are sea-level coral islands, mostly "lagoon islands" or atolls, not more than several meters elevation above mean low-tide level, except for occasional sand dunes and gravel or boulder storm ridges. High islands are mostly volcanoes arising from the ocean floor to above sea level, but the class also includes elevated atolls, mixed islands of old weathered volcanic rock and limestone, and a very few islands of granite (Seychelles) or metamorphic rock (Yap). This difference between high and low islands seems to be of great ecologic and biogeographic significance. The floras (and faunas) of low islands are impoverished and relatively uniform, though by no means identical. They have very few endemic species and are mostly composed of species with very obvious means of dispersal—mostly with propagules adapted for floating or for adherence to birds feathers or feet. Most of the species have very wide distributions and most occur in strand vegetation on high islands and even on tropical continental shores. Such species are *Suriana maritima*, *Scaevola taccada*, and *Tournefortia argentea*. The reasons for this restriction of floras are matters of speculation. Perhaps the most convincing is the suggestion that, during the climatic optimum or post-glacial xerothermic period, several thousand years ago, the sea rose 1–2 m, due to great amounts of melt-water from glaciers and ice-caps, drowning all terrestrial biota on islands of 2 m or less. However, not all marine geologists accept the idea of such a world-wide rise, and subsequent lowering of sea level. It would certainly account for the biogeographic facts about low islands.

It is interesting that even a slight elevation above the 2 m of the typical low coral island, such as that of the "feo" or rough limestone of certain of the Tuamotu islands (Anaa, Niau) or the 5 m or so of Aldabra, in the western Indian Ocean, is accompanied by an increase in numbers of species over nearby lower islands. The greater the elevation, usually the larger the flora and the more endemic species.

In the trade wind belt and other regions of prevailing winds from one direction, one side of a high island is wet and the other dry, with a resulting differentation of two distinct floras.

Certain regions of the oceans have much drier climates than others. For example, the Marshall Archipelago extends from a semi-arid belt in the north to an exceedingly rainy one in the south, with a profound effect on the flora. The northernmost, Pokak, has a flora of nine species, while Ebon the southernmost has perhaps 60 indigenous species (the exact number is not known).

Another important difference in oceanic islands is in volcanic versus limestone substrata. Most oceanic high islands are volcanic, almost all low ones are limestone, some high ones, such as Guam, Rota, Rurutu, and others are part volcanic and part limestone. The floras inhabiting the two kinds of material are largely different. Some species, of course are found on both.

Another floristically important difference or gradient is in isolation. Far away Ducie atoll, between Pitcairn and Easter Island, perhaps the most isolated of all, has a flora of one species of higher plant, *Tournefortia argentea*. The Fiji group, close to the larger continental islands, has several thousand species.

The diversity of habitats that results from the intersection of these environmental and geographical gradients is greater than expected, and provides opportunities for establishment of many of the storm-, water-, or bird-carried waif plant- (and animal-) propagules that, over millions of years of geological history, find their way across the ocean barriers from not just the closest, but from lands at all distances and in all directions. Professor W. A. Setchell once remarked to me that, given enough time, any organism can get anywhere. The geologist Harold E. Palmer said to me, in the same connection, that geology will give you all the thousands of years that are necessary to account for the flora of the Hawaiian Islands, the most isolated of all the major groups of oceanic islands. The same diversity of habitats has provided opportunity for spectacular adaptive radiation of the descendents of the few successful original colonists, providing species and lower taxa to occupy most, if not all, the available ecological niches, at least on the older islands.

Nature and Origin of Oceanic Island Floras

The floras of oceanic islands are not only relatively small in numbers of species and highly endemic but also lacking in the diversity of plant families that would be expected even on areas of comparable size and habitat complexity on continents. I indicated many years ago (Fosberg, 1948) that the entire flowering plant flora of the Hawaiian Islands probably descended and evolved from only about 272 successful colonists, resulting in a total indigenous flora variously estimated at between 1750 and 2350 species and well-marked varieties. Of these I estimated about 97% were endemic, the highest rate of endemism of any island, island group, or continental area of comparable size in the world. Even with such a small original stock, species had evolved to fill almost all existing niches. The original successful colonists came, in different proportions, from almost all possible directions.

There have been countless arguments as to how the progenitors of the species indigenous to oceanic islands got there. There are those who cannot imagine any significant number of plants crossing wide ocean barriers by any other means than dry land. In the total absence of any geological evidence for land connections in any direction from the Hawaiian Islands, for example, one must reject this conclusion. Considering the adaptations possessed by insular and other, plants, for dispersal by wind, water, and birds, and considering the time available—it is not hard to conceive of one colonist on the average of every 20 or 30 thousand years making the journey and landing in a favourable habitat. Carlquist (1974) the recognized authority on island biology and I agree completely on this subject, though we disagree on the relative importance of birds and wind as long-distance dispersal agents.

Granted a continuing, though extremely slow, influx of colonists, let us try to visualize the evolution of an island flora. Given a bare, but finally reasonably stabilized exposure of cinders, ash, and lava above the ocean surface, the first successful colonists must be ecologically pioneer species, possibly halophytes, but at least able to endure the baking effect of the tropical sun on the black volcanic substratum.

To begin with competition is completely lacking—the habitat is "open", and any viable seedling has a good chance to survive. In addition to such nutrients as may leach out of the volcanic material, nitrogen may be fixed by myxophyceae, and nitrate, ammonia, and calcium phosphate may be formed from guano dropped by fish-eating sea-birds. The first pioneer colonists multiply to form a vegetation. This by its very presence broadens the range of different habitats, and the number of possible survivors among the potential colonists that arrive in the form of propagules. This, up to a point, is a self-augmenting process. More and more species with different ecological requirements and tolerances may become established, diversifying the flora. This process then slows down as the vegetation becomes closed and "saturated"—the available niches become filled. Colonization does not stop completely as open habitats are continually created by normal geomorphic process, e.g. erosion, landslides, beach-ridge and dune formation.

Meanwhile, colonization has not been the only process contributing to the development of the flora. Evolution, especially adaptive radiation, has been active, producing forms to fill as yet unoccupied niches (Carlquist 1974). In certain genera a number of new, often bizarre species, evolve, giving the island flora its character. All of these are, at least to begin with, endemic. Other endemics result from the extinction of the continental populations from which indigenous island species arose, leaving the island populations as relict endemics. Extinction also occurs among the island species, even without the aid of man. The supposed equilibrium achieved between immigration (plus evolution?) and extinction of species provides the basis for the recently proposed theory of island biogeography of MacArthur and Wilson (1967). This theory seems to some to be rather a combination of the truism of greater area providing opportunity for more species, with a curious disregard of the role of habitat diversity in determining the numbers of species an island will support. There is no doubt that, on any island of considerable geologic age, a dynamic equilibrium exists in the number and individual abundances of species in its flora. Barring catastrophic events, the floras of older islands were without much doubt in such a state of dynamic equilibrium as MacArthur and Wilson visualized, though probably in few cases were the numbers of species a simple function of island area.

Aboriginal man, arriving on an island affects the flora in at least two ways. The human colonists, if their arrival was intentional, usually brought food plants and other useful species with them. Other, usually weedy, species may have come with them accidentally. Off-setting this augmentation of the flora, the opposite effect may be brought about by conversion of certain habitats to agricultural use and by the increase in competition for native plants from the plant immigrants that accompanied the human immigrants. It is hard to know the exact extent of the effect of aboriginal man on the floras of oceanic islands. Certainly it was real, locally it was probably important, but on the whole, probably few species were lost.

The arrival of European man was an altogether different story. With a plethora of exotic plant species brought purposefully or by accident, came an invasion of herbivorous mammals against which the native plants, evolved in isolation from such creatures, had no defenses whatever. Their browsing and trampling not only destroyed native plants directly, but opened up the stable, closed native vegetation to colonization by a never-ending stream of exotic invaders. Some native species became extinct or very rare, their living space taken by exotics. A few, such as *Sida fallax* may have profited in terms of available habitat, and perhaps, also, in terms of evolutionary opportunity. The availability of new habitats may account for the numbers of new, local varieties or even species of this relationship in Hawaii. Other species, genetically not well isolated, were brought together by break-down of their geographic or ecologic isolation and formed swarms of hybrids obscuring the taxonomic distinction of the species, making the study of floristics in such places difficult and identifications uncertain or even impossible.

The pantropical flora mentioned above, selected and disseminated by man, now dominates large areas, especially in the lowlands, of most tropical high islands. A single member species of this pantropical flora, *Cocos nucifera*, the coconut, dominates most low islands.

In all probability most of the indigenous species do still exist, but some that may have been common are now rare, or even absent from islands they formerly inhabited. Determination of what the original vegetation of tropical oceanic islands originally looked like, and of what species made up its composition, has become increasingly difficult.

State of Knowledge of Tropical Oceanic Island Floras

We have been collecting and studying the floras of tropical oceanic islands for just over 200 years. The explorations of Banks and Solander, sailing with Captain James Cook on the ship H.M.S. *Endeavour*, began the study of Pacific island floras, in 1769. Their own accounts of the plants were never published, but their specimens have been studied by botanists ever since.

One might logically think that with 200 years of work, knowledge of these floras would be complete to the smallest detail. This is far from being the case. The results of the earlier explorations were marred by a lack of appreciation for precision in geographical origin of specimens, lack of interest in the kind of places the plants inhabited, and lack of any attempt to record on the labels any of those characteristics of the plants that disappeared on drying. Most collections made more than 75 years ago yield no more information than that the plant grew on a particular island, and such morphological information as very scrappy fragments showed. Little attempt was made to get all the species on an island. If a plant had been found on other islands in the region it might be ignored. Some of the specimens collected on Captain Beechey's voyage, for example, bear the datum, "coral islands". An assumption seemed to dominate the collectors' minds that the floras of islands were more or less alike.

Illustrative of the limitations of data on the floras of islands are the rather abundant collections from the Marquesas archipelago, on the eastern edge of Polynesia, mostly preserved in the Paris Herbarium. Most have as their data "Iles Marquises" or "Nukuhiva, Iles Marquises", frequently not even the date or the collector's name. In the Marquesas there are 13 principal islands, ranging up to 1260 m in elevation, from semi-desert on the leeward sides to cloud forest and boggy plateaus. Only in the 1920s was any attempt made to record what plants inhabited what island and what kinds of plants occupied what habitats. Even then these attempts were inadequate. Meanwhile the introduction of herbivores (goats, cattle and horses) had greatly blurred the pattern of habitats and the natural ranges of the native species. Even whether some species are or are not indigenous is not now clear.

Of the collectors in the nineteenth century in the Pacific Islands, very few were botanists. Most were surgeons or other officers on naval vessels. Some were general naturalists, such as Chamisso, Darwin, Mertens, Marche and Pickering. Only Dr William Hillebrand (1888) author of the "Flora of the Hawaiian Islands" was a resident. The high quality of his flora shows it, even though he was a physician rather than a botanist by profession.

Many of the collections made on these nineteenth century voyages were never worked up. We are, even now, studying for the first time collections made in Micronesia and Polynesia on the expeditions of such famous explorers as Dumont d'Urville, Freycinet and Marche. Even though most of the endemics of the Marianas archipelago were actually collected by the eighteenth and nineteenth century explorers, many of them waited to be described from later collections by E. D. Merrill and several Japanese botanists in the first half of the twentieth century.

Just how thorough has been the exploration of the tropical oceanic islands from the time of Captain Cook until now? Certainly many specimens have been collected. However, a study of these collections suggests that, with the exception of the work of a few recent explorations and the work of resident botanists in such islands as the Hawaiian group, a few localities readily accessible have been very well collected, and a few remote high peaks are also reasonably known. Mid-slopes, valleys that are not easy to reach, whole islands where there is no regular boat or air service, and peaks, ridges and plateaus that require long and difficult climbing, surmounting dangerous cliffs and crossing deep ravines are not well-known. Discovery of remarkable plants in the last several decades, such as the Malvaceous genus *Lebronnecia*, a tree *Abutilon*, a *Rauwolfia*, and an unknown and totally unexpected shrubby

Oxalis in the Marquesas, a new *Gynotroches* in Palau, an *Oreobolus* in Tahiti, a new *Isoetes* in Hawaii, suggests that our coverage is not yet adequate. We cannot say that even the best known islands are really adequately collected. Hawaii is easily the best explored group of high tropical islands, yet there have been dozens of new species and varieties described from there in the past decade and a half.

What is Being Done Now?

Considering the present ease of access of many islands compared with the difficulties in the past, one might expect that a great surge of activity in island botanical exploration would be in progress. Indeed there was considerable local botanical work on islands in the years immediately following World War II. However, with some exceptions, the activity has not been extraordinary. No concerted effort has been made either to fill the gaps or to bring the level of collections up to any planned comparable standard throughout the widespread oceanic island world. Collecting has been opportunistic, much not done by botanists, many of the smaller islands have never been visited by a botanist and a few have never had a plant specimen collected by anyone.

Recent and current collecting efforts may be noted in the Pacific oceanic islands in the Marquesas, Samoa, Niue, the Cook islands, Palau, the wetlands of the Caroline archipelago, in the Northern Marianas, the Phoenix Islands, the Hawaiian Islands, and a few of the many scattered coral atolls. In the Indian Ocean some collecting is going on in the Mascarenes, and the Seychelles, in the Aldabra Group, the Amirantes and some others of the coral islands of the western part of this ocean. In the Atlantic a resident botanist is working in the Canary Islands. Continuing work is going on in the Lesser Antilles and the Virgin Islands (as well as in some of the continental islands). None of the work listed can be called thorough, or intensive, nor is it aimed toward systematic coverage. Much of it is opportunistic or directed toward specific limited aims however, a good number of specimens are accumulating, some from critically threatened areas.

A number of floras are being written, first mention must be made of A. C. Smith's 40 year effort toward a flora of Fiji (Smith, 1979); a cooperative project on the Mascarene flora is being carried out by British, French and Mauritian institutions; Art Whistler, of Honolulu, is gathering materials for a flora of the Samoan Group; a flora of the Galapagos has recently been published and further work is being done; R. A. Howard and colleagues have made substantial progress toward a flora of the Lesser Antilles; a flora of Aldabra and neighbouring islands by S. A. Renvoize and myself is in press. My colleague Dr Marie-Hélène Sachet has published one section of her flora of the Marquesas Islands. She and I have in progress a flora of Micronesia. We have for some time, also, been bringing together materials for a flora, or series of floras, of low coral atolls in all three oceans.

Before World War II the B. P. Bishop Museum sponsored extensive collecting in Polynesia and one expedition to Micronesia. The results were enormously important, but much of the collecting was on short visits and many of the specimens have general localities and scanty ecological and descriptive data. As one of the participants in the Mangarevan Expedition to Eastern Polynesia, the most extensive effort of that period, I may say that we left each island with the feeling that we needed much more time to do a thorough job. Unfortunately expeditionary visits will be for a long time to come, the principal kind of botany carried out on most tropical oceanic islands. There are very few resident botanists in the islands we are considering.

Tasks for the Future and Most Urgent Needs

(i) Certainly, in view of the rapid degradation of the vegetation of most islands, the most urgent task for the near future is carefully planned and thorough collecting. To use a term paraphrased from archaeology this activity might be called salvage botany. Our chances of acquiring knowledge of insular plants, in their normal habitats and relatively unchanged by human activities, are diminishing daily.

(ii) Perhaps just as urgent, but much more difficult, is the creation of a system of critical habitat reserves protecting an adequate series of natural areas to represent the principal ecosystems on islands, and hence providing refugia where indigenous insular species may survive. This seems an almost hopeless task, considering the number of islands and the scarcity and high cost of land on islands. Yet it must be done if the study of island plants is not to become a branch of paleobotany, studying only "herbarium fossils".

(iii) Studies resulting in revisions of genera represented on islands, or of such parts of these genera as are pertinent to island floras, in understanding of the relationships of insular taxa to continental taxa, in the distinction of relict and recent endemics, and in determining which species are really native on the islands where they are found.

(iv) Production of more reliable and comparable floristic lists for individual islands. Our colleagues in related fields need floristic information and can scarcely be expected to wait until floras that take decades to write are completed. It is a fact that there are many islands for which no lists, or only incomplete or obsolete lists exist.

(v) Production of modern descriptive floras with keys and maximum information content is also important.

It should be understood, though I have not emphasized it, that these five desiderata apply equally to cellular cryptogams and to vascular plants.

Finally, to be hoped for, as current and future studies advance, is the emergence of a more meaningful island (and tropical) phytogeography. At present, the information base on which island phytogeographic patterns and speculations rest is so inadequate that phytogeographic conclusions, except on the broadest questions, are not to be taken seriously. Island floras are so fascinating that they inevitably excite curiosity and stimulate speculation. Answers are hard to come by, though my colleague in island studies, Professor Sherwin Carlquist, has gone farthest to provide some of them. His books, while suggesting and describing what has happened, open the way to far more perceptive studies and far better questions.

References

Carlquist, S. (1974). "Island Biology." Columbia University Press, New York and London.

Fosberg, F. R. (1948). Derivation of the Hawaiian flora. *In* "Insects of Hawaii" (E. C. Zimmerman, ed.) Vol. 1, p. 107–118. University Press, Hawaii.

Fosberg, F. R. and S. A. Renvoize (in press). Flora of Aldabra and neighbouring islands. *Kew Bull. Add. Ser.*

Fosberg, F. R. and M.-H. Sachet (1965). Manual for tropical herbaria, *Regn. Veg.* **39**, 133.

Hillebrand, W. (1888). "Flora of the Hawaiian Islands." Privately published by W. F. Hillebrand.

MacArthur, R. H. and E. O. Wilson (1967). "Theory of Island Biology." Princeton University Press, NJ.

Raven, P. (1976). The destruction of the tropics. *Frontiers* **40**, 22–23.

Smith, A. C. (1979). "Flora Vitiensis Nova" Vol. 1. Pacific Tropical Botanical Garden, Lawai, Kauai, Hawaii. (Vols 2 and 3 in prep.).

III. Regional Phytogeography and Investigation of Tropical Floras

Outline of Ecology and Vegetation
of the Indochinese Peninsula

J. E. VIDAL

Museum National d'Histoire Naturelle, Paris, France

Ecology

The various vegetation-types of Indochina can be understood by the
consideration of three fundamental ecological factors: climate, soil and
man. The climate, or more precisely the bioclimate, results from combination of two main elements: temperature and humidity.

The mean annual temperature varies between 25 and 30° C in the
lowlands. Above 1000 m it is lower than 20° C and consequently the
floristic composition of the vegetation is different and is characterised,
for example, by abundance of Fagaceae and Coniferae.

Humidity, or its opposite, drought can be evaluated by various
methods: annual rainfall, number of dry months during the dry season,
number of dry days during the dry season, that is to say xerothermic
index, and finally saturation deficit index.

Precipitations depend on the geographic situation. In some limited
semi-arid areas (South-East Vietnam, North-East Thailand, Central
Burma) the annual rainfall is lower than 1000 mm. Elsewhere in
lowlands it varies from 1000–2000 mm and can reach 3000–4000 mm
close to montanous ranges (Central Laos, South-East Thailand,
South-West Cambodia). In highlands the annual precipitations are
usually more important and can go up to 5000 mm. (Annamitic Range,
Mt. Bokor in South-West Cambodia).

A better evaluation of the efficient humidity or drought is made by
consideration of number of dry and wet months during the year. But
there is not a common agreement for the definition of these months.
According to Mohr (1933) for Java a month is called dry when the
precipitation is less than 60 mm, and wet when it is over 100 mm.

                              J. E. Vidal

Aubréville (1949), for Africa, thinks that a month less than 30 mm is dry and over 100 mm is wet. According to Gaussen (1953, 1955) a month is dry if the rainfall expressed in millimetres is less than twice the monthly temperature expressed on °C. The "ombrothermic or climatic diagrams" based on this definition are very easy to construct using a scale of 10° C = 20 mm precipitation. These diagrams have been followed by several ecologists including Walter (1955, 1971, 1973) and mainly Walter and Lieth (1967) in their "Klimadiagramm-Weltatlas" (Fig. 1).

According to this definition there is no dry season in equatorial zone (Singapore). In tropical lowlands the number of dry months varies between four and five (Laos, Thailand, Cambodia, South Vietnam). But, in the semi-arid areas previously mentioned, it can go up to six to eight months, and close to some montanous ranges it can go down to zero to three months. Above 1500–2000 m, there are practically no dry months.

Another more elaborate method to measure the strength of the drought is to calculate the number of dry days during the dry season, taking into account the morning fogs or dews and the mean relative humidity. This emended number has been called by Bagnouls and Gaussen (1953) a xerothermic index. A similar index was previously proposed by van Bemmelen (1916): "the number of rainy days during the four driest consecutive months of the year". The value of xerothermic index for Indochina varies from 0 (Annamitic Range) to 74 (Chanthaburi), 108 (Bangkok), 114 (Nha Trang), 151 (Phan Rang).

Other ecologists such as Aubréville (1949) prefer to use the saturation deficit index, that is to say the difference between the maximum water vapour pressure (F) and the real pressure in the atmosphere (f). This index gives a good and simple interpretation of the evaporation capacity important for the life of plants. As an example the value of deficit saturation index is 2·5 mm Hg for Xieng Khouang at an altitude of ± 1200 m and 8·5 mm Hg for Paksé in lowland. Several other sophisticated methods have been proposed by Thornthwaite, Holdridge and others. But it is not the place to develop them here.

From this brief outline of the bioclimatic conditions of Indochinese area it is evident that nearly everywhere, except in some limited areas, temperature and humidity are relatively high. Such conditions are convenient for the development of evergreen or semi-evergreen forests. But the soil conditions and human activity very often prevent the existence of these climatic vegetation-types.

On shallow or superficial soils with rocky outcrops (sandstone, laterite, limestone) we found only open deciduous forests. The soils

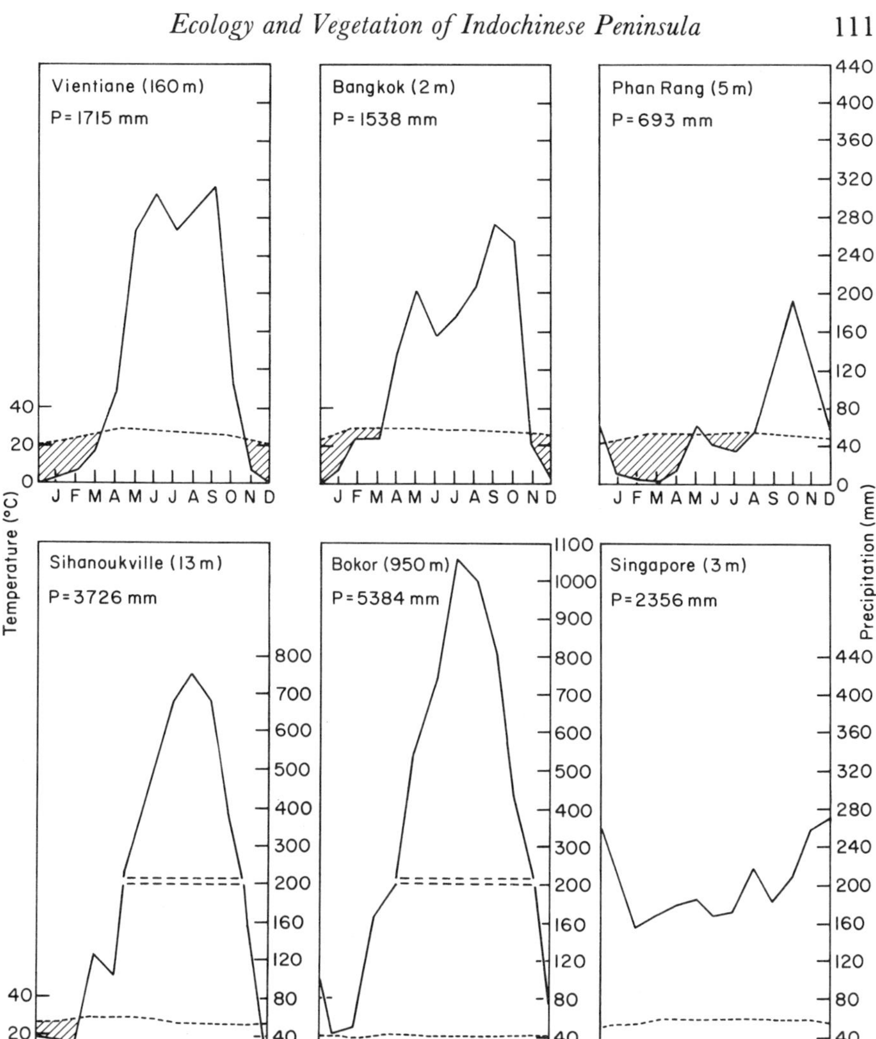

Fig. 1. Climatic diagrams in South-East Asia. The variation in temperature is represented by the broken line, that of precipitation the unbroken line. The scale for temperature is twice that of the scale for precipitation.

degenerated by burning and by shifting cultivation, in montanous areas in particular, bear either savannahs or grasslands. The permanent cultivation areas (rice-fields and various plantations) have in many cases been established in forest sites.

In the Indochinese peninsula, therefore, the types of vegetation are

        J. E. Vidal

various and conditioned mainly either by climatic, edaphic or anthropic influences. Table 1 shows the distribution of the main formations according to these ecological factors.

Table 1. Ecological distribution of main Indochinese vegetation-types.

	Climate	Soil	Man
Wetland formations			
Mangroves	*	**	
Rear mangroves	*	**	
Swamp forests	*	**	
Fresh-water formations	*	**	*
Dryland formations			
Beach vegetation	*	**	
Evergreen forests	**	*	
Semi-evergreen forests	**	*	
Secondary formations	*		**
Mixed deciduous forests	*	*	*
Deciduous Dipterocarp forests	*	**	**
Lowland savannah forests and grasslands	*	**	**
Montane forests	**	*	
Open coniferous forests	*	**	**
Highland savannahs and grasslands	*	**	**

Vegetation

Wetland Formations

Mangroves

Mangrove forests occur on muddy tidal swamps at the mouth of rivers (Menam, Mekong, Red River) and along the sea coasts (Gulf of Siam, South China Sea, Gulf of Tonkin). The trees of these forests are usually arranged by zonal formations from the sea to the land: *Avicennia-Sonneratia* zone, *Rhizophora* zone, *Bruguiera-Kandelia-Ceriops* zone and *Lumnitzera-Xylocarpus-Bruguiera* zone.

The floristic composition is relatively poor compared to that of the rain forest. The greater part of trees belong to the family Rhizophoraceae: *Rhizophora mucronata*, *R. apiculata*; *Bruguiera gymnorhiza*,

B. sexangula, B. parviflora; *Ceriops tagal*, *C. decandra*, *Kandelia candel*. The other families represented are: Avicenniaceae (*Avicennia officinalis*, *A. marina*); Sonneratiaceae (*Sonneratia caseolaris*, *S. alba*); Combretaceae (*Lumnitzera littorea*, *L. racemosa*); Meliaceae (*Xylocarpus granatum*, *X. moluccensis*).

The ground flora is very poor and represented in higher places mostly by *Acanthus ilicifolius* (Acanthaceae) and *Acrostichum aureum* (Pteridaceae).

The peculiar morphological and biological characters of the mangrove vegetation are well known: stilt roots of *Rhizophora*, pneumatophores of *Avicennia* and *Sonneratia*, viviparous reproduction, etc.

Rear mangroves

Beyond the inland edge of the mangrove where the forest floor is inundated only during very high tide the soil is sandy loam and bears a forest which can be called *Melaleuca leucadendra* formation because of the abundance of this tree of the family Myrtaceae. The other most frequent trees are *Bruguiera gymnorhiza* (Rhizophoraceae), *Xylocarpus moluccensis* (Meliaceae), *Heritiera littoralis* (Sterculiaceae), *Intsia bijuga* (Caesalpiniaceae). The common shrubs are *Aegiceras corniculatum* (Myrsinaceae), *Scyphiphora hydrophyllacea* (Rubiaceae) and the ferns *Acrostichum aureum* and *A. speciosum*. Palms are represented by thorny *Phoenix paludosa*, *Oncosperma tigillaria* and spineless *Nipa fruticans*.

Swamp forests

This type of forest is frequently found along depressions beyond the rear mangrove on alluvial muddy soils.

The forest is usually three storeyed. The top storey is composed of tall trees such as *Dyera costulata* (Apocynaceae), *Palaquium gutta* (Sapotaceae), *Melanorrhoea curtisii* (Anacardiaceae) and *Scaphium lychnophorum* (Sterculiaceae). In the second storey can be observed *Nephelium lappaceum* (Sapindaceae), *Hydnocarpus sumatrana* (Flacourtiaceae), *Hopea pierrei* (Dipterocarpaceae), *Heritiera littoralis* (Sterculiaceae) and *Xanthophyllum glaucum* (Xanthophyllaceae). In the lowest storey are found *Melaleuca leucadendra* (Myrtaceae), *Alstonia spathulata* (Apocynaceae) and several species of *Casearia* (Flacourtiaceae) and *Aglaia* (Meliaceae). Palms are very common and consist of many species of spiny rattans (*Calamus*, *Daemonorops*, *Plectocomia*) and other genera such as *Lucuala*, *Areca*, *Pinanga*, *Oncosperma*. The ground flora is represented mainly by monocots (Zingiberaceae, Marantaceae, Orchidaceae, Commelinaceae). Epiphytes are mostly ferns and orchids.

When the soil is sandy the forest is more open with different and more or less stunted trees. *Fagraea fragrans* (Loganiaceae) is frequent with *Beckia frutescens* (Myrtaceae) and *Licuala spinosa* (Palmae). In some places *Melaleuca leucadendra* forms a pure stand.

Fresh-water formations

In places either permanently or temporarily inundated during the rainy season three main vegetation-types can be distinguished:

(i) Aquatic herbaceous vegetation. The water plants are either rooting (rhizophytes) or floating (pleustophytes). Among the former the genera which are very common are: *Xyris, Typha, Scirpus, Nymphaea, Nelumbo, Nymphoides, Ottelia, Trapa, Hydrilla*. In the second category the most common representatives are: *Azolla, Salvinia, Pistia, Eichhornia, Utricularia*.

(ii) Inundated vegetation. A good example of this vegetation is found in the Tonle Sap area in Cambodia. During the rainy season the water level increases up to 10 m and then shrubs thickets become submerged. Only the crown of trees such as *Barringtonia acutangula* (Lecythidaceae), *Mallotus anisopodus* (Euphorbiaceae) remains above water.

(iii) Stream vegetation (rheophytes). The ligneous plants of the streambeds and streambanks possess a strong root system and typical narrow lanceolate leaves. They belong to various families. The most common ones are: *Homonoia riparia* (Euphorbiaceae), *Anogeissus rivularis* (Combretaceae), *Eugenia fluviatilis* (Myrtaceae), *Artobotrys harmandii* (Annonaceae), *Crataeva nurvala* (Capparidaceae). *Saraca indica* (Caesalpiniaceae) is frequently found in riparian or gallery forest together with *Hopea odorata* (Dipterocarpaceae) and *Xanthophyllum glaucum* (Xanthophyllaceae).

Dry land formations

Beach vegetation

The floristic composition of beach vegetation varies according to the topography. Along progressing sandy coasts the *Ipomoea pes-caprae* formation can be observed with some other typical creeping plants: *Spinifex littoreus* and *Ischaemum muticum* (Gramineae), *Canavalia maritima* (Papilionaceae). Shrubby species are represented by *Scaevola taccada*

(Goodeniaceae) and trees by *Casuarina equisetifolia* (Casuarinaceae). Along receding or rocky sea shore where the sand cannot be accumulated the narrow beach is fringed by *Barringtonia asiatica* formation. Some other representatives are: *Calophyllum inophyllum* (Guttiferae), *Terminalia catappa* (Combretaceae), *Hibiscus tiliaceus* (Malvaceae), *Pandanus tectorius* (Pandanaceae), *Cycas rumphii* (Cycadaceae).

On elevated areas of the sea shore the forest is two to three storeyed. Among the trees one can observe *Dialium* and *Sindora* (Caesalpiniaceae), *Diospyros* (Ebenaceae), *Manilkara* (Sapotaceae), *Lannea* (Anacardiaceae), *Terminalia* (Combretaceae). The lowest storey is composed of shrubs such as *Memecylon* (Melastomataceae), *Murraya* and *Atalantia* (Rutaceae), *Harrisonia* (Simaroubaceae). Spiny vines of the genera *Atalantia*, *Acacia*, *Caesalpinia*, *Smilax* are abundant.

Dune vegetation occurs in South-East Vietnam (Nhatrang-Phanrang semi-arid area). On moving sands treelets or shrubs form scattered thickets between which creeping or erect herbs grow. Some representatives of ligneous vegetation are: Dipterocarpaceae (*Vatica odorata*), Myrtaceae (*Eugenia tinctoria*, *Rhodamnia trinervia*, *Decaspermum paniculatum*), Ochnaceae (*Ochna integerrima*, *Gomphia serrata*), Sapotaceae (*Xantolis maritima*, *Manilkara hexandra*). Among the herbaceous vegetation we found *Polycarpon indicum* (Caryophyllaceae), *Phyllanthus arenarius* (Euphorbiaceae), *Desmodium rubrum* (Papilionaceae) and several species of Cyperaceae and Gramineae.

Stabilized sands bear low forest or shrub thickets composed of various families: Annonaceae (*Miliusa*), Caesalpiniaceae (*Sindora*), Bignoniaceae (*Markhamia*), Sapindaceae (*Euphorbia*). On shallower soils spiny shrubs like *Zizyphus* (Rhamnaceae) and *Pleiospermum* (Rutaceae) are found. In the same area the crystalline rocks of the islands and sea shore are covered by dense thickets where *Euphorbia antiquorum* and *Cycas rumphii* are noteworthy.

In open salt marshes the vegetation is represented by typical Chenopodiaceae: *Sueda maritima* or *S. australis* and *Salicornia brachiata*.

Evergreen forests

The typical evergreen forest (rain forest) in the lowlands is found on deep soils in very wet areas where the dry season is very short or absent, for example, in Southeastern and Peninsular regions of Thailand and in South-West Cambodia.

The floristic composition is very heterogeneous. The upper storey is composed of gigantic trees (over 30 m) mostly of the family Dipterocarpaceae (*Dipterocarpus*, *Shorea*, *Hopea*, *Balanocarpus*, *Anisoptera*) and other families like Anacardiaceae (*Mangifera*, *Melanorrhoea*, *Swintonia*), Sapotaceae (*Palaquium*), Sterculiaceae (*Pterocymbium*, *Scaphium*, *Sterculia*, *Heritiera*), Leguminosae (*Dialium*, *Sindora*, *Kompassia*, *Intsia*, *Adenanthera*), Euphorbiaceae, Meliaceae, Moraceae, etc.

The lower storey is constituted of many trees of medium height and dense thickets of shrubs and palms; rattans and vines are abundant. The ground flora is relatively poor and represented mainly by monocots (Zingiberaceae, Marantaceae, Araceae). Epiphytes are mostly orchids, ferns and *Ficus*.

Semi-evergreen forests (seasonal evergreen forests)

In tropical zone where the dry season is well marked for four to six months on rather deep soils, the forests appear as evergreen but contain a variable proportion of deciduous trees (up to 40% of the species in Laos) among which *Lagerstroemia*, *Spondias* and *Stereospermum* are frequent.

These forests as the evergreen forests are dipterocarp forests, the upper storey being composed mainly by dipterocarps (*Dipterocarpus*, *Anisoptera*, *Shorea*, *Hopea*) together with Leguminosae (*Dialium*, *Dalbergia*, *Ormosia*), Meliaceae (*Aglaia*, *Walsura*), Sapindaceae (*Arytera*, *Schleichera*), Lythraceae (*Lagerstroemia*). The lower ligneous storey has the same evergreen aspect as in the rain forest and is composed of various families: Annonaceae (*Polyalthia*, *Uvaria*), Rubiaceae (*Ixora*, *Randia*, *Rothmannia*) etc. The ground flora, epiphytes and vines have more or less the same representatives as in the evergreen forest.

Secondary formations

When the evergreen or semi-evergreen forests are disturbed by man secondary formations soon appear. In some places a tall herbaceous plant of American origin, *Eupatorium odoratum* (Compositae) is the first to spread over the free space. In a few years it is dominated by shrubs and young trees which progressively restore a secondary forest characterized by trees such as *Tetrameles nudiflora* (Datiscaceae) with large buttresses and other genera like *Mallotus* (Euphorbiaceae), *Ailanthus* (Simaroubaceae), *Anthocephalus* (Rubiaceae), *Melochia* (Sterculiaceae).

In other places various species of tall Gramineae (*Saccharum arundinaceum*, *Thysanolaena maxima*) invade the clearings. Frequently bamboos (*Bambusa*, *Cephalostachyum*, *Dendrocalamus*) form almost pure stands in areas where the original forest has been destroyed. Not rarely thickets of wild bananas are mixed together with bamboos in humid lower places.

Mixed deciduous forests

This type of forest intermediate between semi-evergreen and deciduous forest is rather common in the northern and central part of Thailand and Laos, on loamy or rocky soils, up to 500 m altitude. It is essentially composed of Leguminosae (*Afzelia*, *Adenanthera*, *Albizia*, *Xylia*, *Dalbergia*, *Milletia*, *Pterocarpus*), Lythraceae (*Lagerstroemia*), Combretaceae (*Terminalia*), Bombacaceae (*Bombax*), Dilleniaceae (*Dillenia*) and in North Thailand and North Laos of teak, *Tectona grandis* (Verbenaceae). The percentage of deciduous trees is variable and depends on local soil conditions but it is usually about 80%.

The undergrowth is composed of various shrubs, small palms (*Phoenix*, *Rhapis*) and many kinds of bamboos (*Bambusa*, *Cephalostachyum*, *Dendrocalamus*, *Gigantochloa*, *Oxytenanthera*, *Thyrsostachys*). Owing to the open state of the forest the ground flora is represented by many species of Gramineae and Cyperaceae together with Zingiberaceae and terrestrial orchids. Fires are rather frequent but not very noxious. Sometimes they contribute to the stabilization of the forest by aiding the germination of seeds particularly in the teak forest.

Deciduous dipterocarp forests

The widespread occurrence of deciduous dipterocarp forests in various parts of Burma, Thailand, Cambodia, Laos and Vietnam is mostly due to edaphic rather than climatic factors. They grow effectively on sandy or rocky superficial soils with outcrops of laterite, sandstone or limestone. According to the soil conditions two main types can be distinguished: "scrub dipterocarp forests" on dry and shallow rocky soil and "high dipterocarp forest" where the soil forms a rather deep loam. The former type is characterized by stunted trees like *Shorea siamensis*, *S. obtusa*, *Dipterocarpus intricatus*, *Terminalia tomentosa*. In the latter *Dipterocarpus obtusifolius* and *D. tuberculatus* form almost pure stands of well developed trees up to 20–25 m high.

The lower storey is composed of various genera of shrubs and treelets such as *Strychnos*, *Symplocos*, *Diospyros*, *Aporosa*, *Phyllanthus*, *Holarrhoena*, *Dillenia*. A small bamboo of the genus *Arundinaria* and a dwarf palm, *Phoenix acaulis* are frequent.

The ground flora is very rich (about 50% of the total number of species) and consists mostly of tuber- and rootstock-bearing plants that is an adaptative protection against fire. We can note: Zingiberaceae (*Curcuma*, *Kaempferia*), Orchidaceae (*Habenaria*, *Pecteilis*), Amaryllidaceae (*Crinum*), Araceae (*Pseudodracontium*), Commelinaceae (*Aneilema*) various species of Gramineae, Cyperaceae and Eriocaulonaceae, Menispermaceae (*Stephania*), Malvaceae (*Hibiscus*, *Decaschistia*), Papilionaceae (*Eriosema*). Beside the floristic composition another specific character of these forests is their peculiar phenology. About 95% of trees and 80% of shrubs are deciduous.

The intervention of man by fire is an important factor of stabilization of the deciduous dipterocarp forests which can, therefore, be classified ecologically as an edapho-anthropo-climax.

Lowland savannah forests and grasslands

Savannah forests are actually grasslands where stunted trees very sparsely grow together with thorny shrubs. They are found on sandy or rocky soils sterilized by subsequent fires or in areas where the dry season is very long (six to eight months) and annual precipitation lower than 1000 mm (Central Burma, North-East Thailand, South-East Vietnam).

The most frequent trees and shrubs growing there are of the genera *Careya* (Lecythidaceae), *Pterocarpus* and *Dalbergia* (Papilionaceae), *Acacia* (Mimosaceae), *Mitragyne* (Rubiaceae), *Zizyphus* (Rhamnaceae), *Capparis* and *Niebuhria* (Capparidaceae), *Phyllochlamys* (Moraceae).

The herbaceous vegetation is composed essentially of Gramineae (*Imperata*, *Vetiveria*, *Themeda*, *Eulalia*, *Panicum*). In some places inundated during the rainy season the ligneous vegetation cannot grow and grasses form a pure stand (grassland).

Montane forests

Usually the precipitations increase and temperature decreases with altitude. At elevations of 1000 m or more the rainy season is short (one to three months) or absent (cf. Fig. 1). Then the climatic vegetation is a humid forest. This montane forest is either evergreen or semi-evergreen and composed dominantly by Fagaceae (*Quercus*, *Lithocarpus*, *Pasania*, *Castanopsis*), Lauraceae (*Cinnamonum*, *Litsea*, *Machilus*), Magnoliaceae (*Manglietia*), Theaceae (*Camellia*, *Gordonia*, *Schima*). Gymnospermous families are also present such as Podocarpaceae (*Podocarpus*, *Dacrydium*), and over 2000 m Cupressaceae (*Fokienia*) and Taxodiaceae (*Cunninghania*).

In the lower storey shrubs of various families are abundant. Bamboos are represented by erect (*Gigantochloa*) or climbing species (*Dinochloa*). Palms are relatively few (*Pinanga, Phoenix*). Herbaceous vegetation is composed of numerous ferns, terrestrial orchids, Zingiberaceae, Marantaceae, Commelinaceae, Acanthaceae, Rubiaceae. Epiphytes are abundant and are represented by bearded lichens and mosses, various orchids and ferns.

The presence of some trees like *Betula alnoides* and *Sapium discolor* indicate that the forest is not a primary one but a more or less degraded secondary formation. Those forests are frequently destroyed by hill tribes for cultivation and after the soil is sterilized, replaced by savannahs or grasslands.

Open coniferous forests

The open coniferous forests are found usually on poor soils at elevation 200–1200 m.

Three types can be distinguished:

(i) *Pinus merkusii* forest (two needles pine) is found from 200 up to 800 m, occasionally mixed with *Dipterocarpus obtusifolius*.

(ii) *Keteleeria roullettii* (Pinaceae) forms sometimes pure stands (i.e. in Xieng Khouang area in Laos) and sometimes is mixed together with *Pinus* or *Quercus griffithii*.

(iii) *Pinus kesiya* (three needles pine) is found above 800 m up to 1200 m.

These three coniferous species are able to grow on poor soils and therefore can be considered as pioneers for the edification and reconstruction of the climax forest. Pines are the first to occupy the soils of degraded areas like that of grasslands or other kinds of poor soils. They are followed by *Keteleeria* and broad leaves trees or shrubs such as *Quercus, Schima, Tristania, Pieris*. Pines and *Keteleeria* are progressively eliminated from this mixed forest. A more or less deciduous *Quercus* forest then becomes established and prepares the place for a climax evergreen forest. Unfortunately this evolutive scheme is rarely realized because the intervention of man, by cutting, burning and cultivating, prevents the natural dynamic process of the vegetation.

Highland savannahs and grasslands

In montanous areas, savannahs and grasslands constitute the ultimate form of degradation of the vegetation and are very widespread. *Imperata cylindrica* forms the dominant species among numerous other grasses. A few treelets or shrubs (*Helicia, Phyllanthus, Callicarpa*) are found scattered here and there.

120   *J. E. Vidal*

Savannahs evolve by subsequent fires towards grasslands. These are characterized by shorter grasses and Cyperaceae. In some places (i.e. in Plaine des Jarres in Laos) the grassland becomes pseudosteppic by burning and grazing.

Conclusion

In the course of this review of ecology and vegetation of the Indochinese peninsula one cannot fail to notice that the natural vegetation is badly disturbed by human activity. The solution to this problem is not simple on account of ancestral agricultural practices difficult to change. However certain governments conscious of this difficulty make considerable effort to preserve some endangered areas or species by establishing national parks and reserves where people can learn a great deal about plants and animals and how to protect them. We hope that by means of proper education and also by appropriate government measures the natural vegetation could be rationally exploited for the welfare and harmonious development of the Indochinese countries.

References*

Ashton, P. S. (1967). Climate versus soils in the classification of Southeast Asian tropical lowland vegetation. *J. Ecol.* **55**, 67–68.

Ashton, P. S. (1970). Inventaire forestier des terres basses du versant occidental des Monts Cardamomes (Cambodge): l'écologie sylvicole et la botanique forestière, 88 pp, FAO (FO: SF/CAM 6, Rapport No. 5).

Aubréville, A. (1949). "Forêts, Climat et Désertification de l'Afrique Tropicale" Vol. 1. Soc. Edit. géogr. Marit. colon., Paris.

Bagnouls, F. and H. Gaussen (1953). Saison Sèche et Régime Xérothermique. Documents pour les Cartes des Productions Végétales, Série Généralités, t. 3, Vol. 1, Art 8, 1–47.

Bemmelen, W. van (1916). Verzameling van verhandelingen omtrent hetgeen bekend is aangaande den grond van Nederlandsch–Indië en zijn gebruik in den landbouw ten tijde van het Bodem Congres te Djocjakarta Oct. 1916, Vol. 2.

Davis, J. H. (1964). The forests of Burma. *Sarracenia* **8**, 1–41.

Dy Phon, P. (1970). La végétation du SW Cambodgien. *Ann. Fac. Sci. Phnom Penh* **3**, 1–136.

* Complementary references can be found in the following bibliographies: Hansen (1973), Reed (1969), Vidal (1972a)—see under "Phytogéographie" in the subject index, and Walker (1952).

Dy Phon, P. (1971). La végétation du SW Cambodgien. *Ann. Fac. Sci. Phnom Penh* **4**, 1–77.

Edwards, M. V. (1950). Burma forest types. *Ind. For. Rec. n.s. Silvic.* **7**, 135–173.

Gaussen, H. (1954a). Théorie et classification des climats et microclimats. 8th Int. Bot. Congr. Paris 1954, Sect. 7, 124–130.

Gaussen, H. (1954b). Expression du milieu par des formules écologiques. Les Divisions écologiques du Monde (Colloque Intern. CNRS Paris 1954) pp. 13–23 and *Ann. Biol.* **31**, 257–267.

Gaussen, H. (1955). Détermination des climats par la méthode des courbes ombrothermiques. *C.R. Acad. Sci.* **240**, 642–644.

Hansen, B. (1973). Bibliography of Thai botany. *Nat. Hist. Bull. Siam Soc.* **24**, 319–408.

Hundley, H. G. (1961). The forest types of Burma. *Trop. Ecol.* **2**, 48–76.

Kermode, C. W., U. Thein, U. Tin Htut and U. Aung Din (1957). The forest types of Burma. *Burm. For.* **7**, 6–38.

Küchler, A. W. and J. O. Sawyer (1968). A study of the vegetation near Chiengmai, Thailand. *Trans. Kansas Acad. Sci.* **70**, 281–348.

Legris, P., F. Blasco, O. Dottin, J. Fontanel, J. Migozzi and L. Tichit (1972). Notice de la carte de végétation du Cambodge au 1/1000000. *Trav. Sect. Sci. Tech. Inst. Fr. Pondichéry, h.s.* **11**, 1–238.

Martin, M. A. (1973). Notes on the vegetation of the Cardamon Mountains, Cambodia. *Gard. Bull. Singapore* **26**, 213–222.

Mohr, E. C. J. (1933). De bodem der tropen in het algemeen en die van Nederlansch–Indië in het bijzonder. *Kon. Ver. Kolon. Inst. Amsterdam, Meded.* **31**, 91–110.

Ogawa, H., K. Yoda and T. Kira (1961). A preliminary survey on the vegetation of Thailand. *Nature Life SE Asia* **1**, 21–157.

Ogawa, H., K. Yoda, T. Kira, K. Ogino, T. Shidei, D. Ratanawongse and C. Apasutaya (1965). Comparative ecological studies on three main types of forest vegetation in Thailand. I. Structure and floristic composition. *Nature Life SE Asia* **4**, 13–48.

Reed, C. F. (1969). "Bibliography to Floras of SE Asia" 191 pp. P. M. Harrod Co., Towson.

Robbins, R. G. and T. Smitinand (1966). A botanical ascent of Doi Inthanon. *Nat. Hist. Bull. Siam Soc.* **21**, 205–227.

Rollet, B. (1972). La végétation du Cambodge. *Bois For. Trop.* **144**, 3–15; **145**, 23–38; **146**, 3–20.

Rosayro, R. A. de (1974). Vegetation of humid tropical Asia. Natural resources of humid tropical Asia. *Nat. Res. Res.* **12**, 193–215.

Royal Forest Department (1950). "Types and distribution of Forests of Thailand" 5 pp. Bangkok.

Schmid M. (1974). Végétation du Viet-Nam: le massif sub-annamitique et les régions limitrophes. *Mém. O.R. S.T.O.M.* **74**, 1–243.

Smitinand, T. (1966). The vegetation of Doi Chiengdao a limestone massive in Chiengmai. *Nat. Hist. Bull. Siam Soc.* **21**, 93–128.

Smitinand, T. (1977). A preliminary study of the vegetation of Surin Islands. *Nat. Hist. Bull. Siam Soc.* **26**, 227–246.

Smitinand, T. and C. Phengkhlai (1973). Conditions of the lowland tropical forests in Thailand. Pre-congress conference in Indonesia. Symposium: Planned utilization of the lowland tropical forests (Cipayung, Bogor, Java, 1971), 102–110.

Stadelman, R. C. (1966). "Forests of SE Asia" 245 pp. Winmer Bros., Memphis.

Thai Van Trung (1970). "The Forest Vegetation of Viet-Nam" 302 pp. Ed. Sci. Techn., Hanôi.

Tixier, P. (1965). Les parcs nationaux en Thailande. *Sci. Nat.* **72**, 1–16.

Tixier, P. (1966). "Flore et Végétation orophiles de l'Asie Tropicale: les Epiphytes du Flanc méridional du Massif Sud-annamitique" 240 pp. Sedes, Paris.

Tixier, P. (1972). Coup d'oeil sur le plateau Shan. *Sci. Nat.* **113**, 24–29.

Vidal, J. E. (1956). "La Végétation du Laos, 1ère Partie: Conditions Écologiques" 120 pp. Trav. Lab. For. Toulouse, France.

Vidal, J. E. (1959). Conditions écologiques, groupements végétaux et Flore du Laos. *Mém. Soc. Bot. Fr.* **1958**, 1–41.

Vidal, J. E. (1960a). "La Végétation du Laos, 2ème Partie; Groupements Végétaux et Flore" 455 pp. Trav. Lab. For. Toulouse, France.

Vidal, J. E. (1960b). Les forêts du Laos. *Bois For. Trop.* **70**, 5–21.

Vidal, J. E. (1961). Application au Laos de deux méthodes bioclimatiques. *C.R. Som. Biogéog.* **38**, 53–64.

Vidal, J. E. (1967a). Aspects biogéographiques du SE Asiatique. *C.R. Soc. Biogéog.* **43**, 130–140.

Vidal, J. E. (1967b). Paysages végétaux et fleurs d'Asie tropicale. *Sci. Nat.* **83**, 10–21.

Vidal, J. E. (1972a). Bibliographie botanique indochinoise de 1955 à 1969. *Bull. Soc. Etudes Indoch. n.s.* **47**, 655–749.

Vidal, J. E. (1972b). Bibliographical survey of taxonomic and phytogeographic studies related to Indochina from 1955 to 1969. Pre-cong. 12th Pac. Sci. Cong. (Cipayung, 1971), 111–117.

Vidal, J. E. (1973). Voyage botanique en Asie du SE. *Sci. Nat.* **120**, 27–37.

Vidal, J. E. (1976). Etude phyto-écologique autour du barrage de la Nam Ngum (Laos). Bull. Soc. Bot. Fr. **123**, 391–414.

Vu Van Cuong (1974). Flore et Végétation Hydrophytique du S Viet-Nam; 2 fasc. polygraph., 410 pp. Thesis, Paris.

Walker, E. H. (1952). A contribution toward a bibliography of Thai Botany. *Nat. Hist. Bull. Siam Soc.* **15**, 28–88.

Walter, H. (1955). Die Klimadiagramme als Mittel zur Beurteilung der Klimaverhältnisse für ökologische vegetationskundliche und landwirtschaftliche Zwecke. *Ber. D. Bot. Gesellsch.* **68**, 331–344.

Walter, H. (1971). "Ecology of Tropical and Subtropical Vegetation" 539 pp. Reinhold; New York

Walter, H. (1973). "Vegetation of the Earth" 237 pp. Springer-Verlag, New York.

Walter, H. and H. Lieth (1967). "Klimadiagramm-Weltatlas." Fischer Verlag, Jena.

Whitmore, T. C. (1975). "Tropical Rain Forests of the Far East" 282 pp. Oxford University Press, Oxford.

Williams, L. (1965). "Vegetation of SE Asia. Studies of Forest Types" I-IX, 1–302. US Dep. Agric., Washington, DC.

Exploration of the Flora of Thailand

K. LARSEN

University of Aarhus, Denmark

Botanical exploration of Thailand has mainly taken place in the twentieth century, and thus much later than exploration of the neighbouring countries. The main reason for this is that while nearly all tropical Asia shared fate with so many other parts of the tropics and became divided among the main colonial powers, England, Holland and France, Thailand remained an independent kingdom. In the following survey three periods have been recognized.

Period (1690–1900)—Early Botanists

The first botanist known to visit Thailand is E. Kaempfer, who did so in 1690. He does not, however, seem to have made any collections. Nearly a hundred years passed before the next botanical visit. This time the visitor was J. G. Koenig, who in 1778–1779 collected in various parts, but mainly in the Malay Peninsula. Most of his collections were destroyed by the bad weather conditions during his travels, and as Kerr (1939) expresses it: "It may be said that not a single specimen of Koenig's indubitably from Siam, is known, and probably not more than half a dozen of his from the West coast of the Malay Peninsula". With nearly the same meagre result, in spite of great efforts, some other early botanists travelled, mainly in the southern part of the country, viz. G. Finlayson 1821–1822, J. W. Helfer 1837–1839, Ch. S. P. Parish 1860 and Joh. E. Teysmann 1862. Very few specimens of these collectors from Thailand are existing today. A little more was left for posterity by Robert Schomburgk, who stayed in Thailand from 1857–1868. About 240 specimens from these years are deposited in Kew. It seems that the first most extensive collection to come out of the country was that of C. Thorel, who in 1867 travelled along the Me Kong river. It is difficult to see whether a plant is from Thailand or from Cambodia or Laos, but the result of his expedition was about 2000 specimens now deposited in Paris, approximately half of these may be from Thailand. Some other French botanists were among the pioneers in Thailand, viz. J. B. L. Pierre, who during an expedition in 1868 brought home about 300 items, mainly from the Ratburi province, and J. Harmand, who collected about 100 specimens in 1877 on a tour to the Surin province. The Dane C. A. Feilberg travelled in the country from 1868–1869 and collected about 400 items, now deposited in Copenhagen.

In the decade 1889–1899 C. Curtis paid several visits to Thailand. His collections, about 200 items, are now deposited in Singapore. Another important collection to be found in Singapore is that of A. Keith, who collected about 500 specimens in Bangtapan and its vicinity.

Far away from other early collections from Thailand are those of Mr and Mrs Bradley dating from 1890; about 200 items are now to be found in the University of California.

It should also be mentioned that F. H. Smiles in the years 1891–1894 made large collections within the Thai frontiers of those days. Most of his localities, however, are today to be found in the area which later on was included in Laos and, of the approximately 1000 dried specimens sent back to Kew, only about half a dozen originates from present-day Thailand.

A number of occasional visitors collected in Thailand at the end of the nineteenth century. W. Fox *c.* 1879, H. Kunstler 1881, H. J. Murtor 1881–1882, Henry of Orleans 1892, C. Goldham 1895 and E. Haase 1893. All in all these collectors have brought no more than 400 specimens to various herbaria. It should be noted that most of the plants sent to Berlin as the collections of Haase were lost during the Second World War.

The Second Period 1900–1940

At the turn of the century the botanical exploration of Thailand entered a new phase because of the appearance of some well-known professional botanists.

In 1899 D. T. Gwynne-Vaughan, at that time a member of the Skeat Expedition, collected in peninsular Thailand. His collections, about 450 specimens, are now found at Kew and Cambridge.

A collaboration between Danish botanists and Thai authorities was initiated in 1899 by the Danish Scientific Expedition 1899–1900, whose botanist was Johs. Schmidt. This expedition worked in the Chantaburi province and mainly on the island of Koh Chang. More than 1500 specimens were sent to the Botanical Museum in Copenhagen. For the first time thallophytes were collected in Thailand, 95 fungi, 95 lichens and 669 algae. Mosses are represented by 61 collections.

The activities eradiating from Singapore is also best placed in this period. In 1897 H. N. Ridley paid his first visit to the country and in 1900 M. Hanif started collecting in peninsular Thailand. Hanif was a plant collector trained in Penang. His Thai collections are nearly all from the west coast of peninsular Thailand. He was, on some trips, accompanied by H. Nur; their collections are in Singapore.

Two years after the departure of Johs. Schmidt from Thailand a young, 25-year-old medical doctor A. F. G. Kerr arrived in Bangkok, little knowing that he should be the great pioneer in the exploration of the flora of Thailand. Kerr's great work justifies a few more lines about him than about the earlier collectors; a detailed account is found in Jacobs (1962). Kerr was born in 1877 in Eire, he studied medicine and took his doctor's degree in 1901. That year he went to Australia and the next year to Thailand, where he worked until 1932 when he left for England. He died in 1942.

Kerr came to Thailand as a medical doctor, he settled in Chiengmai and soon became interested in orchids. On a leave to England he was encouraged to collect plants by the orchid specialist at Kew, R. A. Rolfe, to whom he showed his sketches of orchids made in Chiengmai. Also D. Prain and O. Stapf at Kew urged him to collect, and supplied him with equipment for that purpose—the snowball had been set in motion. Starting slowly it began to increase and during the next 25 years Kerr collected all over Thailand, which was nearly virgin country for a plant collector. In 1920 he was appointed Government Botanist and the Botanical Section was established under the Ministry of Commerce; Kerr was appointed Director of this first botanical institute in Thailand. During his stay in Thailand Kerr collected about 25,000 items, the largest collection by a single person made in the country. Most of the numbers are represented in Aberdeen, Bangkok (BK), Kew, and the British Museum, but as Kerr generally collected plenty of material, it was possible to distribute it to different herbaria.

It may be appropriate here to mention that A. F. G. Kerr had a younger brother, F. H. W. Kerr, who held a degree in Divinity. He, too, visited Thailand, where he stayed in Chiengmai in 1921 and collected about 455 items.

The first half of the twentieth century can be generally characterized by a growing interest in the study of the rich flora of Thailand. Detailed information is to be found in Bruun (1961).

C. C. Hosseus, a German botanist, who later worked in South America, visited Thailand in 1904–1905, and collected more than 800 specimens, the main set of which is in Munich. Another German, W. Credner, who was a geographer, visited Thailand in 1927–1929. It has not been possible to locate his collections, which may not have been very large, they are neither in Kiel nor in Munich; probably they were in Berlin. A small collection (about 180 items) of Zimmermann from 1901 is deposited in Munich.

A number of persons closely connected with the Royal Forest Department in Bangkok have done a good deal of collecting. Their names are encountered again and again and they deserve mention here. Some have contributed much, others with only a few collections. One in the first category is Mrs D. J. Collins who stayed in Thailand from 1877–1912. Most of her collections, about 2500 specimens in all, originate from the Sriracha-Chantaburi area. Her plants are in the British Museum.

Another important collector is H. B. G. Garrett, who was associated with the Royal Forest Department from 1899 to the time of his death in 1959. His collections, more than 1500 items, are of the highest quality and rich in duplicates.

A young Danish botanist G. Seidenfaden stayed for some time in Thailand in 1934–1935 and during this period collected about 550 items. There are few duplicates and the material is deposited in Copenhagen. After the war he came back as Danish ambassador to Thailand and initiated the fruitful Thai–Danish cooperation. Some other Danes collected some hundred items, now deposited in Copenhagen. Their names and collections are as follows: E. Lindhard 1900–1904 who collected mainly in the Tak area (Raheng on the labels) 81 items; E. Nielsen 1924, 167 items; C. W. Franck 1929–1930, and 1937 about 600 items, A. P. N. Vesterdal 1935–1936 about 500 items. Important collections were also made by Lakshnakara, who was Inspector General of the Ministry of Agriculture, Bangkok. He collected more than 1500 items mainly deposited in Bangkok and Kew.

One of Kerr's plant collectors, Put, has provided an important contribution to our knowledge of the Thai flora, as more than 4500 items were collected between 1926 and 1931. His plants are in Bangkok. Another of Kerr's assistants, Rabil Bunnag (Rabil) also joined the Botanical Section of the Forest Department in 1926. His contribution to BKF is about 360 items. Also Wanapruek Phichara (Vanpruk) was connected with the Forest Department, Bangkok. He collected about

1200 items, mainly in Northern Thailand. Nai Noe Israngkura (Noe) was another assistant of Kerr's. He joined the Department in 1920. His collections were included in Kerr's herbarium. Probably there are not much more than 300 specimens; also included in Kerr's herbarium we find about 90 specimens of Anuwat Wanaraks ("Anuwat") and some collections of D. R. S. Bourke-Borrowe. Finally Winit Wanandorn (Winit) should be mentioned as one of the more important associates of the Forest Department. He has contributed about 2000 items, also mainly from North Thailand. A small collection by Young (57 items) is in the British Museum.

A very good friend of Kerr's was A. Marcan, a mining engineer. He undertook several journeys with Kerr, particularly during Kerr's later years. He collected about 2800 items which were distributed to various herbaria. Most of the specimens may be found at Kew.

In this period the first Japanese collections were made in Thailand, as B. Hayata, on his way to Indochina, interrupted his journey in Thailand and collected about 1000 items now deposited in Tokyo (TI).

To end the list of collectors of this period we may add E. Smith, a medical doctor. She travelled in tropical Asia, and later concentrated on ferns. Her plants are in the British Museum. Also a zoologist, H. McCormick Smith, made plant collections in Thailand in the twenties; about 650 specimens are now in the US National Herbarium. Here we also find the collections of J. Fr. Ch. Rock (J. F. Rock), who collected about 1900 items in Thailand, mainly during three trips to Chiengmai (1919–1921).

A few other names are met with, chiefly in the herbarium in the Royal Forest Department, viz. T. N. Annandale, W. R. S. Ladell, R. M. De Schauncee and D. O. Witt; there is not much information to be found about these collectors and the number of specimens they contributed.

Period (1940–1979)

The war discontinued the botanical work in Thailand during the first years of this period. The first botanical post-war expedition was made in 1946 by S. Bloembergen, G. Den Hold and A. J. G. H. Kostermans in the Menam Kwae Noi Basin from Kanchanaburi (Kanburi), passing Sai Yok and up to the Burmese frontier at the Three Pagodas Pass. About 850 specimens of Phanerogams and 150 Cryptogams were collected, the first set is in Leiden. Later in the year Kostermans continued collecting work alone in the same region; about 450 specimens were collected, the first set of which is in the Arnold Arboretum, USA, while a nearly complete set is in Leiden.

From 1950 collecting activities increased considerably. All in all ≈40 000 collections existed from Thailand at that time. Today, 28 years later, ≈150 000 collections exist.

The most important contributions to our knowledge of the flora of Thailand have come from the following countries outside Thailand: Denmark, Japan and the Netherlands. Below I have tried to survey recent collectors and collections in these countries and in Thailand. Only collectors with more than 500 numbers are reviewed.

The Bangkok herbaria, BK and BKF, each have 50 000/60 000 specimens from Thailand, a large number of which are duplicates from the expeditions from the above-mentioned countries.

The herbarium of the Department of Agriculture (BK) has a nearly complete set of Kerr's collection, i.e. ≈23 000 specimens; furthermore it has several of the Thai collectors from before 1940. Staff members have also collected a considerable number of specimens: I. Badakorn 700, Ch. Chermisirivatana (= Chandraprasong) 2300, P. Sangkhachand 2000, S. Suthisorn 4500 and U. Yongboonkird 500. J. F. Maxwell who was associated with the herbarium 1972–1976, collected several thousand numbers (first and second duplicate sets are in AAU and L).

The herbarium of the Forest Department (BKF) has received a first duplicate set of all expeditions visiting Thailand since 1950. Staff members have taken part in many of these, besides carrying out a large collecting work of their own. By far the most important collector is T. Smitinand with 13 000 specimens. Other collectors are: B. Nimanong 1000, C. Phengklai 4000, L. Phuphatnapong 500, S. Phusomsaeng 1000, B. Sangkhachand 3000, T. Santisuk 1600 and P. Suvarnokoses 2000.

The Dutch collections are all deposited in L. duplicate sets have been distributed to AAU, C, E, K and P. Sleumer (1963) 500, Hennipman and Touw (1965) 1000 pteridophytes and 4000 mosses, Nooteboon (1968) 500, V. Beusekom (1968–1969) 3200, and Geesink (1971–1975) 5000.

Several Japanese expeditions have visited Thailand since 1958. Most of these have been sent out by the University of Kyoto, fewer from Tokyo and Osaka. The expeditions have collected all plant groups. Detailed information has not been available to me concerning the single collectors, but the expeditions from the University of Kyoto have used a special number series T1–T17 868. T. Tuyama collected 700 (TI) while Osaka City University expedition collected 3000 (KYO and TI).

A more detailed list of Danish collectors is shown in Table 1. The Danish expeditions have as a rule collected seven duplicate sets. These are distributed with a first set of BKF. The other sets have mainly been sent to E, K, KYO, L, and P.

Table 1. Danish collections in Thailand since establishment of the Thai–Danish botanical exploration. B—bryophytes, C—carpological collections, F—mainly flowering plants and ferns, O—orchid collected in spirit, W—wood samples, T—thallophytes.

Collectors		No.	Main set deposited	Material
Sørensen, Larsen and Hansen	(1958)	1–6700	C	F
Sørensen, Larsen and Hansen	(1958)	6701–6800	C	T
Sørensen, Larsen and Hansen	(1958)	6801–7250	C	F
Sørensen, Larsen and Hansen	(1958)	7251–7300	C	C, T, W
Sørensen, Larsen and Hansen	(1958)	7841–8000	C	C, T, W
Floto	(1959)	7310–7840	C	F
Larsen	(1961)	8001–9062	C	F
Larsen	(1962)	9063–9687	C	F
Larsen	(1963)	9733–10 727	C	F
Hansen, Seidenfaden and Smitinand	(1964)	10 770–11 719	C	F
Hansen and Smitinand	(1966)	11 720–12 569	C	F
Larsen, Smitinand and Warncke	(1966)	1–1873	AAU	B, F
Larsen and Warncke	(1968)	1874–3391	AAU	B, F
Hansen and Smitinand	(1968)	12 600–13 327	C	F
Charoenphol, Larsen and Warncke	(1970)	3392–5151	AAU	B, F
Larsen, Larsen and Nielsen	(1972)	30 600–32 508	AAU	F
Larsen and Larsen	(1974)	32 600–34 524	AAU	F
Løjtnant and Niyomdham	(1978)	1–227	AAU	F
Seidenfaden	(1957)	7000	private	0

The question that arises is whether Thailand is to be regarded as an undercollected area. Several of the easy accessable nature reserves e.g. Khao Yai, Phu Kradung and Doi Suthep are visited frequently by botanists and certainly "overcollected", while remote areas e.g. the border-jungles in the southern part of the peninsula, harbours a large number of unrecorded and even underscribed species.

A comparison with the area covered by flora Malesiana will elucidate the situation. Unfortunately we have to use the information from 1950 from the Malesian area and compare it with the present situation in Thailand. Table 2 shows the density index (average number collected per 100 square kilometres).

Table 2. Density of index of Thailand 1950 and 1978 compared with the Malesian area 1950.

	Surface	Coll. No.	Index
Java and islands	132 474	247 522	187
Malay Peninsula	132 604	191 055	145
Philippine Islands	290 235	180 090	62
Moluccas	63 575	27 525	43
Thailand 1950	514 000	40 000	8
Thailand 1978		140 000	27
Lesser Sunda Islands	98 625	24 545	25
Celebes	182 870	32 530	18
Sumatra and islands	479 513	87 900	18
Borneo and islands	739 175	91 550	12

The conclusion must be that Thailand is still a country which needs intense exploration, especially so as the forests are rapidly disappearing.

Flora of the Thailand Project

The seeds of this project were sown during the Thai–Danish botanical expedition 1958–1959 of which I was leader. Together with Tem Smitinand several expeditions were later carried out and eventually the plan for a Thai flora took form. In 1964 I arranged a meeting in Kew in order to investigate the possibilities of establishing a board of experienced taxonomists, and in 1967 the editorial board for the flora of Thailand was finally established with representatives from Thailand (BKF), Denmark (AAU and C), the Netherlands (L), France (P), UK (K and E), later also Japan (KYO) became represented. Board meetings have been held in Paris in 1972, Aarhus 1975, and in Kyoto in 1979.

The flora of Thailand is estimated to have ≈12 000 species of pteridophytes, gymnosperms and angiosperms. The pteridophytes comprise ≈400 species, and the gymnosperms are represented by only 25 species.

The following angiosperm families have been treated (the two numerals following the family name represent genera and species respectively): Actinidiaceae 2:3, Apostasiaceae 2:5, Balanophoraceae 1:5, Bonnetiaceae 1:1, Cardiopteridaceae 1:1, Centrolepidaceae 1:1, Conneraceae 6:16, Dilleniaceae 3:15, Flagellariaceae 1:1, Goodeniaceae

2:2, Haloragaceae 2:3, Hanguanaceae 1:1, Icacinaceae 13:18, Illiciaceae 1:2, Juncaceae 1:3, Lowiaceae 1:1, Magnoliaceae 7:13, Ochnaceae 4:5, Oxalidaceae 3:8, Portulacaceae 2:5, Rafflesiaceae 2:2, Restionaceae 1:1, Rhizophoraceae 7:14, Rosaceae 7:14, Saurauiaceae 1:4, Schizandraceae 1:3, Smilacaceae 2:27, Sphenocleaceae 1:1, Stylidiaceae 1:3, Theaceae 9:21, Triuridaceae 1:1, i.e. 31 families, 102 genera and 257 species.

Of the following families the following manuscripts have been accepted by the board: Caesalpiniaceae 20:114, Cannabaceae 1:1, Casuarinaceae 1:2, Droseraceae 1:3, Elaeocarpaceae 2:19, Epacridaceae 1:1, Hippocastanaceae 1:1, Irvingiaceae 1:1, Nyssaceae 1:1, Philydraceae 1:1, Pontederiaceae 2:3, Proteaceae 2:8, Salicaceae 1:2, Simaroubaceae 5:7, Symplocaceae 1:18, Valerianaceae 1:1, i.e. 16 families, 42 genera and 183 species.

Manuscripts are nearly complete for the following families: Bignoniaceae, Campanulaceae, Caprifoliaceae, Cyperaceae, Ebenaceae, Mimosaceae, Papilionaceae, Phaseoleae, Podostemonaceae and Xyridaceae. Several other families have been sent to specialists.

In spite of the great efforts of the board members, it must be realized that only 10% of the flora of Thailand have been treated during a period of 10 years. An acceleration of this procedure can only be obtained through new grants from international and national funds, thus enabling research centres in Thailand to enlarge their staff and encourage foreign specialists to take a more active part in revision work.

References

Bruun, A. F. (1961). Danish naturalists in Thailand 1900–1960. *Nat. Hist. Bull. Siam Soc.* **20**, 71–80

Jacobs, M. (1962). Reliquiae Kerrianae. *Blumea* **11**, 427–493

Kerr, A. F. G. (1939). Early botanists in Thailand. *J. Thailand Res. Soc. Nat. Hist. Suppl.* **12**, 1–27.

Dispersal and Distribution of Malesian Angiosperms

C. KALKMAN

Rijksherbarium, Leiden, Netherlands

Some time ago I had to give two lectures in a course on floristic plant geography. One was on dispersal types and I talked about the customary subjects, the vectors like wind and animals, and about the structural properties associated with the different kinds of dispersal. The second lecture was to treat the subject of areogenesis, and I put down before the students the several factors influencing the enlarging or diminishing, the shifting or disappearance of an area of distribution. Possibly under the influence of the just-held lecture on dispersal I wanted to elaborate on the relationship between dispersal type and area type, but on this point the literature I consulted let me down.

It has been said before that the distribution of a species is not a simple function of its dispersal abilities, and even that dispersal, apart from being a prerequisite for establishment, plays no role in areogenesis. A number of evident observations lay behind this scepticism: there are narrow endemics with dust seeds which appear to be well suited for long-distance dispersal, everyone can find species with similar areas but with their dispersal possibilities looking very different in efficiency.

Nevertheless, it might be expected that if one leaves the anecdotic field and takes a large enough sample of a large flora, correlations between area and dispersal become apparent. So I searched for literature where this connection would be elucidated, more or less expecting or at least hoping to find a paper in which for a certain region or plant group the maximum or average dimensions of the areas occupied by anemochores, exozoochores, myrmecochores, etc. would be given. I did not find it.

Certainly, several dispersal spectra have been published, with percentages of the dispersal types for either a certain vegetation type, a geographic area, or a certain area type. But this was not what I searched for and I decided to do some further exploring.

The choice of the Malesian angiosperms was of course guided by the presence close at my elbow of some "Flora Malesiana" volumes with recent data on the distribution of a few thousands well-revised species. What I want to present here, is not the result of many years of research, it is merely the result of a literature search, leaving—as we will see—a lot of questions.

There are two terms in my equation: dispersal and distribution. We start with dispersal, the most difficult one. What I had feared soon became apparent: very little is known about the actual dispersal methods in most Malesian angiosperm groups. This is not too strongly put. Actual observations on birds or other animals eating a certain kind of berry, on the wind taking winged fruits over such and such a distance, are very rare indeed. Dispersal is a very neglected field. Some monographers do not even mention the subject in their revisions, it is very rare if one finds an observation recorded on a herbarium label, and directed research in the field is virtually absent at least in Malesia. It may be better in other parts of the tropics, but judging from the floras and revisions I have seen, I doubt it.

I would like, therefore, to make two recommendations. First to the residents of the tropics and to travellers. There is need for a lot of observations about dispersal methods and their efficiency. This research must of course go distinctly deeper than the above-indicated relations "bird eats berry" and "wind carries seed". Numbers, percentages, distances, that is what we want to know. This kind of research must mainly come from people living in the area since the duration of an expedition is too short and the other duties will be such that only haphazard observations can be made. However, these are welcome too. Many botanists are not very observant towards animals, but the remark on the herbarium label: "birds eating the fruits" is better than nothing, even when the kind of bird remains a mystery. Botanists living in the tropics, however, can contribute a lot more than that. Not many of them are listening here, but I hope that the proceedings will be read and that my plea will therefore be heard. Virtually every plant is all right to start with, but wild plants from more or less natural habitats would be best.

The second recommendation is directed to taxonomists. Please, do not forget to record in your revisions what you know or suppose about dispersal, either from the labels or from literature, from correspondence, traveller's stories or from your own observation in the field. If floras would more generally transmit this kind of information, the users of the flora might be made more dispersal-minded.

Talking about floras, I would like to make a few remarks. Floras there are of several kinds and one can classify them in several ways. One possibility is to divide them into identification books and information books. Most floras of tropical regions (and other regions too, for that matter) are of the identification type: they lead you to the name of a species, with the aid of keys and descriptions. Other information is restricted, but data on habitat and distribution within or outside the area are always presented.

The information type flora aims at giving much more information about the plants. Hegi's "Flora von Mittel-Europa" is a very good, probably even the best example, and it is one of the very few floras to which one can confidently turn if one wants to know something of a more general nature about a plant. Naturally also the information flora cannot print all information theoretically available, and every editor and author has to make his choice. Will he mention the pollen type, the chromosome number, the technological properties of the timber, the insects responsible for pollination, the type of dispersal?

"Flora Malesiana" is a flora of the "information type", and in most of the family revisions some indication was given about the little that is known about dispersal in the family. There were also families, however, where the word dispersal did not occur in the treatment. Mostly I had to rely on the properties of fruits and seeds, in combination with habit and habitat, to make an educated guess about the most probable dispersal type. These guesses are based for a large part on insights gathered in Europe, and they are certainly not infallible. Not all fruits with a more or less fleshy outer layer will be eaten by animals and if so, not all of them will be effectively dispersed in that way. But most of them will, I think, and the fleshy fruits and the seeds with fleshy or juicy parts (sarcotesta, arillus) were, consequently, considered to be endozoochorous.

I classified the several dispersal types in only three classes, depending on the normal dispersal ranges:

(i) The diaspore is fit for covering small distances only, distances which can be measured in metres. To this class belong the barochores (just falling down), ballistochores (seeds scattered out of the capsules), autochores (with explosion mechanisms), myrmecochores (dispersed by ants), anemochores with heavy diaspores or inefficient specializations, and the species dispersed by rain-wash. It must be realized that, although the distance normally covered is small, there will be circumstances under which that distance becomes substantially larger. A

Lithocarpus fruit will normally fall not far from the tree, but sometimes some of them will fall in a brooklet at a time when there is more water in it than usual, and the fruits will be washed ashore some way down. A *Viola* seed, equipped for transport over some metres by ants, may be swept by rainwash down a hill, become part of a landslide and end up hundreds of metres from its parent plant.

(ii) The diaspores are fit to cover larger distances to be measured in hundreds of metres. To this class belong the anemochores with better specializations and/or lighter in weight than those in (i). In the present class are also placed the species relying on water dispersal by rivers and brooks. In these cases there is often no specialization in the diaspores but the kind of dispersal is indicated by the place where the plant grows: in or near water, in marshes etc. To this class I also reckoned the epi- and endozoochores, since normally the distance covered by animals will not be very large.

(iii) The diaspores are fit to cover large distances, measured in kilometres. To this class I only reckoned the species with dust seed and those which by their habitat and structure are fit for sea-dispersal.

Also in the classes (ii) and (iii) there is a normal range and an exceptional range. Floods and heavy storms may carry a diaspore, or a bird carrying a diaspore, over distances not normally covered.

The ranges of the three classes are summarized in Table 1, which, although is appears very mathematical, is no more than a rough approximation.

Table 1. Approximate range of dispersal of seeds in the three classes.

Distance	Normal range	Exceptional range
Short	0–100 m	1 km
Medium	100–1000 m	10 km
Long	1–10 km	100 km

We turn to the second term of the equation, the distribution. The dimensions of an area can of course be measured in kilometres. Since Malesia is so conveniently divided into a number of islands and island groups, I preferred to distinguish three classes in a less exact manner:

(i) small area—endemic to one island or island group,

(ii) medium area—occurring in two or three neighbouring islands

(iii) large areas—covering more than three islands or present in only two or three but these not neighbouring.

The middle class has, of course, to be distinctly smaller than the entire area. Four adjoining islands may in some cases cover too large a part of the region and that is why I took three as the limit.

For each species treated in "Flora Malesiana" up till now, as far as not obviously introduced or dispersed by man, I defined its area class (small, medium or large) and its dispersal class (short, medium or long distance). Some species had to be left out when the recorded facts about fruit and seed structure were too meagre for my imagination.

I am certain that listening to the story, you have collected in your minds a number of objections. I want your permission, nevertheless, to present my results, summarized in Table 2.

Table 2. Number of species in each of the dispersal and area (small, medium and large) classes.

Dispersal	Small	Medium	Large	Total
Short	222	122	195	539
Medium	1431	349	737	2517
Long	13	7	52	72
Total	1666	478	984	3128

The total number of slightly more than 3000 species covers about 13% of the total number of species in the area. See Tables 3 and 4, and Fig. 1.

Table 3. Area classes in percentages for each of the three dispersal classes and for the entire sample.

Areas	All 3128 spp.	539 spp. short distance	2517 spp. medium distance	72 spp. long distance
Small	53	41	57	18
Medium	15	23	14	10
Large	31	36	29	72

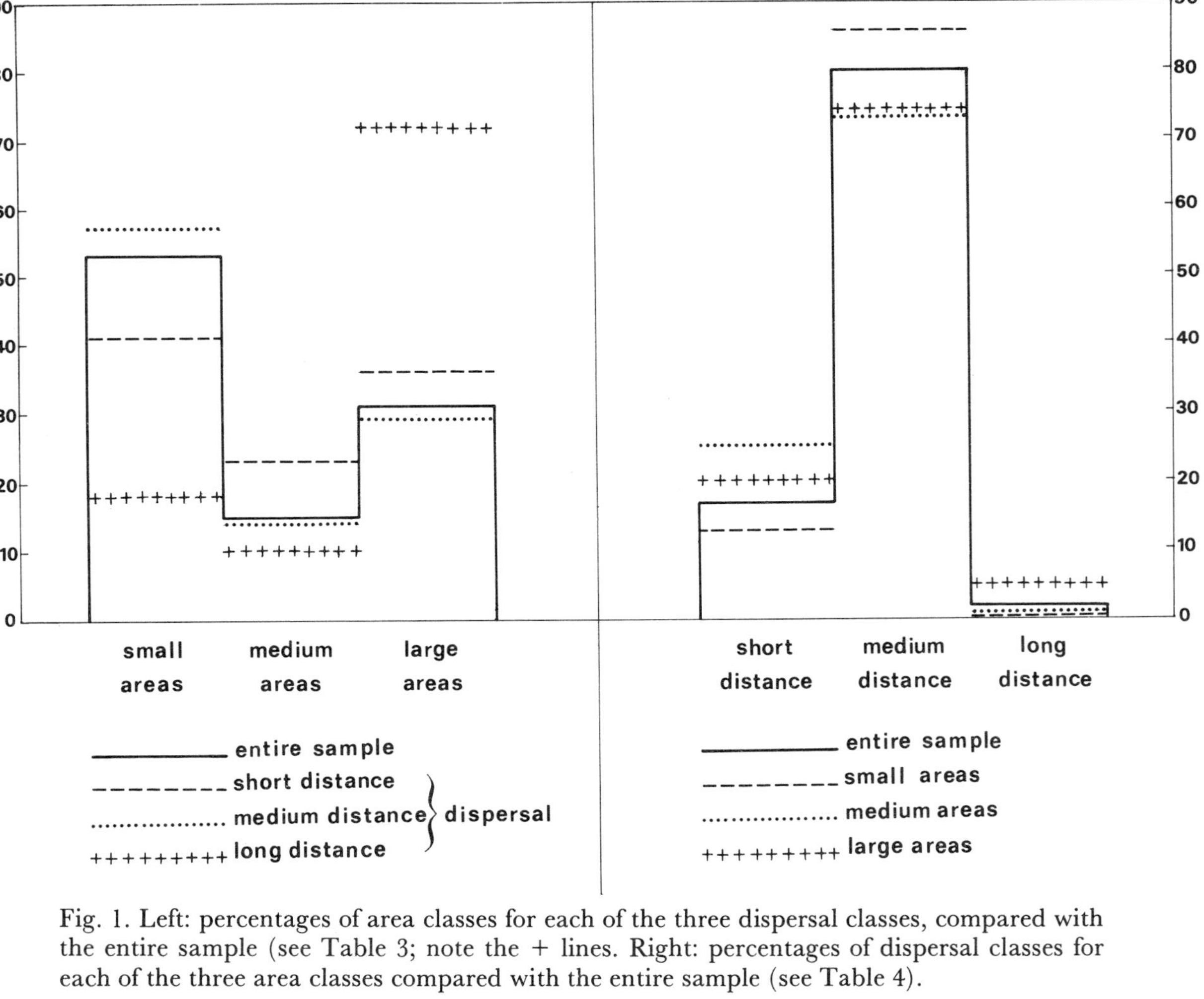

Fig. 1. Left: percentages of area classes for each of the three dispersal classes, compared with the entire sample (see Table 3; note the + lines. Right: percentages of dispersal classes for each of the three area classes compared with the entire sample (see Table 4).

Table 4. Dispersal classes in percentages for each of the three area classes and for the entire sample.

Distance	All 3128 spp.	1666 spp. small areas	478 spp. medium areas	984 spp. large areas
Short	17	13	25	20
Medium	80	86	73	75
Long	2	<1	<2	5

It seems to me that this material is loaded with too many botanical uncertainties to be evaluated statistically. Significant is that the long-distance disseminators have more often a large area of distribution than the representatives of the other two dispersal classes and less often a small area; but their number is small.

Apart from this correlation the conclusion must be, at least for the time being, that the dispersal method is of subordinate importance for areogenesis as compared with other factors. This agrees with the findings of population ecology which indicate that it is the period between germination and establishment which is decisive for survival.

On the other hand, I still cannot quite accept that dispersal efficiency, to be measured in numbers of diaspores per unit of time, and in the average distance covered by them, would not be mirrored in the areas covered by the species if the sample taken is large enough. So in my mind the question arises whether the picture would change when more species are entered, area classes are made smaller, and dispersal classes are made more precise.

The first two adjustments are relatively easy. The sample will grow as more "Flora Malesiana" revisions are made. On the other hand, in a study long ago on the flora of the Lesser Sunda Islands I could show that a sample of ~10% of the species gave a reliable floristic picture.

The third adjustment of course is the crux of the matter. It is not merely a statistical refinement, but it calls for more and better botanical knowledge. And so we are back where I started: too little is known about the realities of dispersal in Malesian Angiosperms.

Maybe it is not altogether inappropriate to have this plea for more knowledge at the beginning of this symposium. I would not say that it is the most urgent field which has to be better explored in tropical botany, but it is certainly one of them.

Phytogeography of the Tropical Flora
of the Indian Desert

M. M. BHANDARI

University of Jodhpur, India

Introduction

In order to discuss the phytogeography of the Indian (Thar) Desert, it is necessary to give a brief survey of its paleogeography. However, in the short time available I will not be able to do more than touch on a few salient phytogeographical features and problems. This land which occupies an area of more than 30 000 sq km, is phytogeographically heterogeneous, since it is a meeting place of western as well as eastern plant taxa.

The Indian Desert is the eastern extension of the great Saharo-Sindian Desert and includes the arid and semi-arid regions of the Indian States of Gujarat, Rajasthan, Haryana and Punjab and West Punjab of Pakistan. It is approximately 100 metres above mean sea level and is situated nearly 300 km west of Aravallies and 600 km north of Kutch. Most of the area consists of dry undulating plains of hardened sand, and the remaining region is a rolling plain of loose sand forming shifting sand-dunes of longitudinal and transverse types. The whole area is part of "maroosthali"—the region of death. A large part of the region in the past is supposed to have remained under the Indian Ocean. Indeed in many parts of Jaisalmer there still remains the illusion of a receded tide which has left large stretches of parched land, punctuated by sand-dunes in systematized curves. Biota of the Thar Desert has western as well as eastern affinities.

Besides the desert's biogeographically important situation, its origin is still debated, although the arid condition is said to be mainly due to climatic oscillation which took place in the last phase of glaciation during the Pliestocene period. The edge of the south-west monsoon shifted eastwards, the extra-peninsular rivers shifted westwards and there was a slow elevation of the Himalayas. With the westerly shift of rivers and easterly shift of monsoon, the forests were cleared and the place which was once the cradle of civilization of Mahinjodaro and Harappa was transformed into desert. It has been estimated to be 4000–10 000 years old (Krishnan, 1952; Wadia, 1960).

These conclusions are based on the archaeological evidences that the Sindo-Rajasthani (Tharian) Desert was well-forested, and swampy conditions existed there in the recent past. Wadia (1960) who has discussed the evolution of the Asian Deserts, has pointed out that before Mid Pleistocene, North India was inhabited by a very large mammalian population, including higher orders. The present shrubby vegetation hardly gives an idea of the thick forests of 4000 B.C.

The present deteriorating conditions, therefore, seems to be due mainly to the change of climate, but man with his sharp axe and his ever hungry herd of goats and camels, is an equally important factor in creating the present conditions. Due to the warmer and drier climate people took to nomadic life, and forests were cleared for firewood much beyond their natural reparative power. Continuous grazing has almost completely prevented natural regeneration. Fossil records certainly play an important role in deciding the origin and evolution of the desert, but the poor fossil material available for studies can only be based on an intimate knowledge of the taxonomy and geographical distribution of the present day flora.

Distribution of the biota throw some light on the origin and the present desertic conditions in general. Blatter and Hallberg (1918–1921) and Biswas and Rao (1953) supported Drude (1890, 1913), who showed a line of demarcation between the "Indo–Malayan" flora and "Perso–Arabian" region from the Gulf of Combay northwards along the Aravallis. Zoogeographical studies (Pocock, 1939–1941; Prakash, 1963) further supported these conclusions. However, Nair and Kanodia (1959) and Vyas (1962) took exceptions to such difference. It has further been pointed out by Vyas (1964) that in Alwar (in eastern Rajasthan), the Perso–Arabian elements predominate over the Indo–Malayan elements. Subsequent to these studies considerable data on the biology of this desert have accumulated (Bhandari, 1954, 1963a,b, 1964a,b, 1965; Prakash, 1962, 1963, 1963–1964) and it has, therefore, been considered proper to review the phytogeography of the Indian Desert in the present communication.

Phytogeography

The plants included are based on the "Flora of Indian Desert" (Bhandari, 1978). Only indigenous taxa have been considered. It may, however, be pointed out that the term "element" often used by authors (e.g. Blatter and Hallberg, 1918–1921; Eig, 1931; Meher-Homji, 1965) has not been employed here, since it seems pedantic to use this term as its meaning differs from author to author and also upon the principle used in such groupings. Moreover, it is desirable that the division of taxa into elements should be based on geography, genetics, history, ecology and dynamics of vegetation. In the present communication, when a taxon has been designated as say Afro–Tharian, it does not mean that the taxon in question has migrated from Africa to Thar, but only that one of the two limits of the taxon in question is Africa and that the other is Thar. It could as well be designated as Tharo–African.

Taxa which have Migrated to the Rajasthan Desert from the West

Taxa Extending up to Thar

Afro–Thariam (Africa to Thar Desert)

Abutilon ramosum (Cav.) Guill. and Perr., *Anticharis senegalensis* (Walp.) Bhandari, *Barleria hochstetteri* Nees, *Cenchrus prieurii* (Kunth) Maire, *Corchorus depressus* (L.) Christensen, *Convolvulus microphyllus* Sieb. ex Spreng., *Dactyliandra welwitschii* Hook.f., *Indigofera hochstetteri* Baker, *Latipes senegalensis* Kunth., *Neurada procumbens* L., *Pegolettia senegalensis* Cass., *Polygala irregularis* Boiss., *Solanum albicaule* Kotschy ex Dunal.

Sharo–Tharian (Sahara to Thar Desert)

Barleria acanthoides Vahl, *Dignathia hirtella* Stapf, *Dipterygium glaucum* Decne., *Leptadenia pyrotechnica* (Forsk.) Decne.

Irano–Tharian (Iran to Thar)

Aristida hystricula Edgew., *Calligonum polygonoides* L., *Commiphora wightii* (Arn.) Bhandari, *Dipcadi erythraeum* Webb. et Berth., *Enneapogon persicus* Boiss., *Glossonena varians* (Stocks) Benth., *Lycium barbarum* L. ex Hook.f., *Tamarix aphylla* (L.) L. Karst.

Taxa Extending Beyond Thar

Afro–Indian (whole of Africa and India)

Actinopteris radiata (Swartz) Link., *Althaea ludwigii* L., *Aristolochia bracteolata* Lamk., *Barleria prionitis* L., *Bergia suffruticosa* (Delile) Fenzl., *Cocculus pendulus* (J. R. G. Forst) Diels., *Cordia gharaf* (Forsk.) Ehrenb. and Asch., *Cyperus conglomeratus* Rottl., *Heliotropium bacciferum* Forsk., *Hibiscus lobatus* (Murr.) O. Ktze., *Hygrophila auriculata* (Schum.) Heine, *Ipomoea dichroa* (Roem. and Schult.) Choisy, *Salsola baryosma* (Roem. and Schult.) Dandy, *Striga gesneroides* (Willd.) Vatke, *Urginea indica* (Roxb.) Kunth., *Vicoa indica* (L.) DC.

Afro–Deccanian (Africa to peninsular India)

Balanites aegyptica (L.) Del., *Peganum harmala* L.

Saharo–Indian (Sahara to India)

Arnebia hispidissima (Lehm.) DC., *Limeum indicum* Stocks ex T. Anders.

Irano–Indian (Iran to India)

Convolvulus rottlerianus Choisy, *Haloxylon recurvum* (Moq.) Bunge ex Boiss., *Pergularia daemia* (Forsk.) Chiov., *Tecomella undulata* (Sm.) Seems.

Irano–Deccanian (Iran to peninsular India)

Erodium cicutarium (L.) L'Herit. ex Ait.

Afro–Malayan (Africa to Malaya)

Under this category examples have been included on the basis of the distribution of taxa in Africa as well as in Malayan/Australian regions which shows their probable direction of migration. *Bidens biternata* (Lour.) Merrill, *Indigofera linifolius* (L.f.) Retz., *Ipomoea coptica* (L.) Roth., *Nothosaerva brachtiata* L., *Trigonella corniculata* L.

Afro–Australasian (Africa to Australia)

Abutilon glaucum (Cav.) Sweet, *Ammannia baccifera* L., *Apluda mutica* L.,

Blumea lacera (Burm.f.) DC., *Corchorus aestuans* L., *Cyperus rotundus* L., *Dentella repens* (L.) Forst., *Glossostigma diandra* (L.) O. Ktze., *Indigofera cordifolia* Heyne ex Roth., *Ipomoea eriocarpa* R. Br., *Peplidium maritimum* (L.f.) Aschers, *Sehima nervosum* (Rottl.) Stapf.

Taxa which have Migrated to the Rajasthan Desert from the East

Taxa extending up to Thar (Indian)

Alysicarpus procumbens (Roxb.) Schindl., *Andrographis echioides* (L.) Nees, *Barleria dichotoma* (Roxb.) Prain, *Clerodendrum phlomoides* L.f., *Cassia auriculata* L., *Echinops echinatus* Roxb., *Grewia tenax* (Forsk.) Fiori, *Kickxia ramosissima* (Wall.) Janchen., *Leucas cephalotes* (Roth) Spreng., *Lindenbergia indica* (L.) O. Ktze., *Mimosa hamata* Willd., *Pithecolobium dulce* (Roxb.) Benth., *Potamogeton pectinatus* L., *Rhus mysorensis* Heyne ex W. and A., *Wrightia tinctoria* R.Br.

Indo–Malayan

Borreria articularis (L.f.) Will., *Grewia abutifolia* Vent. ex Juss., *Momordica dioica* Roxb. ex Willd., *Scirpus littoralis* Schrad.

Sino–Indian

Cuscuta chinensis L., *Cyperus michellanus* (L.) Link., *Heliotropium ellipticum* Ledeb., *Najas australis* Borry ex Chamiss., *Polypogon monsplensis* (L.) Desf.

General Elements

Tharian (Endemics in the Thar Region)

Macro–endemics

Abutilon cornutum Dalz. ex Cooke, *Acacia jacquemontii* Benth., *Aerva pseudo-tomentosa* Blatt. and Hallb., *Ammania desertorum* Blatt. and Hallb., *Anogeissus rotundifolia* Blatt. and Hallb., *Aristida royleana* Trin. et Rupr., *Cadaba fruticosa* (L.) Druce., *Convolvulus scindicus* Stocks, *Corallocarpus conocarpus* (Dalz. ex Dalz. et Gibs.) C. B. Clarke., *Euphorbia jodhpurensis* Blatt. and Hallb., *Glossocordia setosa* Blatt. and Hallb., *Haloxylon multiflorum* (Moq.) Bunge, *Hibiscus punctatus* Dalz., *Ipomoea sindica* Stapf, *Lasiurus ecaudatus* Stay. et Shank., *Malhania magnifolia* Blatt. and Hallb., *Pluchea wallichiana* DC., *Psoralea odorata* Blatt. and Hallb., *Pulicaria rajputanae* Blatt. and Hallb., *Rhynchosia rhombifolia* Blatt. and Hallb., *Sida tiagii* Bhandari, *Tephrosia falciformis* Ramas., *Tribulus rajasthanensis* Bhandari et Sharma, *Zizyphus truncata* Blatt. and Hallb.

Micro—endemics

Abutilon indicum var. *major* Blatt. and Hallb., *A. fruticosum* var. *chrysocarpa* Blatt. and Hallb. *Anticharis glandulosa* var. *caerula* Blatt. and Hallb. ex Santapau, *Aristida adscensionis* var. *pumila* (Decne.) Coss. et Dur., *Barleria prionitis* var. *dichantha* Blatt. and Hallb., *Cleome gynandra* var. *nana* (Blatt. and Hallb.) Bhandari, *Cucumis melo* var. *momoridica* Cogn., *Eragrostis ciliaris* var. *brachystachya* Boiss., *Dipteracanthus patulus* var. *alba* (Saxton) Bhandari, *Pavonia arabica* var. *glutinosa* Blatt. and Hallb., *Polygonum plebejum* var. *sindicum* Hook. f., *Trianthema pentendra* var. *flava* Blatt. and Hallb.

Besides these elements, there are several other taxa which are grouped under paleotropical, neotropical and cosmopolitan (See Table 1).

Discussion

The phytogeographical patterns of taxa in western Rajasthan indicate that the flora of desert is an admixture of western as well as eastern elements. 42·3% of taxa have migrated from west to east, 21·9% from east to west, and general elements constitute 35·8%. The western element in the flora is therefore predominant (see Table 1).

When the above figures are compared with those given by Blatter and Hallberg (1918–1921) for the flora of desert region, the difference seems to be appreciable, the reason being that the earlier authors included the Indian tropics of the old world, tropics in general, all warm countries, endemics, cosmopolitan, temperate and subtropical taxa in one group under the general elements, while in the present study the groupings are different. It is, however, certain that the number of taxa which have migrated from the west is higher than those which have migrated from the east. Recent studies (Vyas, 1962, 1963; Biswas and Rao 1963; Nair and Kanodia, 1959; Nair and Nathawat, 1957) in other parts of Rajasthan have shown that the proportion of the western taxa (elements) gradually decrease as one proceeds away from the Aravallis towards the south-east.

Table 1. The percentage of taxa from different regions.

Taxa	Taxa %
Western elements	
Afro–Tharian	8·7
Saharo–Tharian	2·0
Irano–Tharian	3·5
Afro–Indian	17·0
Afro–Deccanian	0·7
Saharo–Indian	1·0
Irano–Indian	1·6
Irano–Deccanian	0·2
Afro–Malayan	2·3
Afro–Australian	5·3
Eastern elements	
Up to Thar (Indian)	15·2
Indo–Malayan	2·0
Sino–Indian	2·0
Australasian–Indian	2·7
General elements	
Tharian	6·4
Macroendemics	4·26
Microendemics	2·13
Palaeotropical	6·9
Noetropical	13·7
Cosmopolitan	8·8

The distribution pattern of the desert flora also indicates the migratory routes for the desert plants in the past. It seems that west to east migration should have started from Sahara and Iran. Some of the plants, like *Barleria acanthoides*, *Dipterygium glaucum*, *Dignathia hirtella*, *Leptadenia pyrotechnica*, etc. are restricted up to Thar while a few like *Arnebia hispidissima*, *Tecomella undulata*, *Limeum indicum* and *Orobanche cernua* var. *nepalensis* have extended into most parts of Indian. The distribution of this pattern is 8·7%. A similar pattern of migration and ditribution is found within the forms speculated to have migrated from the Iranian region.

While plants like *Aristida hystricula*, *A. pogonoptila*, *Blepharis sindica*, *Calligonum polygonoides*, *Dipcadi erythraeum*, *Haloxylon salicornicum*, *Lycium barbarum* and *Tamarix aphylla* etc. find their eastern limits in Thar, some like *Haloxylon recurvum*, *Corallocarpus epigaeus*, *Leucas uriticaefolia*, *Zizyphus nummularia*, *Pergularia daemia* and *Tecomella undulata* extend to other semi-arid regions of India. It is, however, remarkable that the number of Afro–Tharian elements in the flora of western Rajasthan are appreciably high, which indicates that in the Indian Desert migration has mostly been from the African continent. This is further supported by the fact that out of a total of 549 indigenous taxa in the Indian Desert, 200 plants are common to India and Africa, of these 60 are distributed up to Thar while 149 extend to the other parts of the Indian subcontinent, the number reaching peninsular India being very low (10). Thirteen and 30 plants reach Malaya and Australia respectively. From the eastern side of the Thar 118 plants are found up to Thar; the number from Malayan, Sindo–Indian and Australasian–Indian side is 12, 11 and 15 respectively.

It may, however, be pointed out that recent studies on the flora of the Indian desert have apparently shown a few instances of disjunct distribution e.g. *Dactyliandra welwitschii* (Bhandari and Singh, 1964). Burtt (1971) regards this as one of the examples of widely disjunct components of an old tropical or subtropical arid flora. Two possibilities may be responsible for discontinuous distribution of such taxa. Firstly the flora of the intermediate regions might not have been fully explored, or some of the species in the intermediate areas might have been mistaken or confused with some other species, and ultimately the gaps in the distribution may have been reduced.

It has been shown that high endemics and the occurrence of a large number of local subspecies provide evidence of high phylogenetic plasticity (Mani and Singh, 1963; Roonwal and Chhotani, 1965) and intense speciation on account of the extreme dynamics of the ecological conditions, recent geological, physiographic, topographical and altitudinal changes. Recently Wadia (1960), Krishnam (1952), Randhawa (1945) and Seth (1962) have produced interesting evidences to show that desiccation in northern India set in about 3000–10 000 years ago, and even the areas which are desert today were covered with a wet tropical evergreen vegetation similar to the Indo–Malayan regions. This has been supported by the discovery of fossil leaves from Tertiary beds of Rajasthan (Lakhanpal and Bose, 1951) which have affinities with *Mesua*, *Garcinia* and *Calophyllum* of the Guttiferae. The presence of animals of the swamps, like elephants and rhinoceros in Sind and West Punjab is proved from the seals recovered from Mohinjodaro and Harappa which date from 3250–2750 B.C. Sind and West Punjab are now practically desert. Of the 36 plant taxa which are endemic to Thar, 24 are species while 12 are subspecies, 6·4% of the total flora. It is possible however, that this number may not be final, and a detailed analysis of the biota would add more species. When these figures are compared with those in the flora of the Sahara (Quezel, 1965) the results are interesting, since 3–5% endemics has been reported from this region as compared to 6·4% in the Indian Desert.

A high percentage of endemic species, specially when the age of the Indian Desert is estimated to be only 10 000 years old is therefore difficult to explain. The evolutionary process for a species or subspecies is a long term phenomenon and cannot be completed in such a short period of time. It is, however, probable that either the age of the desert from biological evidence, is more than that estimated by geologists and archaeologists, or the process of speciation has been activated due to rapid ecological changes in the near past. The third speculative reasoning could be that aridity established itself much earlier and the biota started evolving gradually. With the present knowledge on the subject, it would be difficult to give conclusive reasoning in favour or against any speculation.

References

Bhandari, M. M. (1954). On the occurrence of *Ephedra* in the Indian Desert. *J. Bombay Nat. Hist. Soc.* **52**, 10–13.

Bhandari, M. M. (1963a). Notes on India Desert plants. I. New records for the N. W. Indian Desert. *Proc. Raj. Acad. Sci.* **10**, 40–50.

Bhandari, M. M. (1963b). Notes on Indian Desert plants. II. The identity and nomenclature of *Talinum portulacifolium. Ann. Arid. Zone.* **1**, 176–180.

Bhandari, M. M. (1964a). Notes on Indian Desert plants. III. Critical notes on some recently reported new records for N.W. Rajasthan. *Ann. Arid. Zone* **2**, 181–184.

Bhandari, M. M. (1964b). Notes on Indian Desert plants. IV. New names and combinations. *Bull. Bot. Surv. India* **6**, 327–328.

Bhandari, M. M. and D. Singh, (1964). *Dactyliandra welwitschii* (Hook. f.) Hook. f.—*Cucurbitaceous* genus new to Indian Flora. *Kew Bull.* **19**, 332–334.

Bhandari, M. M. (1965). A note on the identification of some unrecorded Desert plants from Kutch. *J. Bombay Nat. Hist. Soc.* **62**, 332–334.

Bhandari, M. M. (1978). "Flora of the Indian Desert." Scientific Publishers, Jodhpur.

Biswas, K. and R. S. Rao (1953). Rajputana desert vegetation. *Proc. Nat. Inst. Sci. India* **19**, 134–138.

Blatter, E. and F. Hallberg (1918–1921). The flora of Indian Desert. *J. Bombay Nat. Hist. Soc.* **26**, 218–246, 525–551, 811–818; **27**, 40–47, 270–279, 506–519.

Burtt, B. L. (1971). From the South: an African view of the Floras of Western Asia. *In* "Plant Life of South-west Asia" (P. H. Davis, P. C. Harper and I. C. Hedge, eds) p. 146. Botany Society, Edinburgh.

Drude, O. (1890). "Handbuch der Pflanzangeographie." Stuttgart.

Drude, O. (1913). Die Oekologie der Pflanzen. *Die Wiss.* **50**, 308.

Eig, A. (1931). Les elements et les groupes phytogeographiques auxiliaries dans la flora Palestinienne. *Fedde Rep. Sp. Novarum Regni Vegetablis. Beih.* **63**, 201.

Krishnan, M. S. (1952). Geological history of Rajasthan and its relation to present day conditions. *Bull. Nat. Inst. Sci. India* **I**, 19–31.

Lakhanpal, R. N. and M. N. Bose (1951). Some tertiary leaves and fruits of the Guttiferae from Rajasthan. *J. Indian Bot. Soc.* **30**, 132–136.

Mani, M. S. and S. Singh (1963). Entomological survey of Himalaya. Pt. 26. A contribution to our knowledge of the geography of high altitude insects of the nival zones from the north-west Himalaya. Pt. 6. *J. Bombay Nat. Hist. Soc.* **60**, 160–172.

Meher-Homji, V. M. (1965). On the Sudano-Deccanian Floral element. *J. Bombay Nat. Hist. Soc.* **62**, 15–18.

Nair, N. C. and K. C. Kanodia (1959). Study of the vegetation of Ajit Sagar Bundh, Rajasthan. *J. Bombay Nat. Hist. Soc.* **56**, 523–557.

Nair, N. C. and G. S. Nathawat (1957). Vegetation of Harshnath, Aravalli Hills. *J. Bombay Nat. Hist. Soc.* **54**, 281–301.

Oldham, C. F. (1874). Lost river of the Indian Desert. *Calcutta Rev.* **59**, 1–27.

Pocock, R. I. (1939–1941) "Fauna of British India, Mammalia" Vol. 1. Taylor and Francis, London.

Prakash, I. (1962). Ecology of the gerbils of the Rajasthan Desert, India. *Mammalia* **26**, 311–331.

Prakash, I. (1963). Zoogeography and evolution of the Mammalian fauna of Rajasthan Desert, India. *Mammalia* **27**, 342–351.

Prakash, I. (1963–1964). Taxonomical and ecological account of the mammals of the Rajasthan Desert. Ann. Arid Zone. 1, 142–161; 2, 150–161.

Quezel, O. (1965). Contribution a l'etudede l'endemism chez les Phanerogames Sahariens. *C. R. Somm. Séanc. Soc. Biogeog.* **359**, 89–103.

Randhawa, M. S. (1945). Progressive dessication of nothern India in historical times. *J. Bombay Nat. Hist. Soc.* **45**, 558–565.

Roonwal, M. L. and O. B. Chhotani (1965). Zoogeography of termites of Assam region, India with remarks on speciation. *J. Bombay Nat. Hist. Soc.* **62**, 19–31.

Seth, S. K. (1962). A review of evidence concerning changes of climate in India during the Proto-historical and historical period. *Indian For.* **88**, 2–18.

Vyas, L. N. (1962). Vegetation of Jaisamand Lake Alwar. *Proc. Raj. Acad. Sci.* **9**, 45–57.

Vyas, L. N. (1963). Studies on the phytogeographical affinities of the flora of north-east Rajasthan. *Indian For.* **90**, 535–538.

Vyas, L. N. (1964). Vegetation of Alwar and its relationship with the north-eastern Rajasthan. *J. Indian Bot. Soc.* **43**, 322–333.

Wadia, D. N. (1960). The post-glacial dessication of Central Asia: evolution of the arid zone of Asia. Monogr. Nat. Inst. Sci. India, Delhi.

Indian Plant Taxonomy

ROLLA. S. RAO

Andhra University, Waltair, India

The scientific study of plants and animals began as early as 4000 B.C. in several ancient cultures of Babylonia, Egypt, Greece, China and India. Ancient Indian scriptures on plants, mainly in relation to medicine, agriculture and horticulture, e.g. Charaka Samhita, Susruta Samhita and Ashtangahridaya Samhita, along with the references in these and other associated works, form the main basis of critical analytical studies on Indian plants, which number about 700 in all. The absence of a workable morphological description of plants is due mainly to the gurukula system* of those days, and the translations and interpretations (nighantu writers) during the last few centuries, with their indiscriminate use of various descriptive terms and vernacular names of their own choice, which made confusion even worse. With the publication of botanical work in Portuguese by Garcia d'Orta in 1565, Indian medicinal plants were introduced to the European countries. This was followed in the seventeenth century by *Hortus malabaricus*, and subsequently the binomial system of nomenclature by Linneaus which was introduced to India through such pupils as Koenig during the eighteenth century.

With the establishment of the then Royal Botanic Garden (now the Indian Botanic Garden) by Robert Kyd in 1787, and with the undertaking of various floristic investigations, explorations and publications in the first half of the nineteenth century by Roxburgh, Wallich, Griffith, Royle, Wight, Hooker, and finally with the publication of the "Flora of British India" (1872–1897), new vistas for Indian plant taxonomy and floristics were developed. Establishment of the Botanical Survey of India by King in 1890 at the Calcutta Botanic Garden,

---

*The gurukula system offered close contact of teacher and pupil, both in theoretical and practical studies, by intensive field work and observation of plants in nature.

and the contributions and the floristic works as regional floras by King, Clarke, Watt, Prain, Burkill, Barber, Cooke, Duthie, Gamble, Lawson, Haines, and several others, marked the highlights of floristic studies and botanical exploration work during the closing of the nineteenth century and the early part of the twentieth century.

Scientific activities of the Botanical Survey of India were kept in abeyance after World War I, and with the virtual abolition of the Botanical Survey of India in 1937, floristic and taxonomic studies were practically discontinued. Reorganization of the Botanical Survey of India in 1954 revived opportunities for botanical exploration and taxonomic and floristic studies. A few contributions during the period 1920–1954, though useful, are quite meagre for a subcontinent of this size. Further, the omission of serious taxonomic studies in the university curricula and the stopping of recruitment of posts connected with systematic botany even in important herbaria including the Calcutta Herbarium and the Botanical Survey of India, had the unfortunate effect of depressing still further the descending spiral of the training of taxonomists. At this stage recruitment of various workers of different disciplines of botany, besides a few trained earlier by the Botanical Survey of India, into the floristic and taxonomic studies of Botanical Survey of India were adopted. Thus the organization had to face a wide gap of a few generations as the traditional background of the classic taxonomy and floristics built by earlier workers of the Botanical Survey of India and university professors like Cooke, Fyson, Blatter, Barnes, Kashyap and forest officers like Talbot, Gamble, Champion was completely lost. However several young workers enthusiastically started botanical exploration and identification work utilising the valuable collections of the Herbaria at Calcutta, Coimbattore, Poona, Shillong and Dehradun, even though most of them were then lying in a moribund condition without adequate staff and facilities.

At this stage of rapid growth, with four important regional circles covering the whole of India, with headquarters at Shillong, Dehradun, Coimbattore and Poona and a central laboratory and main headquarters at Calcutta, there should have been leadership by competent taxonomists, suitable training programmes for serious taxonomic and floristic work and close liaison with the Kew and Edinburgh Herbaria. Such streamlined programming did not take place, with the result that the country is now facing an acute shortage of competent taxonomists to prepare the floras and to revise the families.

Botanical explorations and collections covering different states of India during the period 1955–1978 by the existing seven circles of Botanical Survey of India have been encouraging, and enormous collections need further study. Even though a few areas in some states are still underexplored, and large tracts of a few border states like Arunachal Pradesh are considered as unexplored, it is now felt that due to considerable delay, immediate action is necessary to launch the programme of preparation and publication of district floras and the State Flora Analysis (an elaborate check-list model) so as to make the data on vegetable wealth of the country available for all concerned. Publication of such data would also help the revision work of the families for the Flora of India Project.

An analysis of underexplored and unexplored areas in different states of India are outlined below, and by critical study of the data (see also Fig. 1) it is evident that quite large areas have been well explored. The unexplored regions are valuable areas which need careful and urgent studies, and a serious effort should be made immediately to plan intensive exploration work.

Unexplored States

Andaman and Nicobar—Andaman and Nicobar Islands (many of the islands); Himachal Pradesh—interior Himalayan Ranges of Himachal Pradesh: Rampur, Kalpa, Dankar, Bara Lacha Pass, Kyelang and Dharwas; Jammu and Kashmir—Himalayan Ranges bordering Jammu and Kashmir and Pakistan along Pir Panjal and adjoining ranges along Kotli, Punch, Muzaffarabad, Chilas and Yasin, northern border along Kara Koram Range bordering Jammu and Kashmir and China, Ladakh, Zaskar Ranges along Mintak Pass, Kara Koram Pass, Chushul, Chumar and Char Region; Tripura—southern Tripura; Mizoram—mountain region along Burma border; Manipur—mountain region along Burma border; Nagaland—mountain region along Burma border and Saramati Range; Arunachal Pradesh—interior ranges and borders of Tibet and Subansiri District of Kameng District, most of the ranges of Subansiri, Siang, Lohit and Tirap Districts and borders of Tibet and Burma.

Underexplored States

Kerala—Anamalai Hills, Cardomom Hills, Silent Valley; Tamil Nadu—Nilgiri Hills to Silent Valley (Ootacamund to Palghat), Tirunelveli Hills; Karnataka—Coorg District, Bababudan Hills in Chikmaglur

                    R. S. Rao

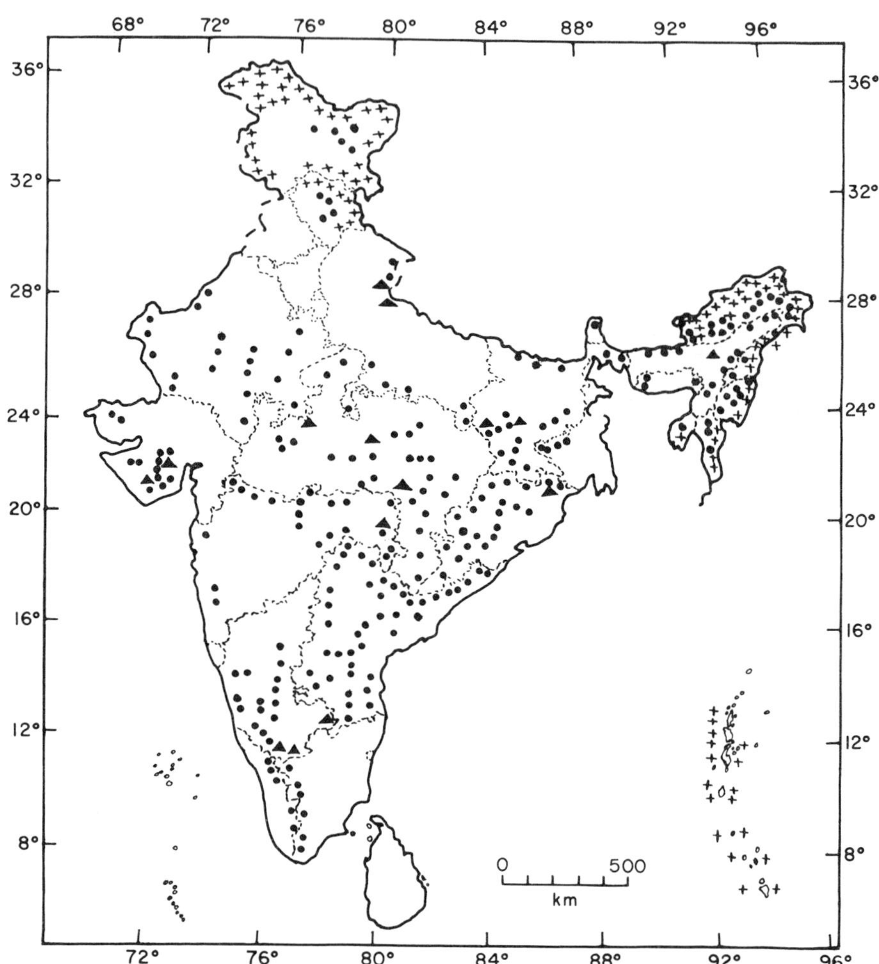

Fig. 1. Map of India showing botanically unexplored (+) and underexplored
(●) regions, and important national parks (▲).

District, Chitradurga to Hospet Range, Sirsi, Garosappa Falls area,
Biligirirangan Hills, Bandipur, Nagarhole and Bannerghatta National
Parks; Andhra Pradesh—Eastern Ghat Ranges, Deccan Plateau, Rayala-
seema, Krishna Valley with Srisailam Ranges, Tirupati Hills and
Seshachalam Hills of Nallamalais; Maharashtra—Satpura Ranges,
Mahadeo Hills, Balaghat Ranges, Chanda District, Taroba National
Park, Nanded and Buldhana areas; Madhya Pradesh—Shivpuri,

Bandhabgarh and Kanha National Parks, Mahadeo Hills, Maikal Bhanrer Ranges, Shahdol–Shahpura area, Raipur Uplands, Malwa Plateau and Morena–Bhind area; Gujarat—Gir and Velavadar National Parks, Mandav and Girnar Hills and a few ranges north-west of Kutch; Rajasthan—Aravali Range a few points between Gogunda–Beawar; Kekri–Shahpura–Bilwara hilly area, Jhalawar–Kota–Sawai Madhupur–Dholpur hilly area (East Rajasthan), a few points along the India–Pakistan border and Munabao–Tanot–Bahla points; Himachal Pradesh—Mandi–Kulu–Manali–Dharmasala area; Jammu and Kashmir—Leh area; Uttar Pradesh—Pithorgarh–Khela area along the borders of Nepal and Uttar Pradesh, Corbett and Dudhua National Parks, Rihand Dam area, Dudhi–Robertsgam area; Banda–Man Ranipur–Samthar; Behar—Raxaul–Sitamarki–Forbesganj region, Nepal border area, Hazaribagh and Palamau National Parks, interior parts of the Rajmahal Hills, Ranchi Hills–Basia–Parahat area; Orissa—Malakanagiri–Jeypore–Kotapad–Nowranga pur hilly area, Nowrangapur–Raigharh–Bhawanipatna–Kotgar–Phulbani hilly area, Borasambar–Sundargarh–Bonaigarh–Joshi-pur–Meghasani hilly area and Simlipal Hills National Park; West Bengal—Midnapur–Behar border, Jalpaiguri, Coochbehar and Ali-purduar Forests; Sikkim—interiors of northern Sikkim; Assam—borders of Assam and Bhutan Forest hill area, Assam and Nagaland, Assam and Jaintia Hills district of Meghalaya, and Assam and the eastern Arunachal Pradesh foot hill area and the Khaziranga National Park; Meghalaya—Garo Hills district, eastern Jaintia Hills district; Tripura—Northern Tripura; Mizoram—Aizawl and Lunglei sur-roundings; Manipur—Ukhrul, Karong, Tamenglong and Kala Naga hilly regions; Nagaland—Nagar hilly regions of Tuensang–Mokok-chung, Dimapur–Kohima hilly ranges and the Barail ranges of Assam and Manipur borders; Arunachal Pradesh—borders of Bhutan, a few interior parts and foot hill area of the Kameng District, around Zero and Itanagar of the Subansiri District, outer ranges and along the track from Along to Ninguing of the Siang District and the outer ranges of hit (Daphabum) and Tirap Districts.

A Field Station has been established at Itanagar, the new capital township of Arunachal Pradesh, and a Regional Circle at Port Blair, the capital of the Andamans and Nicobars, although the facilities available are not yet sufficient to carry out intensive exploration of floristic data. It is therefore necessary to open Field Stations, one in each district of Arunachal Pradesh, Nagaland, Manipur and Mizoram, and also in the Himalayan districts of Himachal Pradesh and Jammu and Kashmir; to improve the facilities of transport to different island groups in the

Andamans Circle and remote corners of the Himalayas in the north-western and the north-eastern zones. Similarly several interesting hilly areas along the Eastern Ghats of the Andhra Pradesh and Orissa can be explored by establishing a Regional Circle of Botanical Survey of India either at Hyderabad or at Visakhapatnam of the Andhra Pradesh State.

Arunachal Pradesh with its five districts covering an area of about 84 000 sq km presents a very rich and interesting flora. During the period 1955–1977, the Botanical Survey of India has organized 25 collection tours, gathering about 3000 species belonging to 1155 genera and about 188 families. Of these, Kameng District (immediately east of Bhutan), has been fairly well explored (Rao *et al.*, in prep.). In the Andamans and Nicobars more than 500 species have been added to the known flora of 1600 species during the last five years, since the establishment of the Regional Circle at Port Blair in 1972. Twenty-one new species have been described for the islands during general collection studies in the past five years. Of the 230 forested islands, collections have only been made from about 85. There is urgent need for conservation of these forests and for collection and presentation of floristic and fauna data as the islands are now subjected to human resettlement.

Special attention has been given to the members of family Orchidaceae with more than 1200 species in India, by way of establishing National Orchidaria, one at Shillong in the Khasia Hills, Meghalaya in 1959 and the other at Yercaud in the Shevaroy Hills, Tamil Nadu in 1963. More than 5000 plants covering about 350 species are under cultivation at Shillong, and about 170 species at Yercaud. Eastern India is very rich in orchid flora, and the distribution of some orchids of South Indian Hills is quite interesting, extending to the Himalayas in the north and to Malesia in the south-east. The study of the orchid flora is a life-long subject and the workers engaged in the study of the family Orchidaceae should be given the opportunity of continuing their work uninterrupted. Unless a suitable policy is adopted this important study will continue to suffer for years to come. Furthermore exploitation of wild orchid material by the various nurseries for commercial purposes is being carried on at alarming rate, in spite of severe restrictions, and it may be too late to get suitable material for proper taxonomic assessment of the family and other allied aspects. Proper arrangements for the study, conservation and protection of Orchidaceae must be made.

Similarly, taxonomic studies on the lower groups of plants such as algae, fungi, lichens, liverworts, mosses and ferns need investigation. Good collections are being maintained by the Botanical Survey of India, with a few workers solely devoted to the Cryptogamic section, but such an effort is not sufficient to prepare the necessary taxonomic works. Collections specially arranged for these groups by the Botanical Survey of India and specialists, and subsequent study in specialized centres abroad, would provide the necessary background to prepare such taxonomic data.

The importance of conservation and protection of endangered flora has been fully realized throughout the world. Promulgation of several acts by different countries and organization of committees by world bodies for nature conservancy and protection, reflect a timely world-wide awakening on the subject (e.g. United States Endangered Species Act, 1973; U.K. Wild Creatures and Wild Plants Act, 1975; Threatened Plants Committee of the International Union for Conservation of Nature and Natural Resources-IUCN). Even regional and specialist groups on tree ferns, cycads, orchids, palms, succulents, etc. have been formed by IUCN. The tropics, with more than 170 000 species of flowering plants, presenting a general estimate of over 20 000 species as endangered and threatened, are still inadequately served with floras and monographs. In India more than 1500 species of angiosperms and gymnosperms should be conserved and protected, but this is not happening, despite the Wild Life (Protection) Act, 1972.

Whilst carrying out floristic studies, the importance of locating various populations among the species, particularly those yielding useful alkaloids, has been realized, and interesting findings have been made. I am closely connected with locating *Iphigenia stellata* Blatter, growing in restricted areas near Mahabaleswar, along the Western Ghats in Maharashtra State. This plant yields high percentages (1·2–1·9) of colchicine. This new plant source and its extraction process were subsequently registered as a patent (No. 127743 CSIR) and given to industry (Ansari and Rao, 1973). The area of such germ-plasm material has been recommended for protection, and enclosed drug-farms have been developed by the Maharashtra State Forest Department. Further programmes have to be developed, with extensions of the systematic cultivation of this species along the various valleys surrounding Mahabaleswar, for the economic benefit of the local population. All such applied aspects of taxonomy are being developed whilst carrying out basic taxonomic work on *Iphigenia*. Indeed recently two more species in the genus were identified.

Similar variations in the yield of alkaloids among various populations of a few species are very frequently observed, e.g. *Costus speciosus* (Koenig) Smith, a polyploid species with 0·38–3% diosgenine (Sarin *et al.*, 1977). Further analysis of such data by biosystematic studies would be useful to locate polymorphism in such taxa, and to isolate useful germ-plasm.

It has been realised by Botanists in India that it is very necessary for properly planned collaboration between the Universities, the Botanical Survey of India, and a few other research institutes like the National Botanical Research Institute, Lucknow in order to achieve success in the Flora of India Project. But to achieve success a methodology to prepare the necessary infra-structure in the country has to be developed immediately in order to repair the damage caused earlier this century. Genuine interest in conservation, protection and utilization of the natural plant resources has to be instilled in students in school, college and university by reorientation of the syllabus, and the introduction of courses on floristics, plant geography and biosystematics.

To achieve such objectives and to produce competent taxonomists for the Flora of India Project, a plan of action at the Andhra Pradesh State level has been prepared and submitted by me to the Government of India*. This involves all six Universities in the State, both to train personnel by way of Ph.D. Thesis programme, and to prepare the 23 districts floras, and subsequently the State Flora as a prerequisite for the Flora of India Programme.

Suitable centres for such work in each district are to be selected either in a college, P. G. Centre or University. With a suitable guide the areas of collection in each district are determined in relation to the forest ranges in the districts. Thus a herbarium, small museum and photo-library would be developed as the data for the district flora is prepared. Such a centre in each district would form the nucleus for developing small herbaria in different colleges, and if possible in high schools, for that district and thus develop the necessary infra-structure for plant conservation programmes. Such district flora data would be further utilized for the preparation of State Flora which would form the main foundation for Flora of India. Every aspect of the conservation programme would depend on the effectiveness of earlier publications on accurate taxonomic studies with important data, brief discriptions and detailed keys to illustrate the district floras in a concise form . Such a pattern has been considered suitable by the majority of workers in the field, as well as the Botanical Survey of India. This district flora pattern is outlined below:

(i) General account: Introduction (general note, historical note, reasons for undertaking the present work and its importance); Past and present work (earlier work, present work with materials and methods

---

*The Department of Science and Technology of the Government of India wants to proceed slowly on such a project by approving two districts at a time because there is such an urgent need for data.

as followed in the field and herbarium and plan of the flora); Topography and general features (covering geology, soil and climate); Forest biodata; General vegetation types (with a few photographs); Floristic analysis; Plants of medicinal, economic and horticultural importance; Interesting plants of botanical value; Endangered plants; Germ plasm material; Game sanctuaries /national parks; Abbreviations.

(ii) Systematic treatment.

(iii) Selected bibiliography.

The systematic treatment under each family covers the following pattern: Dichotomous indented key to families/genera; Genus with author; Detailed key to the species (including exotics if naturalized); Name of species with original citation and citation of latest taxonomic work if any; Citation of change in name or of basionym wherever necessary; Reference to the Hooker's "Flora of British India and the Regional Flora"; Local names in inverted commas at the end of this paragraph; Brief description with primary characters not reflected in the key; Flowering and fruiting period; Reference to good illustration if any; Distribution and frequency (using symbols and numbers with reference to forest ranges to pin-point the localities) (only rare taxa can be cited); Notes if any.

(iv) Cultivated plants, extotics etc. and their reference at the end of each family.

(v) Standardized base map for the district, showing forest ranges and divisions, national reserves, game sanctuaries and important plant-based industries.

(vi) Line drawings of rare, endangered, endemic or other interesting botanical species.

As indicated earlier, data accumulated since the reorganization of the Botanical Survey of India in 1954, together with earlier data available in the old herbaria in India including the Central National Herbarium, Howrah, can immediately be processed. A programme for publication of various district floras is being followed up for the Jowai and Garo Hills (Meghalaya State), Sidhi, Bilaspur, Bastar and Raipur (Madhya Pradesh State), Mirzapur, Kumaon, Pithorgarh, Pilibhit, Chamoli (Uttar Pradesh State), Tonk and Banswara (Rajasthan State), Poona, Thana, Chandrapur (Maharashtra State), Kanyakumari (Tamil Nadu State), Idiki (Kerala State), Carnicobar (Andaman and Nicobar Islands) and Kameng (Arunachal Pradesh State) and, now in press, Goa, Diu, Daman, Dadra and Nagarhaveli. A few districts of West Bengal, Behar and Orissa will also be added after assessing available data.

Check-lists should also be drawn up to bring to light available data on distribution, etc. by way of well-scrutinized and nomenclaturally corrected taxa. These may then be utilized for further flora writing, analysis and presentation of material to the Flora of India workers, creating facilities for research workers in educational institutions, national and regional research laboratories, industry, rural and district museums.

Such modified check-lists termed as State Flora Analysis is an urgent need and may be published by the Botanical Survey of India within the next two to three years for all States of India. The material for the States of Arunachal Pradesh, Meghalaya, Tripura, West Bengal, Himachal Pradesh, Rajasthan, Gujarat*, Maharashtra, Tamil Nadu, Lakshadweep and Minicoy Islands is already available for publication. Revisions of Coriariaceae†, Paeoniaceae†, Dilleniaceae, Polygalaceae, Annonaceae, Commelinaceae, Pittosporaceae, Fabaceae (selected genera) are also well underway.

It is gratifying to record two special symposia, one on "Floristic Studies in India" organized by the Botanical Survey of India, Howrah in November 1977, and another on "Modern Trends in Plant Taxonomy" by the National Botanical Research Institute, Lucknow during March, 1978. Important aspects of the recommendations made by the symposia, refer to (i) the urgent necessity of establishing botanical gardens and arboreta, (ii) close collaboration with the universities and other research institutes engaged in floristic studies, (iii) organizing work-shops and short term courses for specialized training in taxonomy and floristics, (iv) the opening of advanced centres for taxonomy under special assistance programmes of the university grants commission, (v) creating a few chairs in taxonomy in selected universities, and a few other allied aspects. However, along with the maintenance of proper herbaria, experimental gardens and arboreta and museums, availability of documentary and information services in plant taxonomy and other allied subjects have to be properly maintained by the Botanical Survey of India and other Government Institutes.

Focusing the importance of the studies on floristics and taxonomy on tropical botany at the international level (as at this symposium) is very important in drawing the attention of world bodies like the United Nations Educational Scientific and Cultural Organisation (UNESCO), Food and Agricultural Organisation (FAO), national governments and research councils for encouragement, both financial and otherwise, to the developing and the developed countries to complete the tropical floras as soon as possible. With such cooperation,

*Rao *et al.* (in press). "State Flora Analysis".
†Already published as "Flora of India" Part 1 (1978).

conservation and presentation of data suitable for the utilization of plant wealth of the tropical countries, would be possible to a very satisfactory level within the next two decades.

References

Ansari, M. Y. and R. S. Rao (1973), *Iphigenia stellata* Blatter (Liliaceae)—its identity and economic importance. *Bull. Bot. Surv. India* **15**, 118–122.
Sarin, Y. K., A. Singh and S. K. Kapur (1977). Symp. Prod. Util. Veg. Raw Mat. Ster. Horm. Oral Contracept. Abst. Proc. 36.

The Vegetation Types of Ecuador—
A Brief Survey

G. HARLING

University of Göteborg, Sweden

Introduction

The first serious attempts to carry out a phytogeographical survey of Ecuador were made by Sodiro (1874) and Wolf (1892). They divided the country into three natural main regions: (i) the western coastland, (ii) the Andean highland (this region being further subdivided by Sodiro, 1874), and (iii) the eastern lowland, "El Oriente", belonging to the Amazon Basin.

Later phytogeographical works have also been based on this natural but coarse division, e.g. Mille (1918), Diels (1937), Acosta-Solís (1966, 1977), Sparre (1968). All these authors recognize within the three main regions a varying number of more natural formations of the vegetation. The most elaborate of these works are those of Acosta-Solís (1966, 1977); in his very interesting papers he distinguishes no fewer than 18 natural main formations.

The attempt at a phytogeographical subdivision of Ecuador that I shall present here is based partly on these papers and partly on my own experiences from more than three years of field work in Ecuador during the period 1946–1977. According to Fig. 1, 16 main vegetational types are distinguished. It must be emphasized that the extent of most of them is still poorly known and that the limits between them are in many cases vague. The boundary lines on the map are therefore often more or less arbitrarily drawn.

Seashore Vegetational Types

Sandy or Stony Beaches

These are not specially marked on the map. They comprise all open and exposed parts of the seashore and are often more or less barren or

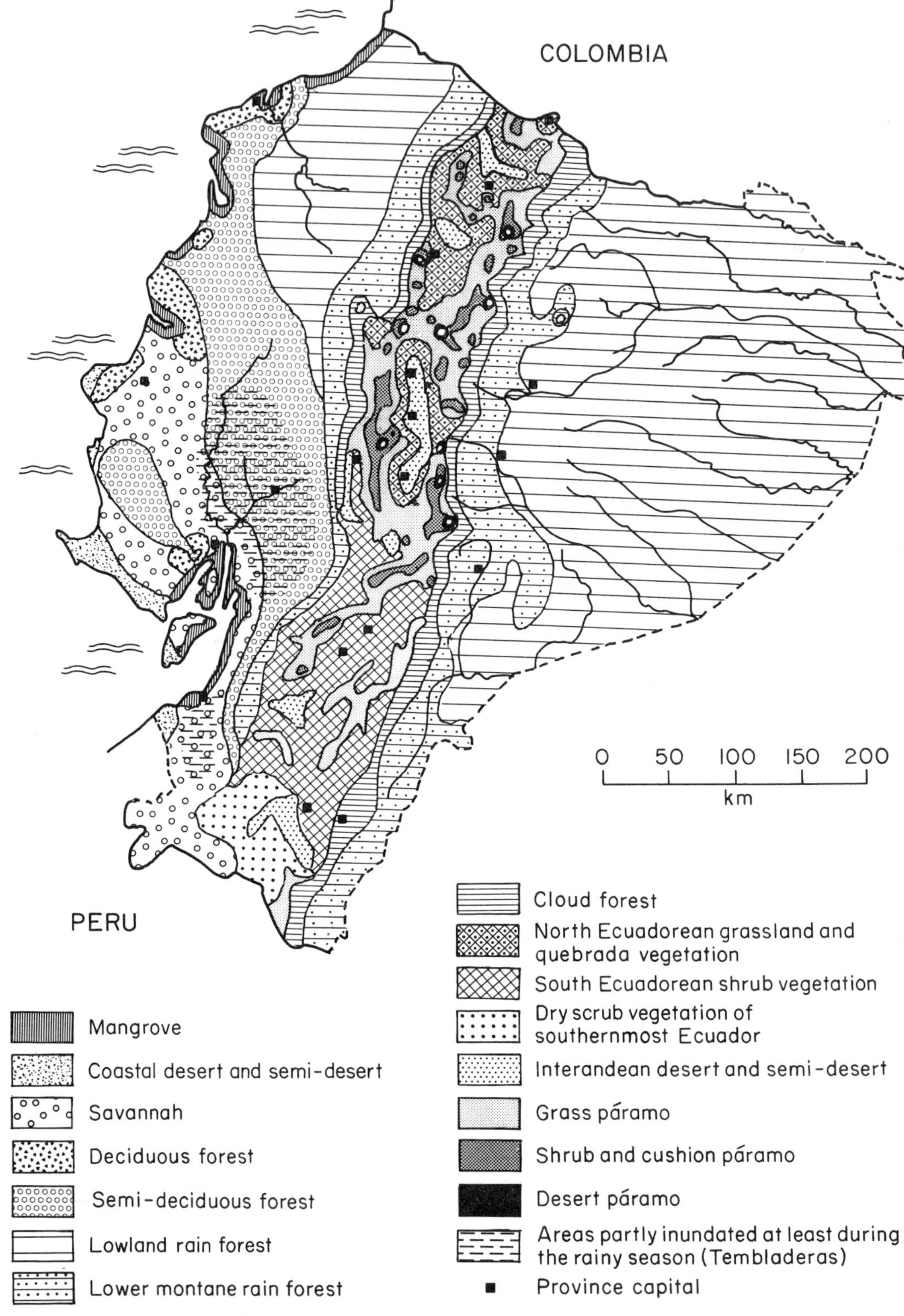

Fig.1 Survey of Ecuadorean vegetation types.

with scattered, low and wind-tortured shrubs of *Maytenus octogona* and *Cryptocarpus pyriformis*. The *Ipomoea pes-caprae* formation dominates in many places, with e.g. *Canavalia maritima* and *Pectis arenaria*. Other common species are *Vallesia glabra*, *Scutia spicata*, *Acacia huarango*, *Cacabus prostratus*, *Sesuvium portulacastrum*, *Trianthema portulacastrum*, *Phyla canescens*, *Coldenia paronychioides*, *Heliotropium curassavicum*, *Ipomoea incarnata*, *Sporobolus virginicus* and *Cenchrus* spp. In a few places, e.g. near Manglaralto, the beach vegetation consists of a border of *Scaevola plumieri*, also found in the Galápagos Islands, and further inland shrubs or small trees of *Hippomane mancinella*. For more detailed descriptions of the beach vegetation see Holm-Nielsen *et al.* (1975).

Mangroves

Mangroves (manglares) are confined to the estuaries and sheltered bays. The most extensive and best developed mangrove is found in northern Esmeraldas, especially between Limones and the Colombian frontier, and in the gulf of Guayaquil. The dominating species is always *Rhizophora mangle* but on the inner side *Avicennia germinans*, *Laguncularia racemosa* and *Conocarpus erecta* are common, in northern Esmeraldas also *Pelliciera rhizophorae*. The large fern *Acrostichum aureum* is characteristic of at least the Esmeraldas mangrove. In open grounds bordering on the innermost mangrove margin halophytes like *Batis maritima* and *Salicornia peruviana* often occur. A detailed description of the Ecuadorean mangrove is given by Acosta-Solís (1959).

Dry Coastal Vegetational Types

Deserts and Semi-deserts

These are found in areas with very low precipitation due to the cold waters of the Humboldt Current. The annual rainfall is normally only about 100–300 mm, which falls from February to April. These arid areas include the Santa Elena Peninsula and most of the Manta region. Characteristic species of this vegetational type are among others the tall *Armatocereus cartwrightianus*, and widely scattered shrubs or small trees of *Cordia lutea*, *Acacia* spp., *Erythrina velutina*, *Cercidium praecox*, *Loxopterygium huasango*, *Carica paniculata*, *Croton rivinifolius*, *Waltheria ovata*, *Wedelia grandiflora*, and the evergreen *Jacquinia pubescens*, the latter species also found in the savannah and in the Guayaquil mangrove. Common grasses are *Chloris radiata* and *C. virgata*, *Anthephora herma-*

phrodita, *Cenchrus pilosus*, *Aristida adscensionis* and *Bouteloua disticha*. In the rainy season there may be a mass flowering of *Leptochiton* (*Hymenocallis*) *quitense* and several annuals. The arid regions of western Ecuador have been described by Svenson (1946).

Savannah

Savannah has a vast extension in southwestern Ecuador. It occurs in areas with higher precipitation, often about 1000 mm, but still with a long and pronounced dry season from May to January. The dominating savannah trees are several Bombacaceae, above all *Ceiba trischistandra* and the species usually but incorrectly called *C. pentandra*. *Eriotheca ruizii*, *Pseudobombax millei* and *P. guayasense* are also locally common. In the Pedro Carbo region and near Guayaquil the ceibas are partly replaced by the beautiful *Cavanillesia platanifolia*. Other common savannah trees and shrubs are *Cochlospermum vitifolium*, *Acacia* and *Prosopis* spp., *Mimosa albida* and *M. pigra*, *Croton rivinifolius*, *Capparis ovata* and *C. ovalifolia*, *Tecoma castanifolia*, and *Tabebuia chrysantha*. The ceibas are leafless during the long dry season but the process of assimilation has been taken over by the green stems and branches as is also the case with *Cercidium praecox*. From February to May, if the rains do not completely fail (as in 1968) the trees, shrubs and most of the ground are covered by cucurbit climbers of the genera *Apodanthera*, *Sicyos*, *Momordica*, *Luffa*, etc. A very common and showy savannah plant is *Ipomoea carnea*, which flowers during most of the year. Common grasses are *Pennisetum purpureum* and *P. occidentale*, *Aristida adscensionis*, several *Panicum* and *Paspalum* spp., *Andropogon bicorne* and *Chloris virgata*. In southernmost Ecuador the *Ceiba* savannah extends as far east as the Macará region.

Deciduous Forest

Deciduous forest (dry shrub), another dry vegetation type, replaces the savannah in many places. Since the climatic conditions are essentially the same as in the savannah zone its occurrence is probably due to different topographic and edaphic factors. Many of the savannah trees and shrubs are found also in the dry forests but generally there is a dominance of thorny leguminous trees like *Prosopis juliflora*, *Acacia huarango*, *A. macracantha*, and *Erythrina velutina*. Another characteristic feature, especially in the dry forests near the coast, is the abundance of cacti, as *Armatocereus cartwrightianus*, *Hylocereus peruvianus* and *Opuntia* spp. Other common species are *Achatocarpus pubescens*, *Muntingia cala-*

bura, *Guazuma ulmifolia* and *Clavija pungens*. Epiphytic bromeliads are sometimes abundant, e.g. *Tillandsia latifolia* and *T. straminea*.

Humid Lowland Forests

Semi-deciduous Forest

Often called seasonal or monsoon forest, semi-deciduous forest has, during the last few decades, largely been destroyed and replaced by cultivation but many remnants still remain. Since this vegetation type is especially poorly known it is an important task to study the remnants before they disappear.

The parts of western Ecuador today or formerly covered with semi-deciduous forest have a much heavier rainfall (up to 2500 mm or more), but with too long a dry season for a true rain forest to occur. During the rainy season the semi-deciduous forest may rather resemble a rain forest but it is lower, composed of fewer species and contains fewer lianas and epiphytes. The ground vegetation is usually particularly well developed. Nevertheless the semi-deciduous forest and the rain forest have many genera and species in common, e.g. *Ficus* and *Cecropia* spp., and several Lauraceae, Myristicaceae and Lecythidaceae. Root-climbing lianas of the Araceae and Cyclanthaceae are represented in the moister semi-deciduous forests, as is the case with giant monocotyledonous herbs like *Heliconia* and *Costus* spp., *Calathea insignis* and *Pleiostachya morlae*. Other characteristic species of this vegetation type are *Myriocarpa stipitata*, *Samanea saman*, *Cordia alliodora*, *Tabebuia chrysantha*, tall bamboos (*Guadua angustifolia*) and several palms, e.g. *Iriartea* and *Geonoma* spp.

In the deforested areas and also in the natural savannah of the inner coastal plain, large parts of the ground are inundated at least during the rainy season (the so-called tembladeras). Here a lot of swamp plants are found: sedges and rushes of various sorts, *Typha*, *Sagittaria*, *Limnocharis*, *Echinodorus*, *Thalia*, *Neptunia*, etc.

Lowland Rain Forest

This covers most of Esmeraldas Province and adjacent parts of Pichincha Province in western Ecuador, and also almost the entire lowland area east of the Andes. Although there are some differences in the floristic composition of the western and eastern rain forests, the general vegetation type is the same, and so I have found no reason for treating them as different formations as do some authors. The western rain forest is a continuation of the Colombian Pacific rain forest, whereas the eastern one belongs to the vast forest system of upper Amazonas. In both cases the forests belong to the mixed type without any dominant tree species and with a very high specific diversity. To the most important tree families are Leguminosae, Moraceae, Lauraceae, Myristicaceae, and Meliaceae.

Like the semi-deciduous forests the Ecuadorean rain forests are now highly threatened by increasing deforestation, usually by the abominable "slash and burn" system. Until recently the rain forests have been just as poorly known as the semi-deciduous ones. However, our knowledge about the western lowland rain forest has been considerably increased through the works of Little (1969), and Dodson and Gentry (1978). Very little has been written about the eastern lowland rain forest except for the useful ecological paper by Grubb *et al.* (1963) on the primary forest near Tena. During the last decade important collections have accumulated in the larger herbaria, particularly from the regions along Río Napo and Río Bobonaza in the north, and Río Upano and Río Zamora in the south. Most of these collections still await identification.

Montane Forests

Lower Montane Rain Forest

Lower montane rain forest gradually replaces the lowland forest at a height of about 700–800 m on both sides of the Andes. It continues to about 2500 m where it changes into cloud forest. The annual rainfall, especially on the eastern slopes of the Andes, seems to be much higher than in the lowlands. In the Mera region (Río Pastaza Valley) it is said to be about 5000 mm. The floristic composition of the montane forest changes slowly with increasing altitude. Typical lowland families, such as Myristicaceae, Lecythidaceae and Vochysiaceae, disappear and are replaced among others by Ericaceae, Melastomataceae, Betulaceae (*Alnus jorullensis*) and Cunoniaceae (*Weinmannia* spp.). In the bamboos *Chusquea* spp. replace the tall *Guadua*. The large woody climbers diminish in number, whereas the epiphytes become very abundant, especially species of orchids, bromeliads, ferns and mosses.

Cloud Forest

Cloud forest (upper montane rain forest or elfin forest) is found between *c.* 2500–3400 m. The Spanish term Ceja de la montaña or Ceja andina is often used for its uppermost part, near the transition to the páramo. It is generally best developed in the eastern cordillera. It is a dense, low, microphyllous forest with the trees overloaded with mosses, filmy ferns and lycopods. The tall tree ferns are abundant. Important families among others, are Rosaceae, Ericaceae, Melastomataceae, Rubiaceae and Compositae. Some of the many characteristic genera are *Podocarpus*, *Hedyosmum*, *Bocconia* (*B. pearcei* and *B. integrifolia*), *Cinchona*, *Ilex*, *Oreopanax*, *Escallonia*, *Miconia*, *Brachyotum*, *Gunnera*, *Fuchsia*, *Sphaeradenia*, *Bomarea* and *Chusquea*.

Interandean Vegetational Types

North Ecuadorean Grassland and Quebrada Vegetation

This vast area of the interandean plateau extending southwards to the Chimborazo Province, was almost entirely destroyed long ago and has been replaced by cultivation. Remnants of the original vegetation are found in ravines, on steep slopes and other less accessible places. These remnants indicate that large parts of the earlier vegetation consisted of a low forest similar to the Ceja andina. The forested areas were probably interspersed with more open, savannah-like land. Nowadays almost the only woods in this zone consist of *Eucalyptus globulus*, introduced from Australia more than a century ago. Characteristic of this region are many Compositae, particularly the genera *Baccharis*, *Diplostephium*, *Eupatorium* (s. lat.), *Gynoxis* and *Barnadesia*, further *Oreopanax mucronulatus*, *Coriaria thymifolia*, *Dalea caerulea*, *Columellia oblonga*, *Siphocampylus giganteus*, *Calceolaria crenata* and *C. hyssopifolia*, *Solanum* and *Cestrum* spp., *Bomarea caldasiana* and *Puya hamata*. Common grasses are e.g. *Cortaderia rudiuscula*, and *Festuca*, *Calamagrostis* and *Stipa* spp.

South Ecuadorean Shrub Vegetation

This is most typically developed in the Cañar, Azuay and Loja Provinces between the altitudes of 2000 and 3000 m. In its general appearance it resembles the preceding formation but it is dryer and dominated by other species. The most important families in this formation are Compositae, Ericaceae, Melastomataceae, Proteaceae and Bromeliaceae. The dominating species in the scrub are among others *Hypericum laricifolium*, *Valeriana hirtella*, *Diplostephium lavandulifolium*, *Miconia* and *Tibouchina* spp., *Oreocallis grandiflora* and *Lomatia hirsuta*.

172 G. Harling

Several *Bejaria* spp. are common, e.g. *B. glauca*, *B. aestuans* and *B. dryanderae*, and so are *Blechnum* (*Lomaria*) *loxense*, *Baccharis genostelloides* (s. lat.), *Barnadesia dombeyana* and *Salvia rugosa*. From the Loja Province the following species may be especially mentioned: *Cantua quercifolia*, *Streptosolen jamesonii*, *Linociera pubescens*, *Stachytarpheta steyermarkii*, *Puya parviflora*, *Sphenostigma spruceanum* and *Porphyrostachys pilifera*. A list of species from Loja Province and the adjacent parts of El Oro Province has been published by Espinosa (1949).

Dry Scrub Vegetation of Southernmost Ecuador

This appears to be much the same as that described by Weberbauer (1945) for extreme northern Peru. It is partly a steppe with low, thorny scrubs and cacti, sometimes (as between Vilcabamba and Yangana) a dry savannah dominated by widely scattered trees or shrubs of *Acacia macracantha*, *Prosopis juliflora*, *Cercidium praecox* and *Erythrina* spp. Here and there stands of *Chorisia insignis*, *Linociera pubescens* and *Cacosmia rugosa* are found. Shrubs of *Croton*, *Duranta*, *Amphilophium paniculatum* var. *molle* and *Stachytarpheta steyermarkii* are frequent.

Interandean Deserts and Semi-deserts

These areas are confined to a number of more or less deep valleys or basins, which lie in the rain shelter of the western cordillera. The most important of these are from north to south: the Chota Valley, the Guayllabamba Valley, the Latacunga—Ambato—Riobamba Basin (to a very large extent irrigated and cultivated), the Chanchán Valley, the Palmira Desert, the Río Jubones—Río León Valley, and the Catamayo Valley. In many places the annual rainfall is probably less than 300 mm but I have seen no exact data. The mostly sparse vegetation is dominated by *Acacia pellacantha*, *Mimosa* aff. *quitensis*, *Caesalpinia tinctoria*, *Croton wagneri*, *Jatropha nudicaulis*, *Buettneria* spp., *Schinus molle* and *Dodonaea viscosa*. The cacti are frequent, e.g. *Trichocereus pachanoi*, *Borzicactus aequatorialis* and *B. sepium*, and *Opuntia* spp. Other characteristic genera are *Ephedra*, *Tillandsia* (both terrestrial and epiphytic spp.), *Furcraea*, *Alternanthera*, *Mentzelia*, *Bryophyllum*, *Echeveria*, *Wissadula*, *Flaveria* and *Onoseris*. The originally introduced *Aloe vera* dominates in many places, especially in the Chota valley.

Andean Vegetational Types

Grass Páramos

Grass páramos (pajonales) occupy most ground between 3400 and 4000 m. Downwards they border on the low ceja scrub or, nowadays, often on cultivated or deforested land. They are dominated by bunch grasses of the genera *Calamagrostis* and *Festuca*, to the south (from Chimborazo Province and southwards) also by *Stipa* spp. The clumps of grass are generally intermingled with herbs and small shrubs, e.g. species of *Ranunculus, Lupinus, Gentiana, Halenia, Castilleja, Valeriana, Baccharis, Oritrophium, Chuquiraga* and *Hypochaeris*. The genus *Espeletia* has its southernmost outposts in the Ecuadorean grass páramos, at an altitude of about 3500–3700 m. *E. hartwegiana* is common in the northern part of Carchi Province and probably also occurs in the Llanganates Mountains (Andrade Marín, 1936).

Shrub and Cushion Páramos

These are mostly confined to altitudes between 4000 and 4500 m. Here the bunch grasses begin to decrease in extent and are largely replaced by shrubs, herbs of various kinds, mats, rosette plants and, especially in the moister páramos, by cushion plants. Small trees belonging to the genera *Polylepis* and *Escallonia* may also occur. In Illiniza there are rather large stands of *Polylepis lanuginosa* at 4200–4300 m. Among the dominant shrubs are several Compositae such as *Chuquiraga jussieui, Diplostephium rupestre*, and *Baccharis* and *Loricaria* spp. Shrubs of *Valeriana* spp. are frequent as are *Calceolaria ericoides* (dry páramos) and *Astragalus geminiflorus*. Very characteristic genera of this zone are among others *Culcitium* (e.g. *C. uniflorum, C. rufescens* and *C. nivale*), *Senecio, Werneria, Oritrophium* (above all *O. pellitum*), *Gentiana, Halenia, Viola, Lachemilla, Draba, Bomarea* (subgenus *Collania*), *Jamesonia* and *Lycopodium*. The cushion plants belong to many different families such as Compositae, Umbelliferae, Ericaceae, Geraniaceae, Plantaginaceae, Cruciferae and Juncaceae. Among the most important cushion-forming species may be mentioned *Azorella pedunculata, A. aretioides* and *A. corymbosa, Plantago rigida, Draba aretioides, Werneria humilis* and *Distichia tolimensis*.

Desert Páramos

Desert páramos (arenales) generally begin at about 4500 m and extend up to the snow limit. Here larger or smaller patches of bare ground alternate with a sparse vegetation of xerophytic grasses, a few herbs and small shrubs, and several mosses and lichens. The most high altitude phanerogams include *Ephedra americana*, *Poa cucullata*, *Rhopalopodium guzmannii*, *Lupinus microphyllus* and *L. smithianus*, *Nototriche pichinchensis*, *Senecio microdon* and *S. comosus*, *Culcitium nivale* and *Werneria rigida*. In some mountains the desert páramo begins at a considerably lower level. The western and southern slopes of Chimborazo are dry and sandy from about 4000 m and upwards (arenal grande) with scattered clumps of *Stipa* and a few shrubs and herbs, e.g. *Calceolaria ericoides*, *Azorella pedunculata* (here forming mats, not cushions), *Calandrinia acaulis*, *Chuquiraga jussieui* and *Hypochaeris sonchoides*.

References

Acosta-Solís, M. (1959). "Los Manglares del Ecuador." Ecuadorian Institute of Natural Sciences, Quito.

Acosta-Solís, M. (1966). Las divisiones fitogeográficas y las formaciones geobotánicas del Ecuador. *Rev. Acad. Colomb.* **12**, 401–447.

Acosta-Solís, M. (1977). "Ecología y Fitoecología." Case de la Cultura Ecuatoriana, Quito.

Andrade Marín, L. (1936). "Viaje a las Misteriosas Montañas de Llanganati." Ecuadorian Institute of Natural Sciences, Quito.

Diels, L. (1937). Beiträge zur Kenntnis der Vegetation und Flora von Ecuador. *Bibl. Bot.* **29**, 1–190.

Dodson, C. and A. Gentry (1978). Flora of the Río Palenque Science Center. *Selbyana* **4**, i–xxx, 1–628.

Espinosa, R. (1949). "Estudios Botánicos en el Sur del Ecuador." Editorial Universitaria, Loja.

Grubb, P. J., J. R. Lloyd, T. D. Pennington and T. C. Whitmore (1963). A comparison of montane and lowland rain forest in Ecuador. *J. Ecol.* **51**, 567–601.

Holm-Nielsen, L., S. Jeppesen, B. Loejtnant and B. Oellgaard (1975). Preliminary Report on the 2nd Danish Botanical Expedition to Ecuador. Botanical Institute, University of Aarhus.

Little, E. L. (1969). "Arboles Comunes de la Provincia de Esmeraldas." FAO/SF: 76 ECU 13, Rome.

Mille, L. (1918). "Nociones de Geografía Botánica Aplicadas al Ecuador." Quito.

Sodiro, A. (1874). "Apuntes Sobre la Vegetación Ecuatoriana." Quito.

Sparre, B. (1968). Ecuador som egen intressesfär. *Fauna Flora* **63**, 137–154.

Svenson, H. K. (1946). Vegetation of the coast of Ecuador and Peru and its relation to the Galápagos Islands. *Am. J. Bot.* **33**, 394–426, 427–498.

Weberbauer, A. (1945). "El Mundo Vegetal de los Andes Peruanos." Ministerio de Agricultura, Lima.

Wolf, Th. (1892). "Geografía y Geología del Ecuador." Leipzig.

The Phytogeographical Position of the Neotropical Vascular Páramo Flora with Special Reference to the Colombian Cordillera Oriental

A. M. CLEEF

State University of Utrecht, Netherlands

Introduction

In the course of our studies of the páramos of the Colombian Andes we became greatly impressed by the richness, in genera as well as in species, of the páramo flora, the vascular flora in particular. A hasty comparison with the cool floras of other tropical and temperate high mountain areas learned that the neotropical páramo probably contains the richest vascular high mountain flora of the world. Checklists of tropical floras are rather scarce, especially of areas rich in species. This is also true for the tropical high mountains.

The first modern compilation of a vascular tropical high mountain flora was published by Cabrera (1958), treating the puna of North-West Argentina near the southern border of the neotropical realm. Additions were given by Ruthsatz (1977). Cabrera (1958) listed about 650 vascular species belonging to 258 genera, including the "prepuna" taxa, which generally occur at lower elevation.

Hedberg (1965) gave a detailed account on the vascular flora of the "afroalpine" belt of the equatorial East African mountains, those of Ethiopia excepted. He distinguished nine genetical flora elements, including in total 278 vascular taxa belonging to 103 genera. Three more genera have become known recently (Hedberg, 1965). The vascular flora of the "tropicalpine" upper reaches of Mt Wilhelm, New Guinea, was recently studied by Smith (1977), who listed nearly 400 native species belonging to 107 genera, pteridophytes excepted.

Together with 16 pteridophyte genera native to this area (Smith, 1977) a total of 123 indigenous vascular genera has been obtained. He distinguished 10 different geographical flora elements, aliens included.

The flora of the neotropical páramos is still incomplete, although in recent years our knowledge has increased considerably. Up till now more than 300 vascular genera are recorded from the páramo area extending from northern Peru up to Venezuela and Costa Rica. The composition of the vascular páramo flora of the Colombian Cordillera Oriental, where most of our studies were carried out, runs as follows: Pteridophyta—20 genera belonging to nine families, Monocotyledonae—62 genera belonging to 12 families, Dicotyledonae—177 genera belonging to 52 families.

The present paper contains a subdivision into seven geographical flora elements of all 259 vascular páramo genera of the Colombian Eastern Cordillera known at this time, thus providing a basis for conclusions on origin and composition of this modern páramo flora.

For phytogeographical comparisons of vascular floras on a worldwide basis the genus level seems most appropriate because taxonomic studies of many vascular taxa still continue at the species level. Data presented here are therefore based on the genus level. To determine the phytogeographical affinities of the páramo flora, the present distribution areas of all composing vascular genera were recorded. Genera with similar areas of spatial distribution were united into geographical flora elements. Only geographical flora elements are described because knowledge is too incomplete yet for distinguishing genetical flora elements (Szafer, 1975; Hedberg, 1965).

Geographical Flora Elements

In the indigenous vascular páramo flora of the Colombian Eastern Cordillera, the following seven geographical flora elements can be recognized (genera marked with *are monotypic).

Páramo Element

Part of the "neotropical element" and consisting of genera limited to the páramos of Costa Rica, Panamá, Venezuela (including the Cordillera de la Costa), Colombia, Ecuador and North-East Peru (Chachapoyas). Also included are genera, which have their main species concentration in the páramos and but a single species on the remote mountains of Guatemala, on the Antilles or on the Guayana shield. Part of the genera is also present in the forest belt at lower elevations. Genera also occurring in the southern páramos along the humid Amazonian slope in the remaining part of Peru and Bolivia are mostly included in the next element. Nineteen genera belong to the páramo element; i.e. 7·3%.

They include: *Aphanactus*, *Aragoa* (Fig. 1), *Blakiella**, *Bucquetia*, *Castratella*, *Espeletia*, *Espeletiopsis*, *Floscaldasia**, *Laestadia*, *Libanothamnus*, *Lourteigia*, *Myrrhidendron*, *Nephopteris**, *Neurolepis*, *Plutarchia*, *Purpurella**, *Swallenochloa*, *Tamania** and *Vesicarex**.

Other Neotropical Elements

These include genera occurring in other neotropical areas as well, but mostly confined to the mountains; 88 genera belong to this group, i.e. 33·8%.

Included are: *Acaulimalva*, *Aciachne**, *Acnistus*, *Aethanthus*, *Ageratina*, *Alonsoa*, *Altensteinia*, *Aphanelytrum**, *Arcytophyllum*, *Arracacia*, *Aulonemia*, *Axonopus*, *Baccharis*, *Befaria*, *Bomarea*, *Brachyotum*, *Calea*, *Cavendishia*, *Centropogon*, *Cestrum*, *Chaetolepis*, *Chaptalia*, *Chromolaena*, *Chusquea*, *Cologania*, *Dendrophtora*, *Diplostephium*, *Disterigma*, *Distichia*, *Ditassa*, *Drymaria*, *Echeveria*, *Eccremis**, *Elleanthus*, *Epidendrum*, *Eriosorus*, *Gaiadendron*, *Geissanthus*, *Grammadenia*, *Gomphichis*, *Greigia*, *Gynoxys*, *Halenia*, *Halimolobus*, *Heliopsis*, *Hesperomeles*, *Jaegeria*, *Jamesonia*, *Lachemilla*, *Loricaria*, *Lorenzochloa**, *Lucilia*, *Lysipomia*, *Macleanea*, *Macrocarpaea*, *Masdevallia*, *Maxillaria*, *Miconia*, *Monnina*, *Monochaetum*, *Moritzia*, *Nierembergia*, *Niphogeton*, *Noticastrum*, *Odontoglossum*, *Oncidium*, *Oreopanax*, *Oritrophium*, *Oxylobus*, *Pachyphyllum*, *Phyllactus*, *Phylloscirpus**, *Plagiocheilus*, *Polylepis*, *Psammisia*, *Pterichis*, *Puya*, *Relbunium*, *Rhizocephalum*, *Sabazia*, *Scaphosepalum*, *Sericotheca*, *Sessea*, *Siphocampylus*, *Stevia*, *Vazquezia*, *Verbesina* and *Werneria*.

Wide Tropical Element

This includes genera widely distributed in the tropics and occurring in at least two different continents: either America–Asia or America–Africa. The former group resembles the austral–antarctic element, but the taxa concerned are confined to the warm ecosystems. Twenty-seven genera belong to the wide tropical element; i.e. 10·4%.

They comprise: *Achyrocline*, *Buddleia*, *Bulbostylis*, *Cheilanthes*, *Clethra*, *Conyza*, *Cosmos*, *Ctenopteris*, *Cyperus*, *Elaphoglossum*, *Eugenia*, *Hymenophyllum*, *Ilex*, *Lippia*, *Paepalanthus*, *Paspalum*, *Peperomia*, *Phytolacca*, *Psychotria*, *Rapanea*, *Siegesbeckia*, *Spilanthes*, *Sporobolus*, *Symplocus*, *Ternstroemia*, *Xiphopteris* and *Xyris*.

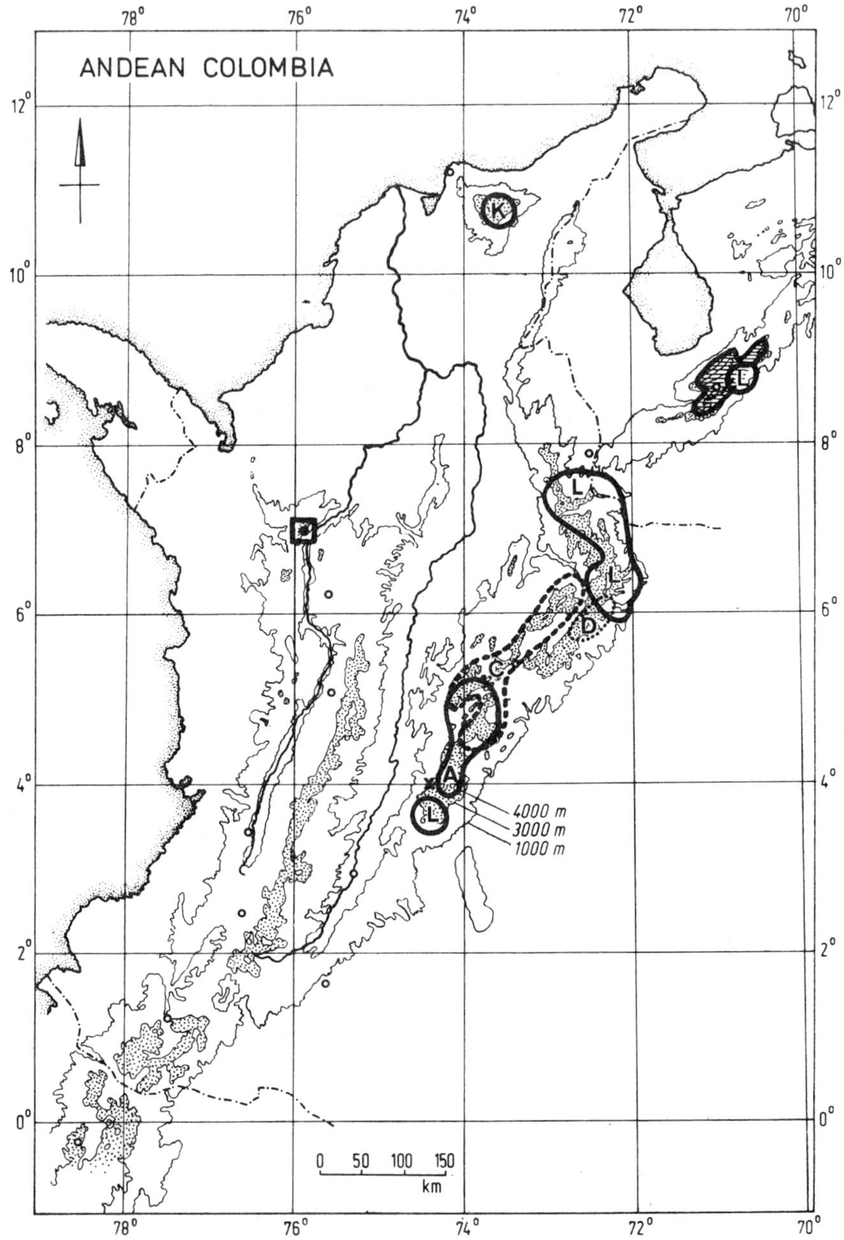

Fig. 1. Distribution of the genus *Aragoa* (Scrophulariaceae). A = *A. abietina* HBK, C = *A. cupressina* HBK, D = *A. dugandii* Romero, K = *A. kogiorum* Romero, ▤ = *A. lucidula* Blake, L = *A. lycopodioides* Benth., ◉ = *A. occidentalis* Pennell, × = *A. perez-arbelaeziana* Romero.

Austral–Antarctic Element

This includes genera occurring in (i) the temperate southernmost part of South America and its shelf islands, including the Juan Fernandez archipelago; (ii) the Subantarctic and Antarctic; (iii) Tasmania and temperate New Zealand and Australia; (iv) the tropical high mountains of South-East Asia and New Guinea, occasionally extending up to China and Japan. The austral–antarctic relationship of the neotropical páramo vegetation was also earlier studied (Troll, 1960; Cleef, 1978). Twenty-four genera belong to the austral–antarctic element; i.e. 9·2%.

Included are: *Acaena, Azorella, Calandrinia, Calceolaria, Cortaderia, Cotula, Desfontainea*, Dysopsis*, Escallonia, Fuchsia, Gaultheria, Lilaea, Lilaeopsis, Mühlenbeckia, Myrteola, Nertera, Oreobolus, Oreomyrrhis, Orthrosanthus, Ourisia, Pernettya, Sisyrinchium, Ugni* and *Uncinia*.

Holarctic Element

Included are genera native to the northern hemispheric temperate and mediterranean floras. It concerns widely distributed genera, which have their main centre of evolution there. Some genera have ranges extending along the cool tropical mountains into the southern hemisphere. Twenty-eight genera belong to the holarctic element; i.e. 11%.

Included are: *Bartsia, Berberis, Castilleja, Cerastium, Cinna, Draba, Erigeron, Gaylussacia, Hackelia, Hypochoeris, Lappula, Lathyrus, Lupinus, Mühlenbergia, Oenothera, Potentilla, Ribes, Salvia, Satureja, Sibthorpia, Spiranthes, Stachys, Thalictrum, Tofieldia, Trifolium, Vaccinium, Verbena* and *Vicia*.

Wide Temperate Element

Included are genera characteristic for temperate and cool areas of both hemispheres. In the tropics this element is limited to the uppermost part of the mountains. 51 genera belong to the wide temperate element; i.e. 19·6%.

Included are: *Agrostis, Alopecurus, Arenaria, Brachypodium, Bromus, Calamagrostis, Callitriche, Carex, Cardamine, Cystopteris, Danthonia, Daucus, Elatine, Epilobium, Equisetum, Eriocaulon, Festuca, Galium, Gentiana, Gentianella* (s.l.), *Geranium, Gnaphalium, Gratiola, Hieracium, Hierochloe, Hypericum, Isoetes, Juncus, Limosella, Luzula, Mimulus, Montia, Myrica, Pilularia, Pinguicula, Plantago, Poa, Polystichum, Potamogeton, Ranunculus, Rubus, Rumex, Sagina, Senecio, Stellaria, Tillaea, Trisetum, Urtica, Valeriana, Veronica* and *Viola*.

                                 A. M. Cleef

Cosmopolitic Element

Included are genera with worldwide or subcosmopolitic distribution as defined by Good (1964). They are well represented in temperate regions; in the tropics they are found from sea level up to the snowcapped mountains. Twenty genera belong to the cosmopolic element; i.e. 7·7%.

Included are: *Asplenium*, *Azolla*, *Bidens*, *Blechnum*, *Eleocharis*, *Eryngium*, *Euphorbia*, *Hydrocotyle*, *Lobelia*, *Lycopodium*, *Myriophyllum*, *Ophioglossum*, *Oxalis*, *Polygala*, *Polypodium*, *Rhynchospora*, *Scirpus*, *Selaginella*, *Solanum* and *Utricularia*.

Unknown Affinity

Two genera do not fit in any of the distinguished geographical flora elements: *Anthericum* (Lil.) and *Coriaria* (Coriar.) If included *Anthericum* and *Coriaria* have most affinity to the wide temperate element.

The procentual subdivision of the vascular páramo flora of the Colombian Cordillera Oriental into geographic flora elements is vizualized in Fig. 2.

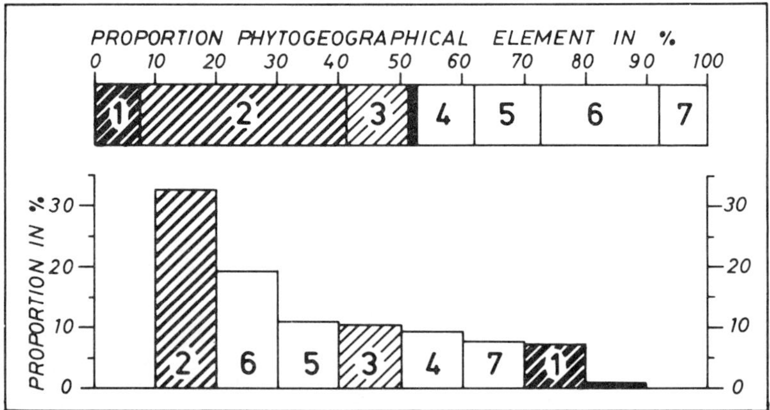

Fig. 2 Percentages of the different geographical flora elements based on vascular genera native to the páramos of the Eastern Cordillera, Colombia. 1 = páramo element, 2 = other neotropical elements, 3 = wide tropical element, 4 = austral–antarctic element, 5 = holarctic element, 6 = wide temperate element, 7 = cosmopolitic element, black = unknown affinity.

It appears that the neotropical component, including the páramo element attains more than 40%, whereas the tropical element in total (first three elements) and the temperate element (last four elements) each contribute almost 50%.

About 30% of the vascular genera occurring in the páramos of the Colombian Eastern Cordillera are also present on the humid high mountains on the Guayana shield. Another 7% exclusively occurs in these páramos and in the drier tropical high Andean puna, extending from Peru towards northern Argentina and Chile. Further genera shared with the puna also grow in other parts of the world.

Discussion and Conclusions

Summarizing it appears that the generic vascular flora of the páramos of the Colombian Cordillera Oriental is basically made up of about half of local tropical elements and half of immigrated taxa.

The neotropical component, as might be expected, is most strongly represented. The proportion of the páramo element is relatively weak at the genus level, however outstandingly high at the species level. Nearly half of the páramo element consists of monotypical genera. Whether they are distinct or not depends on taxonomical appreciation; in this paper they have all been recorded. The group of the other neotropical elements and the austral–antarctic element also contain some monotypic genera. In the temperate component the widely distributed genera are most strongly represented, attaining about 20%. The austral–antarctic and holarctic elments are present with about 10% each. Palynological evidence (van der Hammen, this volume) indicates that in the oldest known Pliocene–Pleistocene páramo flora of the Sabana de Bogotá the austral–antarctic element was much more strongly represented. Northern immigrants were at that time still lacking, apparently because the Andean chain was present and readily accessible before the establishment of the Panamanian connection with the cool holarctic floras.

Sometimes it was hard to decide to which element certain genera belong. *Ilex* for example, treated here as belonging to the wide tropical element, is mainly developed in the tropics, but has species growing in the cool holarctic. On the other hand some genera of holarctic affinity, e.g. *Berberis*, *Erigeron*, *Ribes* and *Vaccinium* penetrate into the cool parts of the southern hemisphere along the tropical high mountains. *Erigeron* even raised new species in humid warm tropical regions (Solbrig, 1962). *Isoetes* is treated as belonging to the wide temperate element, but it has at least one species growing in the warm tropical Amazonian lowlands. Despite of their presence in Fuegia and on the Malvinas Islands, most species of the Composites *Baccharis* and *Oritrophium* are limited to the American tropics; hence these genera are considered to belong to the neotropical element. *Hymenophyllum*, recorded here as a wide tropical fern genus, has a few species growing in the climatological humid parts of the subantarctic zone and the temperate northern hemisphere. Austral–antarctic elements may also penetrate far into the northern hemisphere as examplified by *Lilaeopsis* and *Sisyrinchium*, both found as far north as Alaska.

The method of distinguishing geographical flora elements disregards in some cases genetical affinities, and should therefore carefully be applied. For example certain taxa treated here as neotropical and wide temperate elements exhibit genetic ties with the holarctic flora, e.g. *Befaria*, *Echeveria*, *Eryngium*, *Halenia*, *Myrica*, *Sericotheca* and by *Carex* sect. Paniceae and *Plantago* sect. Leucopsyllium and sect. Oreades. Similarly austral–antarctic affinities are found in *Chaptalia*, *Diplostephium*, *Distichia*, *Gaiadendron*, *Hydrocotyle*, *Laestadia*, *Lysipomia*, *Oritrophium*, *Polylepis* and *Rhizocephalum*, and in *Blechnum* subgenus Lomaria, *Calamagrostis* sect. Deyeuxia, *Geranium* sect. Andicola, *Pinguicula* sect. Temnoceras and *Plantago* sect. Oliganthus. Finally, the páramo species of the genera *Gratiola*, *Limosella* and *Myriophyllum* are widespread in the cool southern hemisphere. These data reinforce considerably the importance of the austral–antarctic element in the neotropical páramo flora.

Phytogeographical analysis at species level of the páramo flora will provide different results, because most species of the genera belonging to the first five elements are endemic to páramos. The relatively few endemic páramo genera and the absence of endemic families account for the geologically short time that the páramo biota came into being.

Admittedly this kind of phytogeographical analysis undoubtedly includes deficiencies and omissions, but at present it seems the most suitable approach. After some hesitation I omitted from the list the following genera, because they were only occasionally found in the páramo belt and generally occur in other plant formations at lower elevations. They include: *Alnus*, *Barnadesia*, *Bocconia*, *Borreria*, *Brugmansia*, *Cynanchum*, *Cuphea*, *Dicranopteris*, *Dodonaea*, *Evolvulus*, *Habenaria*, *Iresine*, *Lantana*, *Laplacea*, *Lepechinia*, *Lepidium*, *Pellaea*, *Pleurothallis*, *Polygonum*, *Psoralea*, *Pteridium*, *Rhamnus*, *Sisymbrium*, *Steiractinia*, *Stelis*, *Stipa*, *Symbolanthus*, *Telipogon*, *Tillandsia*, *Tradescantia* and *Vallea*. Alien genera were also excluded: e.g. *Achillea*, *Anthoxanthum*, *Dactylus*, *Digitalis*, *Ulex*, etc.

With respect to the composition of the vascular páramo flora it appears that the Compositae and Gramineae are the most important families, including the largest number of genera. The same is true for cool temperate floras. The tropical component, however, is very striking in the páramo flora; thus the Orchidaceae are the third family in number of genera. Most tropical families are limited to the subpáramo: e.g. Clethraceae, Myrsinaceae, Symplocaceae and Theaceae. Other tropical families, e.g. Bromeliaceae, Loganiaceae, Loranthaceae, Melastomataceae, Phytolaccaceae, Piperaceae and Xyridaceae, do occur commonly in the lower belts of the neotropical páramos, whereas in the tropical high mountains of East Africa (Hedberg, 1965) and New Guinea (Smith, 1977) they are virtually lacking above the forest limit.

As was shown earlier the vascular páramo flora is the richest in taxa of the cool floras of the high tropical mountains of the world. This is also true if the vascular páramo flora is compared with temperate, arctic and antarctic vascular floras. A detailed comparison is in preparation. In forthcoming papers also attention shall be given to the problem why the neotropical páramo flora is so rich in genera and species.

Acknowledgements

Thanks are due to Dr José Cuatrecasas, Dr S. Rob Gradstein and Dr Tom van der Hammen for valuable additions and constructive criticism leading to the improvement of the manuscript. To Mr H. Rypkema for the drawings, to Mrs E. G. van Bemmel-Henken for typing the manuscript, to the Netherlands Foundation for the Advancement of Tropical Research (WOTRO) for supporting fieldwork and finally to the staff of the Instituto de Ciencias Naturales, Bogotá, for collaboration in identification of large part of the vascular páramo collections.

Addendum

Vascular genera to be added to the páramo flora of the Colombian Cordillera Oriental are: *Colobanthus* (Caryoph.) belonging to the austral–antarctic element and in addition the tropical Andean orchid genera *Aa* and *Myrosmodes*. These new reports do not affect in any substantial way the proportions of the phytogeographical features of the vascular páramo flora outlined in this paper.

References

Cabrera, A. L. (1958). La vegetación de la Puna Argentina. *Rev. Invest. Agr.* **11**, 317–412.

Cleef, A. M. (1978). Characteristics of neotropical páramo vegetation and its Subantarctic relations. *Erdwiss. Forsch.* **1**, 365–390.

Good, R. (1964). "The Geography of Flowering Plants." Longman, London.

Hedberg, O. (1965). Afroalpine flora elements. *Webbia.* **19**, 519–529.

Ruthsatz, B. (1977). Pflanzengeseilschaften und ihre Lebensbedingungen in den Andinen Halbwüsten Nordwest-Argentiniens. *Diss. Bot.* **39**, 168 pp.

Smith, J. M. B. (1977). Origins and ecology of the tropicalpine flora of Mt. Wilhelm, New Guinea. *Biol. J. Linn. Soc.* **9**, 87–131.

Solbrig, O. T. (1962). The South American species of Erigeron. *Contr. Gray Herb. Harvard Univ.* **191**, 3–79.

Szafer, W. (1975). "General Plant Geography." Państowowe Wyndawnictwo Naukowe, Warsaw.

Troll. C. (1960). The relationship between the climates, ecology and plant-geography of the southern cold temperate zone and the tropical high mountains. *Proc. R. Soc. B,* **152**, 529–532.

van der Hammen, T., J. H. Werner and H. van Dommelen (1973). Palynological record of the upheaval of the Northern Andes: a study of the Pliocene and Lower Quaternary of the Colombian Eastern Cordillera and the early evolution of its high-Andean biota. *Rev. Palaeobot. Palynol.* **16**, 1–122.

Plant Refuge and Dispersal Centres in Venezuela: Their Relict and Endemic Element

J. A. STEYERMARK

National Herbarium, Caracas, Venezuela

Introduction

I believe I am safe in stating that less is known of the geographical distribution of neotropical flowering plants and ferns than of the vertebrates. Partly for this reason, the vertebrate zoologist is able to state with somewhat more assurance his concept of dispersal centres, centres of origin, and refuge areas. Any botanist who has worked with collections of neotropical plants, and is engaged in their identification, is keenly aware of the multitudinous areas still awaiting exploration within the neotropics, especially those covered by tall forest or those located in remote or largely inaccessible places. As Müller (1973) has so well stated: "the biographical study of South and Central America is still in the beginning".

Therefore, I am quite cognizant at this time of the limitations surrounding a presentation of such a subject as geographical patterns of plant distribution in Venezuela with reference to the relict and endemic elements of the flora which appear to be associated with plant refuge and dispersal centres. Providing the many areas still awaiting to be explored in Venezuela are still available a century hence, and providing there is an increasing number of plant collectors in the coming century, then, perhaps, by that time a much more adequate and accurate model on this subject can be presented. Moreover, since the solutions to many of our problems of phytogeography still remain to be supplied by a combination of evidence from the fields of future botanical exploration, paleobotany, and palynoloy, I may be excused in expressing some uncertainty with reference to some aspects of the subject at hand. Nevertheless, the data now available, based upon known collections and patterns of phytogeographical distribution of the more than 15 000* presently known Venezuelan taxa, provides us with a basis for the present contribution. A study of the Venezuelan flora shows, as it does in many other tropical regions, a very complicated and diverse phytogeographical pattern. Numerous taxa appear to have originated in many sectors of the country, where they are now isolated and endemic, while countless others seem to have entered the various sectors of Venezuela from many distributional centres, and, to add to the complexity, appear to have entered at different time intervals in the geological past.

It would appear that the primary or first factor to have delimited the present distributional pattern of Venezuelan taxa is the historical, geological, or physiographic factor, i.e. the physical delimitation of placement originally of the species within certain areas by the establishment of topographic barriers erected by the Andes, the Coastal Cordillera, the Guayana Shield, and marine transgressions. Edaphic barriers, such as special soils and outcroppings of rock substrate, along with special climatic conditions, must likewise be considered together with the physiographic factor. A second factor, that of the Pleistocene and Post-Pleistocene changes of climates and their influence on the development of the present pattern of plant distribution, would necessarily be conditioned by the already previously established physiographic, ecologic or edaphic barriers.

In this presentation it will be noted that the major areas of plant refuges, dispersal centres, or of the locations of endemic and relict species, are associated with some part of the Andes, the Coastal Cordillera, the Serrania del Interior and the region of the Guayana Highland or Pantepui. The other refuge areas, mainly of lowland elevations, are selected on the basis of their mostly unique floras, some of which have already been recognized, in part or in whole, by zoologists and botanists, as refuge centres for the preservation of floras or for the differentiation and dispersal of taxa. These essentially lowland centres appear to have been associated with the variations and changes of climate during Quaternary and Recent phases. In the brief time at our disposal, we can present only a limited discussion of these areas.

Perhaps, before entering further into an account of the major areas of Venezuela, it would be well to elucidate the terms of relict and endemic elements, refuge and dispersal centres. An endemic element, as the term is here employed, is one restricted in its present distribution to a

* The number of total taxa of flowering plants and ferns in Venezuela is presently calculated as between 15 000–20 000, with a future number estimated to reach 25 000 when the country has been more intensively explored.

limited area, where it may have originated and is autochthonous, whereas a relict element is found either far removed from the main portion of its natural geographical range, or occurs at the extremes of its natural distribution. We can illustrate this by the example of *Lagenanthus princeps*, an endemic monotypic genus of a limited area of the Venezuelan–Colombian border, and by *Delostoma integrifolium* and *Desfontainea pulchra*, both of the last two representing relicts at the extremes of their natural distributional range. Refuges generally have been applied to those forested areas of usually humid tropical lowland forest, which have remained as forested areas during dry phases of Pleistocene time while most of the surrounding area was covered by savanna. However, refuges may also apply, under an extended definition, to edaphic open areas of special soils or rock substrate, such as are found in various parts of Pantepui. But, as applied to forested areas, the forests served temporarily as refuges during the Pleistocene for the isolation of species.

A dispersal centre, following the definition by Müller (1972), is a special type of forest refuge which has acted as a preservation area of plants and animals during arid phases of the Quaternary cycle. Then climatic changes during the Pleistocene and Post-Pleistocene initiated repeated expansions and fragmentations of these forested areas, the result being that eventually differentiation among populations, which had been previously isolated, occurred due to secondary contact with each other during humid phases of the Holocene, at which time the forests re-expanded. The present complex patterns of distribution, where previously isolated species appear to have come into secondary contact with one another, is the result.

Sometimes the forest refuge is the same as the dispersal centre, but it is not necessarily so. Also, the centre of origin of a species may coincide with its dispersal centre, but, on the other hand, the centre of origin of a species may become widely separated from its centre of dispersal. Finally, it should be noted that dispersal centres may be non-forest as well as forest types.

Venezuelan Refuges and Dispersal Centres

Taking up briefly, then, these selected areas, let us devote the major portion of the remainder of this presentation to those comprising mos of the Venezuelan flora. We find that the principal centres for endemic and relict taxa in Venezuela are situated within the territory now occupied by the Andes, the Coastal Cordillera, the Serranía del

             J. A. Steyermark

Interior (Interior Coastal Range), and the Pantepui Region, which includes the summits and talus slopes of the table mountains of the Guayana Highland, the Gran Sabana, and the edaphic lowland sand savannas and igneous rock outcrops of the Territorio Federal Amazonas (see Fig. 1).

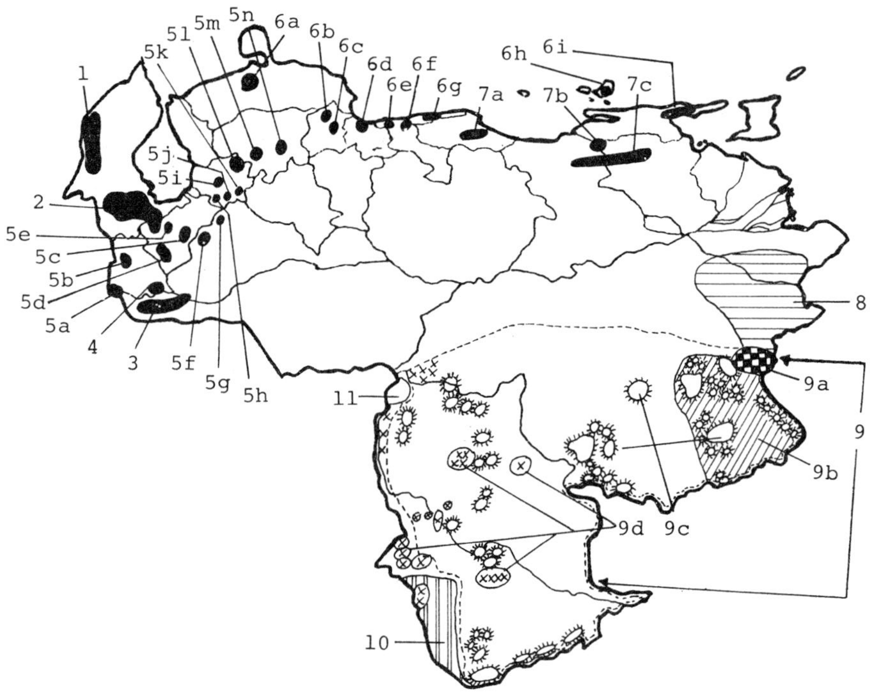

Fig. 1. Plant refuges of Venezuela. 1 = Perija; 2 = Catatumbo; 3 = San Camilo; 4 = Ayari; 5 = Andean: a = Tamá, b = Paramos Tachira, c = Paramos Merida, d = Santa Cruz, e = El Molino, f = Aguada, g = Barinitas, h = Jajo, i = Escuque, j = Guirigay, k = Guaramacal, l = Humocaro, m = Yacambu, n = Terepaima; 6 = Cordillera de la Costa: a = San Luis, b = Aroa, c = Nirgua, d = Borburata, e = Rancho Grande, f = Colonia Tovar, g = Naiguata, h = Margarita, i = Paria; 7 = Serrania Interior: a = Guatopo, b = Turumiquire, c = Guacharo; 8 = Imataca; 9 = Pantpui: a = Venamo, b = Gran Sabana, c = Tepui, d = Amazonas Savannas; 10 = Rio Negro; 11 = Atures.

Coastal Cordillera and Serrania del Interior Versus the Andes—General Remarks

Although the Coastal Cordillera and Serrania del Interior would appear to be eastern extensions of the Andean chain, actually their geological histories have been markedly different. The Venezuelan Andes are relatively more recent than the Coastal Cordillera, having arisen, at the earliest, above the surrounding areas in the Paleocene (van der Hammen, 1961; Simpson, 1975), and continuing into the end of the Pliocene, at which time the Andes attained their greatest height, their principal axes at present directed south-west to north-east. The Cordillera of Perijá, an integral part of the Andes, was uplifted later than the rest of the Venezuelan Andes, having been uplifted as recently as late Pleistocene (Simpson, 1971; Gansser, 1955). The Coastal Cordillera, on the other hand, started to elevate during the upper Cretaceous period, and, with subsequent elevations in Europe and Pliocene, attained its present configuration, with its general orientation west to east. During the stages when both mountain systems were undergoing their tectonic histories, they underwent different erosional stages at diverse times of the uplifts. Moreover, they differ in their component geological formations, and in the relatively higher elevations of the Andes.

The discontinuity of the Andes and the Coastal Cordillera is clearly indicated by the tectonic depression of Turbio–Yaracuy (sometimes termed the Yaracuy Depression), extending from the north-eastern portion of the state of Yaracuy to the south-western part of the state of Lara, thus severing the western end of the Coastal Cordillera from the eastern terminus of the Andes in the south-eastern portion of the state of Lara.

Obviously, the above historical and structural differences in these two major cordilleras have been determining factors which have resulted in the essentially marked divergent floras between them. Although many taxa are common to both the Andes and the Coastal Cordillera, nevertheless each of the systems of mountains possesses its own characteristic and endemic elements. Genera which are found in the Coastal Cordillera but not in the Andes are the following: *Razisea, Lacmellea, Resia, Maripa, Mapania, Croizatia, Fissicalyx, Roucheria, Diolena, Llewelynia, Macrocentrum, Linociera, Cespedesia, Elvasia, Asterogyne* (Aristeyera), *Warrea, Schlimia, Psilochilus, Chysis, Leuddemania, Triphora, Napeanthus, Nautilocalyx, Billbergia, Selenicereus, Diospyros, Platycentrum, Necramium* (Fig. 2), *Glomeropitcairnia, Dichapetalum, Stephanopodium, Dendrobangia, Dussia, Quiina, Dicranostyles, Marila, Gloeospermum,*

Fig. 2. Paria relict *Necranium gigantophyllum*.

Onoseris, Dicranopygium, Schlegelia, Borojoa, Amaioua, Neoblakea, Amphidasya, Froesia, Trichoceros, Platystele, Pterobesleria, Trigonia, Stenostephanus, Cinnamodendron, Macradenia, Houlletia, Eurystyles, Tresanthera, Macropharynx and *Bathysa*.

Genera, on the other hand, found in the Andes, but not in the Coastal Cordillera are given in the following partial list: *Jamesonia, Chevreulia, Diplostephium, Hinterhubera, Oritrophium, Podocoma, Soliva, Axinaea, Castratella, Bucquetia, Lythrum, Lagenanthus, Oreobolus, Fuchsia, Brachtea, Hypochaeris, Jungia, Aetanthus, Chaetolepis, Altensteinia, Llagunoa, Anguloa, Brachionidium, Chondrorhyncha, Cyrtidium, Peristeria, Phragmipedium, Arracacia, Azorella, Niphogeton, Lucilia, Oxybolus, Plagiocheilus, Schkuhria, Vasquesia, Werneria, Hofmeisterella, Polycycnis, Lachemilla, Limosella, Ottoa, Delostoma, Macrocarpaea, Tovomitopsis, Pinguicula, Centronia, Myrtus, Ada, Caucaea, Chrysocycnis, Fernandezia, Pityphyllum* and *Sertifera*. Obviously many of these Andean genera are those which inhabit higher altitudes than exist on any of the summits of the Coastal Cordillera, and for that reason alone are absent from the latter.

Andean Refuges and Dispersal Centres

Having referred to generalities between the Andes and the Coastal Cordillera, we can pass to a discussion of the various plant dispersal centres in the Andean section of Venezuela, 15 of which have been segregated, as follows: Perijá, Tamá, Táchira and Mérida páramos, Santa Cruz, El Molino, Aguada, Barinitas, Jajó, Escuque, Guirigay, Guaramacal, Humocaro, Yacambú and Terepaima. Each of these is considered as a unit from which differentiation and isolation appears evident. In all the centres enumerated, the primary factor in their isolation appears to have been the historic one, i.e. physiographic barrier correlated with the diverse ramifications of the Venezuelan Andes when its axes were modified into disparate branches. As the Andean chain reaches its northernmost South American ramifications, it assumes the configuration of an enormous Y, of which its western branch continues as the Cordillera of Perijá.

Perijá

The mountains of Perijá, which constitute the border between north-eastern Colombia and north-western Venezuela, consist of a highly dissected topography of sandstones, shales, and limestones, with elevations up to 3400 m. A strong endemic element is present and most pronounced in the forested areas above 1100 m, and becomes even more predominant between 2400–3400 m. Some of the endemic species, such as *Miconia limitaris*, *Calea perijaensis* and *Senecio perijaensis* are common to both the Colombian and Venezuelan sides of Perijá. On the other hand, geographically closely related (sympatric) pairs of species maybe isolated in the case of *Calceolaria perijaensis* on the Colombian side and *C. tripartita* subsp. *yuparum* on the Venezuelan side. *Chimarrhis perijaensis* and *Miconia perijensis*, in contrast, are known thus far only on the Venezuelan side, while *Chaetolepis perijensis* is at present restricted to the Colombian portion. Other endemic species include *Pterogastra glabra*, *Baskervillea venezuelana*, *Pleurothallis hypocrita*, *Pterichis latifolia* and *Spermacoce perijaensis*.

Relict taxa which are in the Perijá Mountains consist of Colombian floral elements, such as *Peperomia discilimba* and *Psychotria erythrocephala*. This Colombian element shows its predominance in the forests, beginning at the higher altitudes of about 1800 m, manifest by such species as *Clethra repanda* and *Cecropia telenitida*. The main relationship of the Perijá flora is especially with the Colombian Andes. Widely distributed Andean species, such as *Cinchona pubescens*, *Psila brachylaenoides* var. *ligustrina*, and Andean genera, *Arcytophyllum* and *Espeletia*, are indicative of the Andean relationships. A relationship with the Coastal Cordillera flora is shown by the Coastal Cordilleran endemic, *Psiguria racemosa*, and by the disjunct distributions of *Poulsenia armata*, *Peperomia distachya* and *Solandra grandiflora*. A number of relict elements of the flora, especially those restricted to the lower and middle altitudes of the Perijá Mountains, appear to represent taxa derived from more widely distributed neotropical dispersal centres in Mexico, Central America and the Antilles. Examples of such are evidenced in the distribution of *Miconia schlimii* (Central America to Colombia), *Clidemia purpureoviolacea* (Costa Rica to Colombia) and *Donnellsmithia peucedanoides* (México to Colombia). In general, however, the taxa encountered on the lowest slopes at 250–500 m are commonly encountered species or ones mainly with a wide distribution in other parts of Venezuela, and in other parts of tropical America. Such are the species whose distributional range encompasses Mexico, Central America, the Antilles, South America, and sometimes even the United States. An exception to this generality is noted in *Solanum aquatile*, found only in Colombia and at an altitude of 300 m on the Venezuelan side of Perijá.

Tamá

The eastern branch of the Andes, which starts in Colombia, enters Venezuela in the state of Táchira, but here the Andes are broken in their eastward course by an important geological barrier, the Táchira Depression. This geosyncline, which originated as a tectonic trough, has had a marked influence on the isolation and differentation of adjacent Andean floras. Until this geosyncline became available as a land area at the beginning of the Pleistocene, about a million years ago, a branch of the sea covered this trough, and beginning with the Eocene served as a watercourse allowing the seas to extend between the Maracaibo Basin to the north and the Orinoco Basin to the south in what is at present the western llanos of the Apure. The Depression thus separated the Andes and acted as a barrier. The Tamá dispersal centre, consequently owes its existence and importance to the Táchira Depression of the south-westernmost corner of the Venezuelan Andes. Simpson (1971, 1975) treated this Tamá area only as one of several páramo "island" patches of the northern Andes, but it is actually one of the more significant forest refuges and dispersal centres of Venezuela, even though it has not been generally recognized as such by most biogeographers. Not only does the Tamá centre include the Páramo de Tamá and other páramos of Cobre and Judio lying to the west of the Táchira depression, but involves the surrounding lower forested slopes of the mountains to the west of the Depression north to San Antonio de Táchira, Bramón and Santa Ana.

Actually, no less than 145 species of flowering plants reach their north-eastern limits of dispersal in the Tamá centre, of which 88 or 60%

are known only from portions of the Colombian Andes. These include: *Begonia ferruginea*, *Guzmania confinis*, *Tillandsia stipitata*, *Brunellia elliptica*, *Diplostephium rosmarinifolium*, *Senecio colombianus*, *Lagenanthus princeps* (Fig. 3), *Geranium santanderiense*, *Juglans columbiensis*, *Pinguicula diversifolia*, *Centronia insignis*, *Miconia buxifolia*, *Myrtus oxycoccoides*, *Ada aurantiaca*, *Chrysocycnis schlimii*, *Odontoglossum spectatissimum*, *Palicourea albert-smithii*, *Solanum juglandifolium* and *Hydrocotyle gunnerifolia*. The remainder of the 145 taxa in the Tamá centre are distributed chiefly in various Andean countries from Bolivia, Peru, Ecuador and Colombia north-east to the Tamá centre, and include the following genera at their north-eastern limit in South America: *Delostoma* (Fig. 4), *Oreobolus*, *Tovomitopsis*, *Desfontainea*, *Lythrum*, *Bucquetia*, *Caucaea*, *Pityphyllum*, *Porroglossum* and *Setifera*.

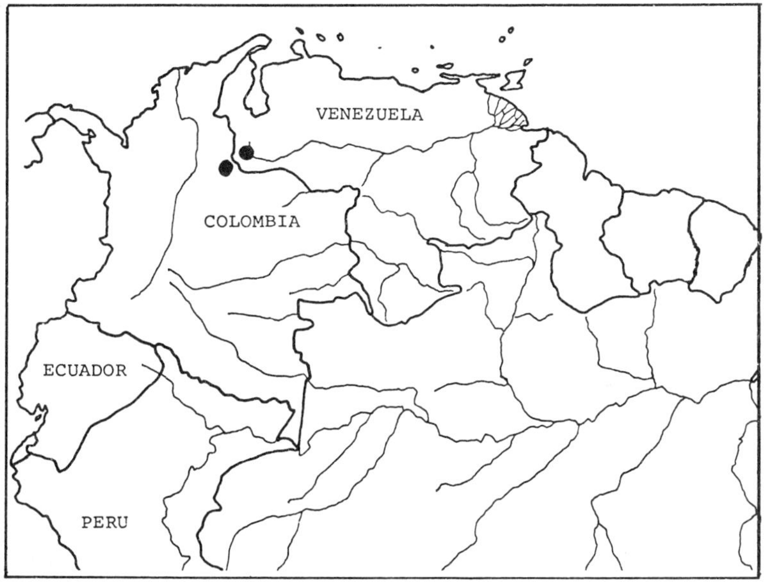

Fig. 3. Tamá endemic *Lagenanthus princeps*.

Eighty-two additional plant species are known to be endemic to the Tamá centre, and such genera as *Castratella* have their centre of origin in this part of the Andes. Data from the avifauna of the Tamá centre, supplied by Phelps (in Steyermark, 1975a) are similarly significant, showing 39 subspecies and six species of birds of Colombia and/or of more southern Andean countries which reach their north-eastern limits in the Tamá centre, and one species and three subspecies as endemics (Phelps and Phelps, 1953, 1955).

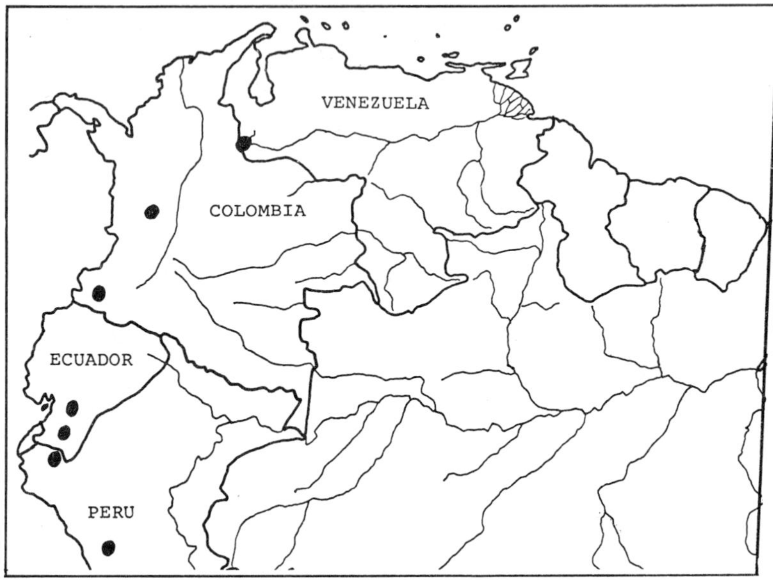

Fig. 4. Tamá relict *Delostoma integrifolium*.

Other Andean Centres

The remainder of the Andes, lying to the east of the Táchira Depression, is divided mainly by the Río Chama into two major portions: (i) the more south-eastern-lying Sierra Nevada de Mérida, which continues northeastward into the Sierra de Santo Domingo, and (ii) the more north-western-lying Sierra del Norte or La Culata. Each of these, with its various ramifications and differences in relief, has served as an historical basis in the initial segregation and isolation of habitats. Since the final uplift of the Venezuelan Andes did not take place until at least the end of the Pliocene, it would not seem possible for the evolution and differentiation of the plant taxa to begin until the Pleistocene (Simpson, 1975). Although the highest altitudes of the Venezuelan Andes are found today in those portions lying within the state of Mérida, where snow-capped stretches and glaciers occur, nevertheless other areas of high altitudes are known with various páramos scattered over the Andean landscapes of the states of Táchira, Trujillo and Lara, culminating north-eastward in the páramos of La Naríz, Las Rosas, Jabón and Cendé. The páramos are separated both vertically and laterally, occurring in patches or "islands" (Simpson, 1975), and, differentiation at the specific level has taken place. Where the páramos are of smaller extent and of lesser altitude above sea level, the number of endemic species is reduced percentage-wise as compared with the flora of the more extensive páramos of higher elevations found in the state of Mérida. Thus, in the Páramo de las Rosas and Páramo Cendé at 3000–3300 m in the Humocaro centre near the eastern Andean sector of the state of Lara are differentiated the endemic *Carex larensis*, *Drosera cendensis*, *Espeletia trujillensis*, *Hinterhubera adenopetala* and *Senecio rigidifolius*; at altitudes of 3200–3500 m in the Guirigay centre of the state of Trujillo occur such endemics as *Greigia aristeguietae*, *Puya venezuelana*, *Espeletia arborea*, *E. frailejonota*, *E. tenorae*, *E. viridis* and *Hydrocotyle aristeguietae*; and at the 2600–2900 m level of the Páramo of Guaramacal in the state of Trujillo are to be found the endemic *Miconia suaveolens*, *M. trujillensis*, *Espeletia griffinii* and *E. lopez-palacii*.

However, the greatest degree of specific differentiation in the Andean centres has taken place in the montane forests, situated below and removed from the páramo habitat, at altitudinal levels from approximately 1500 to nearly 2900 m. Evidence is at hand that during glacial periods of Pleistocene, cooler temperatures caused a lowering in elevation of the vegetational belts, thus increasing the horizontal expanse of the flora which existed at the higher elevations (van der Hammen, 1972a; van der Hammen and González, 1960; Geel and van der Hammen, 1973; Simpson, 1975). The result of the reduced temperatures at that time is believed to have increased the extent of the páramo area, at the same time decreasing the level of the subtropical climate to an altitude in some areas of 500–700 m (van der Hammen and González, 1960; Müller, 1973). Upon changes of climate concomitant with warmer interglacial phases of the Pleistocene, rising warmer temperatures prevailed and led to a shift upward of the montane forest. It would appear, therefore, that as a result of this upward shift, the horizontal expanse of the páramo and the high altitudinal zones and their habitats became more isolated and more restricted (Müller, 1973).

Moreover, as the Andean montane forest elements shifted upward during the warmer interglacial phases, different routes up and over the mountain slopes were followed. These routes became often interrupted and were probably greatly influenced by such physiographic barriers as the numerous Andean valleys, whose generally lower elevations above sea level interrupted the vertical shift of the montane forest flora (Simpson, 1975). In this manner, it seems likely that such shifts and changes which occurred during the cooler and warmer phases of the Pleistocene, together with the physiographic ramifications and discontinuities present in the Venezuelan Andes, combined to act as significant factors in the isolation of the montane forest populations with resultant differentiation of species occupying various dispersal centres (Müller, 1973). Such differentiation has been the basis for the naming of the various dispersal centres in this presentation, indicated for the Andes, in addition to the already discussed Perijá and Tamá, as follows: Santa Cruz, El Molino, Aguada, Barinitas, Jajó, Escuque, Guirigay, Guaramacal, Humocaro, Yacambú, Terepaima and the Táchira and Mérida páramos.

The páramos of Táchira and Mérida, together with those of Jajó, Guaramacal, Humocaro, Santa Cruz, and El Molino correspond to the dispersal centres located in the higher altitudes of the Andes, whereas Aguada, Barinitas, Escuque, Yacambú, and Terepaima more properly belong to the montane forest centres generally situated at lower altitudes between 1000 and 2000 m. The Barinitas, Aguada and Guaramacal centres, located on the south-east-facing slopes of the Andes, are separated from one another by the valleys of the rios Santa Domingo, Canaguá and Boconó, respectively, while the Escuque, Humocaro, Yacambú and Terepaima centres occur on other ramifications at the northern and eastern portions of the Andes, and are separated by the ríos Motatán, Tocuyo, Guanare, Portuguesa, Turén and Turbio. The Andean element of the flora is pronounced throughout the above-mentioned dispersal centres, with the exception of Terepaima, which is the most extreme eastern of the centres and which manifests a strong admixture and influence of the flora from the adjacent dispersal centres to the immediate east in the Coastal Cordillera.

In summary, we find a total of 506 taxa endemic to the various centres enumerated for the Andes. Of this total, the highest concentrations of endemics are found in the montane forests and páramos of the state of Mérida, where 155 or 30·6% of the endemic flora occurs, and secondarily in those endemic taxa whose range is scattered throughout the Venezuelan Andes. The latter group totals 103 taxa, representing 20·4% of the Andean flora. The other large concentration of Andean endemics is found in the Tamá centre, with a total of 82 taxa, or 16·2% of the Andean flora. This is not surprising when it is realized that the Tamá centre was isolated from the remainder of the Venezuelan Andes by the geological Táchira Depression, which acted as a barrier. The remainder of the Andean centres yield a relatively small percentage of endemics, with a combined total of 166 taxa comprising 33% of the Venezuelan Andean endemics (as tabulated): Táchira paramos (34); Molino (22); Humocaro (20); Terepaima (18); Jajó (15); Santa Cruz (14); Perijá (11); Escuque (8); Guirigay (8); Guaramacal (5); Barinitas (5); Aguada (4); Yacambú (2).

Coastal Cordillera Refuges

Turning now to the Coastal Cordillera, it is to be noted that, as already indicated, although a common linkage does exist between the floras of the Coastal Cordillera and those of the Andes, there are many differences between them, influenced partly by their dissimilarities in geological history, geological formations and altitudinal maxima. The Coastal Cordillera (or Cordillera del Norte) extends for approximately 1000 km, from its western extreme by the Depression of Yaracuy–Barquisimeto (or Yaracuy Depression), where it is separated geologically from the Andean chain, to its eastern terminus at the tip of the Península of Paria in the state of Sucre. Its highest elevation is attained on the Pico de Naiguatá above Caracas, with a height of 2765 m above sea level. In its course eastward, the Coastal Cordillera is interrupted by the Depression of Unare in the Gulf of Cariaco on the northern borders of the states of Miranda and Anzoategui before reappearing again east of the Golfo de Cariaco. The Island of Margarita, although now separated, is considered as belonging to the Coastal Cordillera geologically.

To the south of Valencia in the state of Carabobo and lying to the south of the Coastal Cordillera, the Interior Coastal Cordillera or Serranía del Interior is evident at its western terminus. It was formed at a somewhat later geological interval than the northern part of the Coastal Cordillera and has an average lower elevation. In the east it reaches its terminus near Caripito, just west of the Gulf of Paria, in the state of Monagas. Two main portions of the Serranía del Interior are usually recognized: (i) a western sector, comprising the galeras, the morros of San Juan with Cerro Platillón reaching a height of 1930 m above sea level, and the mountains of the Guatopo National Park and El Guapo-Bachiller, the whole area extending east to the Boca de Uchire by the western border of Anzoátegui state, and (ii) an eastern sector, which comprises the mountainous areas east of Bergantín and those around Caripe, with a maximum elevation of 2596 m attained in Cerro Turumiquire.

The summits and uppermost slopes of the Coastal Cordillera are mainly covered by humid cloud forest, which may extend between 900 and 2000 m in the western and central sectors, as in the Sierra de San Luis, Sierra de Aroa, Nirgua, Salom, Borburata, Rancho Grande, Colonia Tovar, Avila, Naiguatá and Península de Paria. In the Serranía del Interior, cloud-forested summits are likewise present and may descend as low as 700 m on the humid summits of cerros del Bachiller of the state of Miranda.

The Paria centre has been generally recognized as a distinct refuge and dispersal centre by those who have given some attention to the subject, although Brown (1975) terms the area "Sucre-Trinidad". The other sectors of the central and western parts of the Coastal Cordillera have been treated either as the "Rancho Grande" refuge (Vanzolini, 1970; Prance, 1973; Brown, 1975), or, as defined by Müller (1973), as "the Venezuelan montane forest centre" with elevations restricted above 1400 m, as contrasted with his "the Venezuelan coastal forest centre", of nearly the same geographical circumscription, except for its north-west extension into coastal Falcon state, but with altitudes below 1400 m.

An analysis of the specific differentiation of the flora of this extensive region, distributed from the western terminus of the Coastal Cordillera in the eastern part of the state of Lara to the tip of the Península of Paria in the state of Sucre, indicates the need of a classification into a much greater number of discrete units that should be considered as refuges and dispersal centres in the same sense and with similar importance as those referred to already under "Rancho Grande" and "Paria".

Although numerous taxa within the Coastal Cordillera manifest a more or less continuous distribution throughout the cloud-forested summits and upper slopes from the Sierra de San Luis in Falcón state to the Peninsula of Paria in Sucre state, there are many others which are restricted and found only in limited portions of one or more mountains, justifying thereby their inclusion into distinct refuge and dispersal centres as endemic or relict elements of the flora. Among examples of generally distributed endemic taxa throughout the Coastal Cordillera may be mentioned the following: *Anthurium longissimum*, *Didymopanax glabratum*, *Begonia sucrensis*, *Maytenus karstenii*, *Henriettella tovarensis*, *Physosiphon lansbergii*, *Bactris setulosa*, *Hoffmannia apodantha*, *Psychotria lucentifolia* (Fig. 5) and *Meliosma pittieri*.

An enumeration of the various centres of the Coastal Cordillera and Serrania del Interior with lists of some of their respective endemic and relict species is presented in the appendix.

A survey of the endemic taxa reveals a total of 247, of which 174 are restricted to the Coastal Cordillera (Cordillera del Norte) and 73 to the more southern Serranía del Interior (Interior Coastal Range). An additional 29 are endemic to the two cordilleras with stations in other Venezuelan dispersal centres. The dispersal centres of the two cordilleras showing the greatest number of endemic species are, in order of rank: Naiguatá (Coastal Cordillera) with 69, Rancho Grande (Coastal Cordillera) with 40, Guácharo (Serranía del Interior) with 38, Turumiquire (Serranía del Interior) with 37, and Paria (Coastal Cordillera) with 29.

Fig. 5. Coastal range endemic *Psychotria lucentifolia*.

As compared with the Andean centres, those of the combined northern Coastal Cordillera and more southern Serranía del Interior have less than half the total number of endemic elements (247 as compared with 506). On the other hand, the flora which at present inhabits the humid cloud-forested summits of especially the northern Coastal Cordillera is one manifesting a significant disjunct distribution which is definitely related to an Amazonian–Guayanan dispersal. Because of this disjunct distribution, the flora of the summits of the Coastal Cordillera would, in particular, indicate a more ancient phytogeographical history than that of the Venezuelan Andes.

The apparently close relationship existing between the flora of the Coastal Cordillera and the Guayana–Amazonian sector, with a disjunction shown of several hundred to a thousand kilometers is in contrast with the Andean element, of which the Guayana–Amazonian flora is much less in evidence. Such genera of the Coastal Cordillera as *Froesia* (Fig. 6), *Macrocentrum* (Fig. 7), *Gloeospermum*, *Stephanopodium*, *Elvasia*, *Cespedesia* and *Roucheria* may be cited as striking examples of this relationship. In some instances, the taxa, obviously relicts, such as *Graffenrieda weddellii* (Fig. 8), *Gloeospermum sphaerocarpum*, *Platycentrum clidemioides* and *Cespedesia spathulata*, isolated on the upper slopes and cloud-forested summits of the Coastal Cordillera, have not differentiated specifically or subspecifically from their dispersal centres several

                    J. A. Steyermark

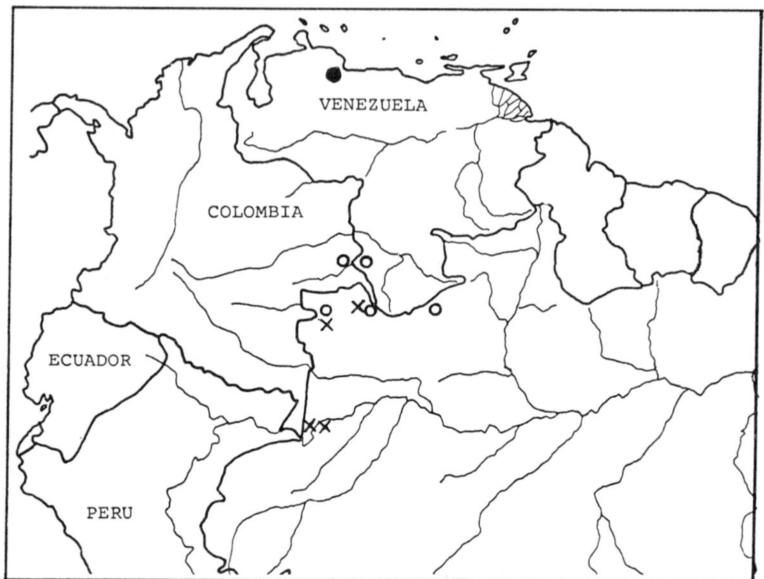

Fig. 6. Coastal range endemic *Froesia venezulensis* (●), *F. tricarpa* (○) and *F. crassiflora* (x) of Amazonian affinity.

hundred to a thousand kilometers to the south. In other cases, such as *Macrocentrum yaracuyense*, the coastal cordilleran species has become differentiated from its closest relative, *M. neblinae*, of Cerro Neblina of the Guayana Highland, where the genus *Macrocentrum* has undoubtedly had its centre of origin. *Froesia*, a genus consisting of three species, has two taxa in Amazonian Brazil and adjacent Colombia and Venezuela, with *F. venezuelensis* differentiated as a third species on the Coastal Cordillera summit of the state of Yaracuy. The evaluation of these relationships will be discussed later.

These isolations on the northern Coastal Cordilleran summits, exemplified by *Macrocentrum yaracuyense* of a genus otherwise limited mainly to the southern sandstone tepuis of the Guayana Highland, or by *Gloeospermum sphaerocarpum* and *Graffenrieda weddellii*, elsewhere dispersed chiefly in Amazonia or other lowland regions, and by *Qualea pittieri*, *Froesia venezuelensis* and *Stephanopodium venezuelanum*, which are differentiated specifically, but far removed geographically from the remainder of the ranges of their respective genera—manifest consistent disjunct patterns which afford evidence of a former, more ancient connection with the flora to the south.

To account for this type of isolation, as well as that of a number of

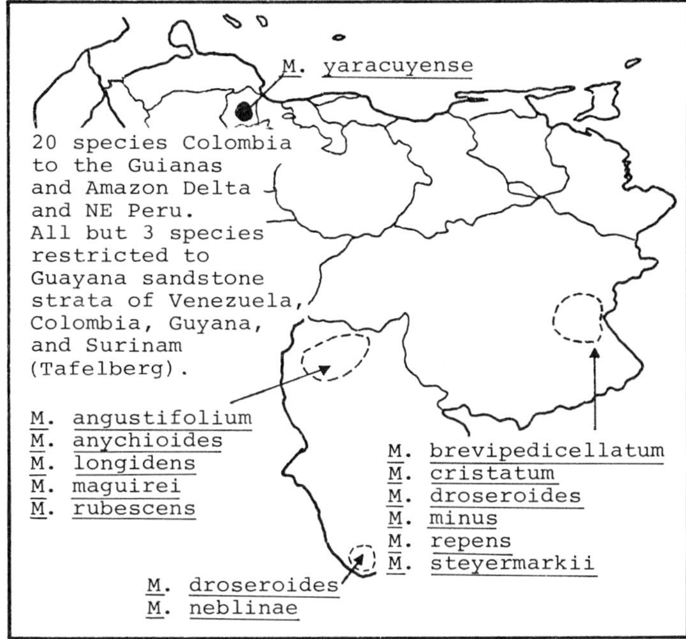

Fig. 7. Coastal range endemic *Macrocentrum* of Guayanan affinity.

Fig. 8. Coastal range relict *Graffenrieda weddellii* of Amazonian affinity.

other similarly distributed taxa,* it would seem evident that at some time in the past, the forest floras of the Coastal Cordillera and Serranía del Interior had been in contact with the forest floras to the south. This contact could conceivably have taken place during one of the humid phases of the Pleistocene by either of one or both of two different operations: (i) tongues and remnants of gallery forests persisting in non-inundated sectors of the llanos and/or of the Orinoco–Río Negro drainage were able to connect with sectors of the Coastal Cordillera, which thereby permitted migration over lowland land bridges; (ii) the alternation of cooler and warmer, and humid and arid phases during the glacial and interglacial phases of the Pleistocene led to vertical shifts of climatic change with corresponding effects on the forest flora. These forest floras, which were displaced downward during the cooler phases of the glacial period, became shifted upward to higher levels during the warmer interglacial stages.

In the cases of *Vriesea scalaris* and *Tillandsia globosa* (Guácharo) and *Vriesea bituminosa* (Rancho Grande), the disjunction is with eastern and southern Brazil. In the examples of *Guzmania cylindrica* (Aroa), *Resia ichthyoides*, and *Rustia longifolia* (Borburata), the disjunction is with the Colombian centres of montane forest. Some species of the avifauna, such as *Pyrrhura leucotis*, *Piprites chloris*, and *Xanthomias virescens*, also show a similarly disrupted range (Phelps, 1966). In their upward migrations, the lowland floras encountered a cooler but similarly humid environment on the cloud-forested summits of the Coastal Cordillera. This humidity factor is probably responsible for the survival of these relict species, which had previously been adapted to the humid, but lower elevations. Apparently, most of the taxa, judging by present-day distributions of the Amazonian–Guayanan flora, were not able to adapt to these climatic changes and became extinct. The

---

* Examples of similar types of disjunction of the taxa of the Coastal Cordillera and Serranía del Interior are to be noted in the following species, whose refuge centre is indicated when only that centre is involved: *Polypodium megalophyllum* (Nirgua), *Roucheria laxiflora* (Nirgua), *Rhynchospera exaltata*, *Pleurostachys puberula* (Naiguatá), *Diospyros guianensis* subsp. *porphyria*; *Aspidosperma fendleri*, *Mapania pycnocephala* subsp. *fluviatilis*, *Dendrobangia boliviana*, *Epidendrum sclerocladium*, *Lankasterella caespitosa*, *Pleurothallis arbuscula* (Nirgua), *Maxillaria imbricata*, *Psilochilus modestus*, *Spiranthes macrantha*, *Vriesea bituminosa* (Rancho Grande), *Vriesea scalaris* (Guácharo), *Tillandsia globosa* (Guácharo), *Chlorophora brasiliensis* (Guácharo), *Sloanea obtusifolia* (Naiguatá), *Playcentrum clidemioides* (Paria), *Cespedezia spathulata* (Paria), *Ilex vacciniifolia* (Turumiquire), *Marcetia taxifolia* (Turumiquire), *Ugni myricoides*, *Laurembergia tetrandra* (Turumiquire), *Stelis maderoi*, *Passiflora bogotensis*, *Doryopteris sagittifolia* (Guácharo), *Macropharynx strigillosa* (Guácharo), *Cordia ucayaliensis* (Guácharo), *Codonanthe uleana* var. *integrifolia* (Guácharo), *Calliandra carbonaria*, *Rustia longifolia* (Borburata), *Resia ichthyoides* and *Guzmania cylindrica* (Aroa).

vertical displacements and adaptations resulted in either the evolution and differentiation of new taxa, such as *Froesia venezuelensis*, *Qualea pittieri* and *Stephanopodium venezuelense*, or the persistence and isolation of some of the previous inhabitants of the lowland flora such as *Graffenrieda weddellii* and *Gloeospermum sphaerocarpum*. In the present-day warmer and relatively drier lowlands at the base of the Cordillera Costanera and Serrania del Interior, the remnants of the Amazonian–Guayanan floral elements have failed to persist and are to be found only on or near the more humid summits.

Examples of genera which have many of their taxa distributed in Amazonian Brazil, Colombia, Peru and the Guianas, but represented by only one or two species in the Coastal Cordillera and/or Serrania del Interior and/or adjacent Trinidad and Tobago are: *Tapirira* (*T. dunstervilleorum*, endemic), *Schlegelia* (*S. parviflora*, not endemic), *Dicranostyles* (*D. costanensis*, endemic), *Dacryodes* (*D. costanensis*, endemic), *Quiina* (*Q. cruegeriana*, not endemic), *Licania* (*L. subrotundata*, endemic), *Anomospermum* (*A. reticulata*, not endemic), *Macrolobium* (*M. obtusum* [endemic], *M. colombianum* [non-endemic], *M. steyermarkii* [endemic], *M. floridum* [endemic]), *Malanea* (*M. sanluisensis*, *M. fendleri*, *M. pariensis*, *M. hirsuta*—all endemic), *Tocoyena* (*T. costanensis*), and *Coussarea* (*C. pittieri*, *C. moritziana* —endemic).

A close relationship has already been noted (Steyermark, 1966, 1974) between the summit flora of Cerro Turumiquire and the Venezuelan Guayana, to the extent of 18 species, or 7·7% of the total flora manifesting this relationship. Of this percentage 12 species or 5·3% of the flora is confined to Cerro Turumiquire and the Venezuelan Guayana, while six other species are found on Cerro Turumiquire, other parts of the Coastal Cordillera and the Venezuelan Guayana. Species characteristic of portions of the Venezuelan Guayana, such as *Ilex vacciniifolia*, *Marcetia taxifolia*, and *Syngonanthus gracilis*, are isolated on the summit of Cerro Turumiquire. Of equal significance is the taxon *Ternstroemia unilocularis*, endemic to Cerro Turumiquire, but most closely related to *T. discoidea* of the Venezuelan Guayana, because of the unilocular ovary. Moreover, the genus *Drosera*, characteristic of sandy and acid soils of the Guayana, is represented on the summit of Cerro Turumuiquire by *D. tenella*, a species known elsewhere in Venezuela only in the lowlands of Territorio Federal Amazonas. Additionally, the Coastal Cordillera is evidenced in the flora of Auyan-tepui of the Guayana Highland to the extent of 14 species, or 5·2% of the flora represented between the elevations of 700–1480 m. Of the total number (698) of taxa known from Auyan-tepui, 32 are represented in the flora of the Coastal Cordillera, or 4·6% of the total phanerogamic flora of

Auyan-tepui (Steyermark, 1967, 1974). Phelps (1966) notes that *Nannopsittaca panychlora*, a monotypic genus and species, is confined to the Peninsula of Paria, and elsewhere is only in Pantepui.

The above data provide supporting evidence for the differences existing between the floras of the Coastal Cordillera and the Andes. The derivative elements of the Amazonian–Guayanan floras are definitely much stronger in the case of the Coastal Cordillera. As has been indicated earlier, these two mountain systems had different geological histories, rock formations and altitudinal maxima. The availability and occupation of the sites in these two major cordilleras must have been affected by the different times and stages to which they were submitted to the forces of erosion and peneplanation. Apparently the Coastal Cordillera offered more accessible contacts and dispersal routes for Amazonian–Guayanan floral elements at some period in the past when Andean areas were either not simultaneously available for occupation, or contacts and migrational routes from the south and east were blocked by physiographic or ecologic barriers.

In this connection the data secured by Phelps (1966) concerning the component elements of the subtropical avifauna in the Coastal Cordillera and Serranía del Interior are instructive. Of the 251 species of birds recorded from the Mérida Andes of an origin from the eastern Andes of Colombia, only slightly more than half (126 species) of these are found in the Coastal Cordillera. Of the remaining 125 species absent from the Coastal Cordillera, 76 or 60% are birds restricted to the higher altitudes of 2800–4500 m, altitudes not found in the Coastal Cordillera. Moreover, as one travels eastward from the mountains above Caracas to Caripe and finally to Paria, there is a decreasing percentage of the birds from the Mérida Andes present. Mountains of smaller extent and less elevation are the reasons given for this decreasing percentage eastward. Data which I have extracted from the numbers of endemic plant species in the Coastal Cordillera and Serranía del Interior also give similar results (69 for Naiguatá, 38 for the Guácharo or Caripe area and 29 for Paria), except for the Turumiquire centre with 37 known endemic species. The Guacharo and Turumiquire centres combined would give a total number of 75 endemic species. Elsewhere in this presentation, it has been suggested that the smaller number of plant taxa in the Coastal Cordillera and Serranía del Interior is partly due to the absence of the higher altitudes prevailing in the Venezuelan Andes. Except for the anomalous situation of the higher altitude of the Cerro Turumiquire area, the remainder of the data from the plants would correlate well with that from the avifauna.

Pantepui Refuges

The third major centre in Venezuela where relict and endemic species are numerous occupies the Pantepui area. I have adopted the term *Pantepui* originally proposed by Mayr and Phelps (1967) and followed by Müller (1973), but changed by Brown (1975) into the "Roraima", "Ventuari", and "Imerí" (in part) refuges. Whereas Müller restricts the Pantepui centre to montane forests of the Guayana Highland, I am employing the use of the *Pantepui* centre in the broadest sense to include, not only the "sandstone tabletop mountains in the Venezuelan Territorio Amazonas and Estado Bolívar and in the adjacent border regions of Brazil and Guyana" (Mayr and Phelps, 1967), but also the Gran Sabana at the base of the eastern Venezuealan tepuis, the lowland edaphic savannas and igneous "laja" formations of the western part of the Territorio Federal Amazonas, and the extreme northeastern sector drained by the Río Venamo and tributaries of the Upper Río Cuyúni. The extensive forest area within circumscribed portion of southern Venezuela is likewise included. By combining the above numerous and heterogeneous floral elements into one centre, greater coherence and relationships between these elements and their derivatives can be brought together within a more unified pattern.

The Pantepui centre has, by far, the greatest number of endemic species within Venezuelan territory. Maguire (1970) calculated for the entire Guayana region an estimated total of 8000 species, of which 75% are confined to the region, including approximately 4000 endemic species and nearly 100 newly described genera. Of the Guayana Highland flora itself, Maguire estimated that "considerably more than 50 percent" of the 8000 species "will be endemic to the over-all area" and that the summit flora of the sandstone table mountains would contain probably some 2000 species, of which 90–95% would be endemic. Considering the long geological history and availability of the Pantepui centre for occupation by the flora, the differential stresses of uplift or peneplanation by erosional forces, concomitant with varying changes of climate concurrent with arid and humid, or warm and cool phases, of glacial and interglacial Pleistocene times, all combined with the important edaphic constituent of the basic structural unit of the highly acid Roraima sandstone—the sum total of these circumstances have probably combined to produce an unusually high level of specific differentiation and endemism.

While the endemicity of the species confined to the actual summits of the Guayana Highland is usually higher than that of the endemicity of the genera, it is surprising to find the endemicity of the generic flora which is confined to the summits of the tepuis in Venezuela so surprisingly low. In an analysis which I have recently completed, of a total of 459 genera occurring on the actual summits of the tepuis, only 39 of the genera, or 8·5% of the summit flora, are in fact strictly endemic to the summits. To this total of 39, an additional 40 genera are found mainly on the summits, but also occur either on some portion of the lower talus slopes or/and the surrounding lower altitudes of the Gran Sabana and contiguous areas. Counting these additional genera, the total number would be elevated to 79, or 17·2% of the summit flora, but the figure would apply then to genera which are mainly but not strictly endemic to the summit. Included within this last category is *Heliamphora*, the well-known South American pitcher plants, as well as the following genera: *Pterozonium*, *Orectanthe*, *Cephalocarpus*, *Everardia*, *Didymiandrium*, *Rhynchocladium*, *Amphiphyllum*, *Kunhardtia*, *Saxofridericia*, *Stegolepis*, *Brocchinia*, *Connellia*, *Cottendorfia*, *Navia*, *Hexapterella*, *Carptotepala*, *Apocaulon*, *Raveniopsis*, *Spathelia*, *Blepharandra*, *Diascidia* (including *Sipapoa*), *Leitgebia*, *Philacra*, *Poecilandra*, *Bonnetia*, *Macrocentrum* (with one species in the Coastal range), *Elaeoluma*, *Chorisepalum*, *Notopora*, *Ledothamnus*, *Velloziella*, *Tylopsacas*, *Gongylolepis*, *Stenopadus*, *Stomatochaeta*, *Maguireothamnus*, *Merumea*, *Neblinathamnus* and *Nietneria*.

A similar analysis of the specific components of the summit flora tends to show a much lower average percentage of endemic elements confined to the summit than the 90–95% estimated by Maguire (1970). On a recent trip to the summit of Ptari-tepui, whose summit flora had been previously unknown, about 90% of the species had already been collected from the lower or upper talus slopes of that table mountain (Steyermark, 1957). Likewise, the summit flora of Cerro Autana showed a surprising number of species characteristic of or associated with much lower altitudes (Steyermark, 1974, 1975). On the recently explored summit of Cerro Kukenan, adjacent to Mount Roraima, only 31·3% of the summit flora was found to be so restricted, while the remaining 68·7% is known to occur on lower talus slopes of other table mountains. On Auyan-tepui (and the adjacent Uaipán-tepui), it has been recorded (Steyermark, 1967) that the total number of vascular plant taxa (375) found only on the summit represents 45% of the total flora (826). However, the total number of endemic vascular species on the summit represents only 20·6% of that total. If the figures are modified to include all the phanerogamic species found from the summit down to the upper slopes above 1500 m, then the percentage of the phanerogamic flora at this level is elevated to 63%. In the avifauna of Pantepui, Mayr and Phelps (1967) found that "not a single one of the 30 endemic Pantepui species is restricted to the summit plateaus", all of them being found also on the talus slopes. Only four subspecies of birds were found to be endemic to the summits.

The fact that the 79 endemic genera (17·2% of the total summit flora) are confined to the summits and talus slopes of the Venezuelan table mountains would indicate that the Pantepui centre has served as the centre of origin and as the original one for the 79 autochthonous genera. From this original centre a few of these endemic genera have dispersed centrifugally into outlying contiguous areas (Maguire, 1970). However, an analysis of the remainder, or 82·8% of the non-endemic genera found on the summits, but not restricted to them would suggest a centripetal dispersal, indicating that the greater part of the generic flora now present on the summits of the table mountains has been derived from other sources, and has attained its present location as a result of past immigrations or dispersals from other areas (Camp, 1947). The two principal sources for this non-endemic generic flora would appear to have been derived from (i) the lowland peripheral areas to the south of the Guayana Highland, mainly from the Amazon basin, and (ii) from genera showing widespread neotropical and/or pantropical distribution. This pattern of immigration is shown by the 17·7% of predominantly lowland genera found on the tepuis which are strongly indicative of an Amazonian Basin dispersal, and by the 48·6% which appear to have speciated from genera having a more widespread neotropical and/or pantropical distribution.

Ornithological data from Pantepui would tend to support the botanical data, as it has been suggested (Mayr and Phelps, 1967) that the table mountains have served as "a minor centre of differentiation which has contributed elements to other parts of South America, yet many elements on Pantepui give the impression of being relics of formerly more widely distributed types".

The relative contrasts in endemicity and speciation in the floras of the table mountains to the east and west of the Caroní River has been documented (Steyermark, 1967; Maguire, 1970; Steyermark and Brewer-Carias, 1976). Brown (1975) has separated the main part of the Guayana Highland region into an eastern "Roraima" refuge, extending from the forested slopes of the Pacaraima and Parima mountains and from Roraima north and west to the upper río Caura in Venezuela and south to the extreme upper Orinoco, and a more western "Ventuari" refuge of the forested area surrounding the río Ventuari in south-central Venezuela. Inasmuch as the flora of the Guayana Highland area appears to be a closely unified, but also differentiated, model, it would seem preferable to maintain it as one large unit dispersed throughout the Roraima sandstone formation, but separable into two major floristic areas, which have developed into principally an eastern and a western division. This division of Pantepui has likewise been adopted for the avifauna (Mayr and Phelps, 1967).

Gran Sabana

This is equivalent to the "Roraima" centre recognized by Müller (1973). Situated in the eastern division of the Guayana Highland or Tepui section of Pantepui, the Gran Sabana can be differentiated as a distinct dispersal centre. It occupies an area within the Roraima formation distinguished by extensive savannas and shrubby vegetation criss-crossed by streams with a diversified gallery forest flora, and overlying successively higher terraces or plateaus at elevations ranging from 350–1300 m above sea level. Circumscribed as such, the Gran Sabana extends in Venezuela from its northern limits in the vicinity of Luepa, Kavanayen, Kamarata and Canaima, south to the vicinity of Santa Elena de Uairén and Icabarú, and west to Urimán and Uonán, in which transect it surrounds or lies at the base of a number of the eastern tepuis. East of Venezuela it extends into adjacent Guyana and southward into the Río Branco of northern Brazil.

A number of species have become differentiated in the Gran Sabana. Some of these endemics include: *Panicum erectifolium, Rhynchospora karuaiana, R. sororopana, Cladium costatum, Carptotepala insolita, Eriocaulon dimorphopetalum, Paepalanthus steyermarkii, Syngonanthus venezuelensis, Brocchinia steyermarkii, Nietneria paniculata, Trimezia fosteriana, Cecropia kavanayensis, Euplassa venezuelana, Roupala minima, Thesium tepuiense, Drosera arenicola, Licania lasseri, Humiria balsamifera* var. *pilosa, Spathelia fruticosa, Dacryodes glabra, Byrsonima bolivarana, Tetrapterys pusilla, Qualea ferruginea, Vochysia ferruginea, V. rubiginosa, Phoradendron semivenosum, Polygala blakeana, Croton kavanayensis, Ilex karuaiana, Maytenus apiculata, Cupania kukenanica, Cupania roraimae, Matayba reducta, Poecilandra pumila, Bonyunia minor, Clusia pusilla, Tovomita angustata, Orthaea crinita, Oxythece steyermarkiana, Cynanchum bolivarense* and *Cynanchum caespitulosum*.

Some of these taxa occur in the open savanna areas, sometimes in swampy, sometimes on dry rocky terrain, or have differentiated in the gallery forests or forested quebradas, which at different altitudinal levels traverse the savanna landscape. The populations which have speciated in this centre are confined principally to the 350–1300 m level, but in some cases have ascended to slightly higher levels of 1500–1600 m on the lower talus slopes or open shoulders of adjacent tepuis, and occasionally have reached the summits of these tepuis, as exemplified by *Heliamphora heterodoxa* and *Trimezia fosteriana*.

Amazonas Savannas

In contrast to the Gran Sabana centre, the areas of Amazonas savannas in the western extreme of Pantepui, have developed at lower altitudes of 100–200 m especially well developed in the vicinity of the Rio Guainía, the lower Ventuari, Pacimoni, Pimichin and Atabapo Rivers. They have originated on special edaphic sandy soils having high acidity and low water-holding capacity, which have given rise to savannas with special types of herbaceous and ligneous vegetation. A large number of endemic species have become differentiated in this centre, including many taxa of the Rapateaceae, Xyridaceae, Eriocaulaceae, Bromeliaceae, Olacaceae, Leguminosae, Humiriaceae, Euphorbiaceae, Ochnaceae, Melastonataceae, Apocynaceae, Bignoniaceae and Tetrameristcaceae. Mention of some of these taxa includes: *Duckea junciformis*, *Rapatea yapacana*, *Schoenocephalium cucullatum*, *Guacamaya superba*, *Curtia quadrifolia*, *Nautilocalyx steyermarkii*, *Syngonanthus vareschii*, *Pachyloma pusillum*, *Sauvagesia nudicaulis*, *Mabea linearifolia*, *Dulacia redmondii*, *Macrolobium savannarum*, *Licania lanceolata* and *L. savannarum*, *Humiria fruticosa*, *Pentamerista neotropica*, *Ambelania lanceolata*, *Galactophora pulchella*, *Humiria fruticosa*, *Lacmellea pygmea*, *Macairea maguirei*, *Sipaneopsis maguirei*, *Dendrosipanea wurdackii* and *Navia gracilis*.

Atures

In the Territorio Federal Amazonas in the vicinity of Puerto Ayacucho, near and bordering the Orinoco River, another edaphic formation, the igneous outcroppings of "lajas", is characterized by a distinctly differentiated flora. These outcrops form part of the basement complex of the Guayana Shield Roraima formation. While many of the outcrops are found in other portions of the Venezuelan Guayana, they attain a more pronounced character in the vicinity of the rapids of the Atures, which lends its name to this centre.

Some of the endemic species in this area are *Borreria pygmea*, *Rudgea maypurensis* and *Tocoyena brevifolia*. On the igneous outcrops elsewhere, such as on Laja Catipán on the Río Yatua, are found *Pitcairnia wurdackii*, and along the same river on other igneous outcrops on the lajas of Cerro Arauicaua there are the endemic *Saxofridericia spongiosa*, *Paradrymonia yatuana* and *Decagonocarpus oppositifolius*. At the base of Cerro Sipapo on igneous outcrops occurs *Navia brocchinioides*, another endemic.

Rio Negro

Contiguous with the south-western portion of Pantepui is a lowland humid forest area encompassing the Rio Negro basin and tributaries in south-western Venezuela, adjacent Colombia and Brazil. The easternmost portion of the Rio Negro refuge may include a small part of the "Imerí" refuge so designated by Brown (1975), Prance (1973), and Haffer (1974), although for the most part there is little or no overlapping.

A large number of species are restricted to this area, and do not enter for the most part, into the contiguous Pantepui sector, although some penetration is evident northward near the mouth of the Río Atabapo in the Caño Cupavén and in the Yapacana sector, both of which areas are in the Ŏrinoco drainage. Of the many endemic species known from the Rio Negro centre, the following few may be mentioned: *Macrolobium truncata, Mouriri angustifolia, Adelobotrys barbata, Miconia truncata, Psychotria spiciflora, Psychotria pacimonica, Tocoyena pendulina, Coussarea grandis* and *Pagamea sessiliflora.*

Venamo

An area of major endemism and relict flora is developed near the north-eastern terminus of the Roraima formation of the Pantepui centre and is contiguous with it. It comprises a highly diversified, rich humid forest overlying the Roraima sandstone as well as the igneous formation in contact with the sandstone, and encompasses the drainage of the Río Venamo and upper Río Cuyuni tributaries. Its centre lies in the Sierra de Lema and Cerro Venamo at the northern edge of the Roraima formation escarpment by the Venezuelan–Guyana border. The eastern portion of the area extends into adjacent Guyana as far as the Pakaraima Mountains and Potaro River drainage. Although manifesting an intimate relationship with the flora of the Guayana Highland, with which it forms an integral part, nevertheless a different pattern of speciation is evident here and not present westward in Pantepui. Elevations extend from approximately 400 m on the lower northern and north-eastern slopes of the escarpment bordering the area to 1650 m on Cerro Venamo, 1300 m on Cerro Uei, and 1700 m on Cerro Uananapán. In general, the forest is more humid, with a greater abundance of epiphytes than in other parts of the Guayana Highland to the west.

Among the numerous endemic species which have become differentiated in this area, the following may be mentioned: *Phainantha myrteoloides, Platycarpum rugosum, Sloanea crassifolia, Caraipa psilocarpa, Rapatea steyermarkii, Sauvagesia longipes, Macrocentrum steyermarkii, Calycophyllum venezuelense, Hillia psammifolia, Psychotria hemicephaelis, Broc-*

chinia micrantha, *Guzmania venamensis*, *Ladenbergia venamoensis*, *Guzmania steyermarkii*, *Cottendorfia gracillima*, *Rhoogeton leeuwenbergianus*, *Chorisepalum acuminatum*, *Alloplectus savannarum*, *Stegolepsis steyermarkii*, *Senefelderopsis venamoensis*, *Habenaria nilssonii*, *Solanum rufistellatum*, *Cyphomandra bolivarensis*, *Solanum puberuloba*, *Matelea coriacea*, *Endlicheria nilssonii*, *Cayaponia botryocarpa*, *Gurania simplicifolia*, *Sipanea wilson-brownei* and *Dunstervillea mirabilis*, this last being the only known endemic genus of Orchidaceae in Venezuela.

Relict species of the Venamo area include *Markea porphyrovaphes*, *Burmannia kalbreyeri Solanum coriaceum* and *Mapaneopsis effusa*.

Remaining Refuge Centres

Imataca

Just north of the Venamo sector of the Pantepui centre is a dispersal area of great phytogeographical significance. It is a mainly densely forested zone of tall humid forest comprising all the elevated Sierra de Imataca and Altiplanicie de Nuria (the summits of which attain elevations from 500–700 m) in the state of Bolívar and adjacent Territorio Delta Amacuro, together with the contiguous lowland forests to the north, east, south and west, delimited on the north-west by the headwaters of the Rio Supamo eastward to the northern tributaries (Yuruari, Corumo, Botanamo) of the Río Cuyuni, and extending to the streams draining north from the Sierra Imataca, and thence continuing east into the Guianas. The name "Imataca" has also been adopted by Brown (1975) and Prance (1973), although the latter limits the Imataca centre to a much smaller sector. This area is also recognized by Müller (1973), but only as a northern extension of his "Guyanan" centre.

Up to the present time only a few endemics are known from the Venezuelan portion of Imataca centre, these being *Picramnia nuriensis*, *Piranhea longipedunculata* (Fig. 9), *Paullinia nuriensis*, *Sloanea megacarpa*, *Dilkea magnifica*, *Passiflora nuriensis* and *Episcia adenosiphon* (Steyermark, 1968; Jablonski, 1967; Leeuwenberg, 1969; Steyermark and Marcano-Berti, 1966). *Miltoniopsis santanaei*, an orchid described as a new species from this area, is also found in Colombia and Ecuador.

The Imataca centre is noteworthy as a large refuge centre for an exceptionally high number of species whose main geographical centre is in the Guianas to the east and with the floras of the states of Pará, Amapá and Amazonas, Brazil, to the south and south-east (Steyermark, 1968). No less than 24 genera and 317 species of flowering plants,

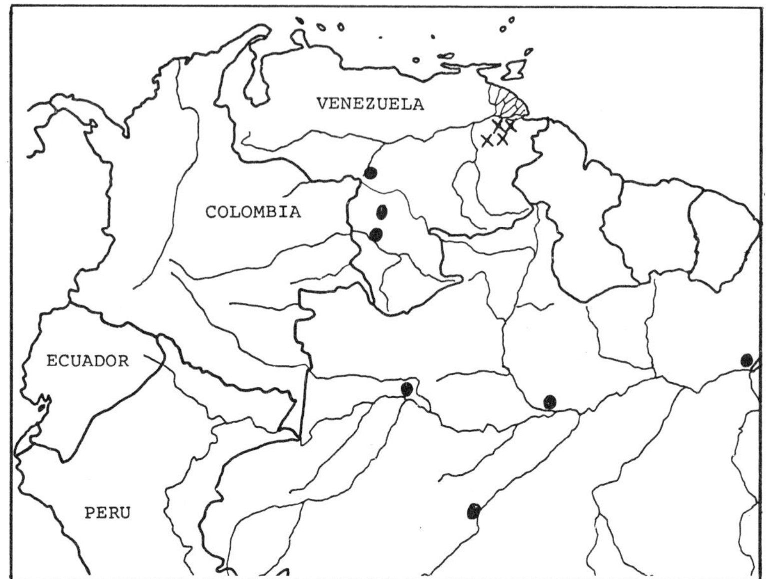

Fig. 9. Imataca endemic *Piranhea trifoliata* (●) and *P. longepedunculata* (x).

previously unrecorded for Venezuela at the time of publication of the Catalogo de la Flora Venezolana (Pittier *et al*, 1945, 1947) occur in the Imataca area and registered as new to the Venezuelan flora. They are known to occur only in this sector. Examples of species in Venezuela restricted entirely or mainly to the Imataca centre are the following: *Parahancornia amapa*, *Systemodaphne mezii*, *Pausandra martinii*, *Qualea dinizii*, *Lecointea amazonica*, *Parinari rodolphii* and *Cayaponia guianensis*.

Relationships similar to those found in the flora of this sector of Venezuela with the Guyanan and northern Brazilian floras are evident in some of the geographical patterns documented for the vertebrates, as indicated by Müller (1973) and his "Guyanan centre" relationships with his "Pará" and "Amazon" centres.

Catabumbo

An important refuge and dispersal centre is found in a region of lowland forest developed in altitudes mainly of less than 1100 m, and generally between 100 and 300 m. It is situated to the south and west of Lake Maracaibo, between the drainage area of the Río Catatumbo on the lower slopes of the Perijá Mountains and the base of the north-west slopes of the Mérida Andes. On the lower forested slopes of the Río Onia of the Andes of Mérida, the relict flora of this area is prominent. The Catatumbo centre of Müller (1973) coincides with that presented here.

A few endemic species, such as: *Anthurium praemontanum*, *Philodendron mesae*, *Rhodospatha perezii*, *Spathiphyllum perezii* and *Besleria ornata*, are known from the Catatumbo centre, but the relict element is especially noteworthy, and consists of a number of taxa otherwise known only from Amazonian Brazil and Peru or Colombia, as well as from the lowland portions of the states of Apure, Barinas and Territorio Federal Amazonas. It is the only sector on the north side of the Andes retaining remnants of the Amazonian flora through the Catatumbo refuge. A good example of this geographical pattern is found in *Faramea capillipes*. Besides the area of the Río Onia in the state of Mérida, where the relict element is predominant, other lowland areas of the Andean states of Táchira, Mérida and Trujillo, as well as occasional eastern localities in Monagas and Bolívar states, are associated with the same geographical pattern. Examples of the Catatumbo sector associated with lowland areas in Colombia are found in *Ochoterenaea colombiana*, *Aphelandra impressa*, *Miconia mocquerysii*, *Palicourea buntingii* and *Vochysia lehmannii*. Some of the relicts of this centre show an influence and contact with the following area of San Camilo.

San Camilo

The lowland forests of San Camilo in the state of Apure and those of the Río Caparo in the state of Barinas contain a considerable quota of floral elements associated with what Prance (1973) refers to as a western Amazonian distribution. Some of these elements show an influence of contact with the Catatumbo centre, discussed above.

The endemic taxa are related to those distributed generally in the lowland Amazonian forests, especially of western Amazonia. A number of such examples are shown by *Bonamia apurensis*, *Besleria multiflora* (most closely related to a Brazilian Amazonian species, *B. gibbosa*), *Besleria ornata* (related to *B. immitis* of Amazonian Colombia and Peru), *Rhynchanthera apurensis* and *Rauvolfia leptophylla* var. *orientalis*, the last taxon showing previous relationships with the lowland flora of Colombia. Other endemics of the San Camilo centre manifest affinities elsewhere, such as *Lorostelma venezolanum*, most closely related to the only other known species of the genus, *L. struthanthus* of eastern Brazil and *Forsteronia apurensis*, related to the Antillean *F. portoricensis*.

Taxa, such as *Capparis sola*, *Licania latifolia*, *Dichapetalum latifolium*, *Henriettella rimosa*, *Leandra aristigera*, *Maxillaria equitans* and *Piper hermannii* are good examples of the western Amazonian geographical pattern of distribution. Other species, such as *Miconia matthaei*, *Miconia pterocaulon* and *Adelobotrys adscendens* have a more widely distributed pattern, ranging into Central America.

Ayari

An area situated at the base of the Andean foothills at low elevations in the state of Táchira in the drainage of the Río Uribante, together with an eastward extension into the adjacent state of Barinas, manifest a distinct, although small, refuge and dispersal centre. Endemic species pertain to *Clidemia steyermarkii* and *Campylocentrum steyermarkii*, while various taxa with affinities to species endemic in the Magdalena and Meta drainage of Colombia are evident in this centre. *Pitcairnia echinata*, *Mimosa bauhinaefolia*, *Mimosa colombiana* and *Miconia magdalense* may be mentioned among these relict taxa.

Final Remarks

Other areas in Venezuela, such as the coastal portions of the states of Miranda, Sucre, Falcon, and Carabobo, indicate patterns of differentiation and speciation of certain elements of flora. This can also be stated with reference to the llanos of Venezuela, where occur such endemic species as *Hymenocallis venezuelensis*, *Stilpnopappus apurensis* and *S. pittieri*, *Eriocaulon rubescens*, *Polygala aristeguietae*, *Cuscuta aristeguietae*, *Vernonia aristeguietae*, *Borreria aristeguietae* and *Limnosipanea ternifolia*. However, these areas have not been included in this presentation as pertaining to refuge or dispersal centres.

The conclusions reached in the present work are based on floristic and phytogeographical data of Venezuelan plants, and, as such, represent the first botanical attempt of a localization of plant refuges and dispersal centres in Venezuela. A comparison with refuge centres which have been proposed by various other workers indicates that there are a greater number of refuge centres and subcentres in Venezuela recognizable botanically than have been generally attributed to the country by zoogeographers. In many cases, it has been found that there is a concurrence between the botanical dispersal centres and refuges in Venezuela and those indicated by zoogeographers.

Appendix

Endemic and relict species in the Coastal Cordillera and Serrania del Interior of Venezuela—among some of the more obvious examples, the following taxa are included within their respective refuge or dispersal centres (* = occurs in other refuges).

COASTAL CORDILLERA

BORBURATA

9 Endemics: *Napeanthus spathulata*, *Pterobesleria aristeguietae*, *Resia ichthyoides**, *Clusia cochlanthera*, *Oedematopus aristeguietae*, *Psittacanthus costanensis*, *Peperomia carabobensis*, *Peperomia tamayoi**, *Cestrum aristeguietae*

Relicts: *Marila grandiflora*, *Macradenia brassavolae*, *Notylia fragrans*, *Rustia longifolia*

NIRGUA

14 Endemics: *Razisea spicata* var. *pulchra**, *Ruellia chrysantha*, *Tillandsia steyermarkii*, *Dichorisandra diederichsanae*, *Phoradendron longiarticulatum*, *Struthanthus granulatus*, *Macrocentrum yaracuyense*, *Froesia venezuelensis*, *Ladenbergia buntingii*, *Psychotria aneurophylloides*, *Psychotria yaracuyensis*, *Rudgea buntingii*, *Trigonia costanensis*, *Aegiphila arcta*

Relicts: *Roucheria laxiflora*, *Polypodium megalophyllum*, *Pleurothallis arbuscula*, *Mapania pycnocephala* subsp. *fluviatilis**

AROA

14 Endemics: *Lastrea aristeguietae*, *Beloperone steyermarkii**, *Justicia leptophylla**, *Justicia pannieri*, *Ruellia pulverulenta**, *Phyllanthus croizatii*, *Peperomia buntingii*, *Peperomia tamayoi**, *Peperomia croizatiana*, *Hoffmannia aroensis*, *Hoffmannia stenocarpa*, *Psychotria aroensis*, *Rondeletia aristeguietae*, *Simira aristeguietae**

Relicts: *Dennstaedtia dissecta*, *Hypolepis stübelii*, *Trichipteris microdonta*, *Trichipteris procera*, *Prestonia surinamensis*, *Guzmania cylindrica*, *Eupatorium macrocephalum*, *Peperomia percuneata*

SAN LUIS

10 Endemics: *Mandevilla lata*, *Asketanthera steyermarkii**, *Alloplectus congestus*, *Lepanthopsis steyermarkii*, *Pleurothallis nubensis*, *Stelis sanluisensis*, *Epidendrum garcianum*, *Geonoma paraguanensis*, *Malanea sanluisensis*, *Psychotria sanluisensis*, *Eupatorium santensis**

Relicts: *Osmunda cinnamomea**, *Trichomanes trigonum**, *Philodendron holtonianum*, *Peritassa granulata*, *Masdevallia ecaudata*, *Masdevallia striatella*, *Pleurothallis longipedicellata*, *Restrepia nittiorhyncha*, *Stelis flexuosa*. Although the San Luis centre is completely detached from both the Cordillera de la Costa and the Andes, and has no geological connections with either, the flora manifests a closer relationship with that of the Coastal Cordillera to the east than with the Andes. This is evidenced by the large number of endemic Coastal Cordilleran species which reach their western limits in the Sierra de San Luis (Steyermark, 1975). The Cerro Santa Ana is included within this San Luis centre.

COLONIA TOVAR

5 Endemics: *Elaphoglossum reichenbachii, Mikania hexagona, Passiflora stellata, Stelis pittieri, Masdevallia rechingeriana**

Relics: *Asplenium dissectum, Didymoglossum pusillum, Diplazium celtidifolium, Hymenophyllum dependens, Hymenophyllum valvatum, Hypolepis viscosa, Lastrea linkiana, Polybotrya canaliculata, Trichipteris elegans, Liabum sessiliflorum, Hemiangium excelsum, Miconia miconioides, Epidendrum diffusum, Liparis brachystalix, Spiranthes tovarensis, Stelis lutea, Passiflora mollis, Piper crassinervium*

RANCHO GRANDE

40 Endemics: *Gunnera pittieriana, Lepanthes longiracemosa, Lepanthes mimetica, Piper regresivanum, Arachnothrix venezuelensis, Manettia badilloi, Costus venezuelensis, Alsophila swartziana, Hymenophyllum maxonii, Selaginella bombycina, Aphelandra steyermarkii, Justicia stipitata*, Anthurium paradisiacum, Monstera henry-pittieri, Philodendron aristeguietae, Philodendron henry-pittieri, Philodendron rhodaxis, Rhodospatha badilloi, Aechmea cymoso-paniculata, Aechmea lasseri, Guzmania hedychioides, Guzmania virescens, Protium araguensis, Eupatorium clavisetum, Mikania araguensis, Dichapetalum steyermarkii, Sloanea grossa*, Clidemia farinasii, Meriania subumbellata, Dorstenia aristeguietae, Pourouma venezuelensis, Peperomia fragrans, Piper tovarense, Amaioua brevidentata, Borojoa venezuelensis, Coussarea pittieri, Faramea garciae, Psychotria agostinii, Psychotria araguensis, Schradera glabriflora.*

Relics: *Miconia rubens, Asplenium sanguineolentum, Dennstaedtia cicutaria, Diplazium plantaginifolium, Lastrea pachyrachis, Pteris altissima, Pteris laciniata, Stigmatopteris tijuccana, Vittaria scabrida, Vriesea cowelii, Vriesea bituminosa, Ichthyothere scandens, Eupatorium nemorosum, Mikania lindleyana, Odontoglossum crocidipterum*

NAIGUATA

This area mainly includes the Avila and Naiguatá Mountains. Some of the taxa are also found on the other side of the Caracas Valley to the south in the Serranía del Interior. A subcentre may be differentiated as *Petaquire*, situated on the seaward humid slopes to the west of Avila, between Colonia Tovar and Junquito, with altitudes varying from 1300–2100 m. The taxa limited to the *Petaquire* subcentre are indicated by†.

69 Endemics: *Stigmatopteris alloeoptera, Justicia lindaviana, Guatteria venezuelana*, Lacmellia costanensis, Peltastes manarae** (on the other side of the Caracas valley), *Philodendron danteanum, Oreopanax reticulatus†, Marsdenia naiguatensis, Begonia boucheana, Pitcairnia microcalyx* var. *microcalyx, Achyrocline flavida, Aspilia avilensis, Mikania allartii, Verbesina laevifolia, Weinmannia lansbergiana, Elateriopsis caracasana, Sloanea lasiocarpa, Vaccinium crematum, Croizatia naiguatensis, Besleria labiosa*, Arthrostylidium naiguatense, Arthrostylidium subpectinatum, Calamagrostis naiguatensis, Clusia articulata* subsp. *mirandensis** (also on other side of Caracas Valley), *Anaectocalyx manarae, Henrietella manarae, Heliconia villosa*, Myrica pubescens, Cybianthus pittieri*, Heisteria olivae, Linociera avilensis, Bulbophyllum manarae, Comparettia venezuelana, Cryptophoranthus sarcophyllus, Epidendrum kermesianum, Epidendrum manarae, Epidendrum platyotis, Habenaria ernstii, Habenaria galipanensis, Habenaria gollmeri, Maxillaria proboscidea, Odontoglossum wageneri, Oncidium liminghei, Oncidium pardalis, Oncidium picturatum, Pogonia nana, Stelis gutturosa, Peperomia aristeguietae, Piper catucheanum, Piper naiguatanum,*

*Licania subrotundata, Bathysa pittieri** (also on opposite side of Caracas Valley), *Psychotria avilensis, Psychotria sanmartensis* var. *costanensis, Talisia morilloi, Calceolaria macropoda, Symplocos lasseri, Rinorea oraria, Macrolobium steyermarkii†, Blakea longibracteata†, Clidemia saltuensis†, Lepanthopsis steyermarkii†, Piper pseudoeucalyptifolium†, Piptocarpha steyermarkii†, Rudgea costanensis†, Markea costanensis†.*

Relicts: *Talauma dodecopetala*, Asplenium dimidiatum, Pleurostachys puberula, Grammitis limula, Hymenophyllum sericeum, Lycopodium subulatum, Plagiogyria semicordata, Pteris gigantea, Pteris spinosa, Trichipteris leucolepis, Prestonia mucronata, Guzmania sanguinea, Espeletia neriifolia*, Sloanea obtusifolia, Graffenrieda weddellii*, Eltroplectris (Centrogenium) roseo-album, Epidendrum mojandae, Epidendrum polyanthum, Epidendrum radiatum, Malaxis histionantha*, Malaxis ventricosa, Malaxis wendlandii, Notylia punctata, Oncidium anomalum, Oncidium sessile, Spiranthes tenuis, Stelis angustifolia, Stelis micrantha, Stelis ophioglossoides, Warrea warreana, Passiflora kalbreyeri*, Ixora ferrea.*

PARIA

29 Endemics: *Cnemidaria amabilis*, Cnemidaria consimilis, Elaphoglossum agostinii, Anthurium pariense, Xanthosoma pariense, Begonia verruculosa, Guzmania membranacea, Aspilia sucrensis, Asplundia pariensis, Dicranopygium macrophyllum, Stephanopodium venezuelanum, Besleria mortoniana, Episcia membranacea, Besleria hirsutissima, Tovomita coriacea, Dendrophthora filiformis, Henriettella bracteosa, Miconia lucida* subsp. *pariensis, Tococa broadwayi, Topobea steyermarkii, Elvasia steyermarkii, Epidendrum dunstervilleorum, Piper parianum, Ixora agostiniana, Malanea pariensis, Psychotria humensis, Psychotria pariensis, Psychotria sucrensis, Cestrum pariense.*

Relicts: *Asplenium pulchellum, Cochlidium seminudum*, Dryopteris leprieurii, Hymenophyllum semiglabrum, Nephelia imrayana, Polypodium blepharolepis, Polypodium nematorhizon, Selaginella hartii, Trichomanes fimbriatum, Trichomanes osmundoides, Mendoncia flava, Stemmadenia cerea*, Anthurium aripense, Begonia glandulifera, Cordia panamensis, Aechmea aripensis, Aechmea dichlamydea* var. *trinitensis, Aechmea gigantea, Glomeropitcairnia erectiflora, Maytenus sieberiana*, Aspilia verbesinoides, Pollalesta condensata, Vernonia trinitatis, Connarus patrisii, Diospyros ierensis, Alchornea latifolia*, Marila grandiflora*, Dussia martinicensis*, Swartzia simplex, Dendrobangia boliviana*, Aniba megaphylla, Ocotea trinitadensis*, Eschweilera trinitensis*, Clidemia cruegeriana*, Necramium gigantophyllum*, Platycentrum clidemioides*, Notylia wullschlaegeliana*, Trichilia roraimana, Cespedesia spathulata, Lepanthes ovalis, Sobralia ciliata, Triphora gentianoides, Euterpe pubigera, Peperomia panamensis, Quiina cruegeriana, Schradera clusiaefolia, Tresanthera pauciflora, Pouteria mammosa, Cyphomandra tobagensis, Gloeospermum sphaerocarpum.*

The Paria refuge is noteworthy for the large number of species known only from Trinidad and/or Tobago, and for the relict Amazonian–Guayanan elements isolated on the summits of Cerro de Humo and Cerro Patao of the Peninsula de Paria (Steyermark and Agostini, 1965; Steyermark, 1973, 1974).

MARGARITA

This island, geologically, is part of the Coastal Cordillera.

5 Endemics: *Ctenitis ampla*, Mikania johnstonii, Blakea monticola, Epidendrum johnstonii, Pleurothallis johnstonii*.*

218   *J. A. Steyermark*

Relics: *Cochlidium seminudum*, Cyathea arborea, Dennstaedtia ordinata, Goniopteris megalodus*, Lycopodium struthioloides*, Polypodium nematorhizon*, Trichomanes alatum, Vaccinium latifolium*, Pleurothallis floribunda, Spiranthes adnata, Rinorea melanodonta**.

SERRANIA DEL INTERIOR

GUATOPO

This centre includes the Cerros del Bachiller to the east of the Guatopo National Park; The Cerros del Bachiller have a number of relics and endemic species not found in Guatopo National Park.

8 Endemics: *Justicia oxypages, Tococa perclara, Heliconia rodriguensis, Asterogyne (Aristeyera) spicata, Piper guatopoense, Piper patulipilum, Borojoa universitatis, Tresanthera thyrsiflora*.

Relics: *Danaea elliptica, Elaphoglossum dussii, Pteridanetium citrifolium, Trichomanes spicatum, Besleria longipes*, Maxillaria caespitifica, Ornithocephalus bonplandii*

TURUMIQUIRE

This centre includes adjacent mountains with altitudes of mainly 1500 m and above. Mountains so included are Cerro Peonía, Cerro Negro and Cerro San José, in addition to Cerro Turumiquire itself.

37 Endemics: *Elaphoglossum aristeguietae, Hymenophyllum gollmeri*, Ilex vesparum, Cynanchum caudigerum, Cynanchum sucrense, Begonia laxa, Begonia otophylla, Tournefortia gracilipes, Capparis guaguaensis, Stellaria venezuelana*, Tradescantia venezuelensis, Baccharis erectifolia, Gnaphalium caeruleocanum, Oyedaea maculata, Pollalesta vernonioides, Senecio badilloi, Verbesina tatei, Viguiera leptodonta, Carex culmenicola, Carex turumiquirensis, Befaria steyermarkii, Gaultheria tatei, Croton turumiquirensis, Alloplectus ichthyoderma* var. *pallidus, Besleria conformis, Panicum pilatum, Stylogyne turumiquirensis, Plinia fruticosa, Maxillaria patula, Monnina steyermarkii, Monnina tatei, Relbunium glaberrimum, Oxythece rigidiopsis, Styrax costanus, Ternstroemia steyermarkii, Ternstroemia unilocularis, Pilea tatei*.

Relics: *Blechnum columbiense* var. *bogotense*, Ctenitis platyloba, Elaphoglossum glossophyllum, Osmunda cinnamomea*, Sticherus rubiginosus, Ilex vaccinifolia, Rhynchospora dissitiflora, Drosera tenella, Syngonanthus gracilis*, Gesneria onacaensis, Juncus marginatus* var. *setosus, Eubrachion ambiguum*, Phoradendron emarginatum, Marcetia taxifolia, Ugni myricoides, Laurembergia tetrandra*, Stelis maderoi, Passiflora bogotensis, Meliosma herbertii, Calceolaria tripartita*, Styrax hypargyreus, Symplocos umbellata*, Xyris acutifolia**.

GUACHARO

This centre includes the mountains of lower altitudes, mainly between 300 and 1400 m, in the vicinity of Caripe, Guácharo and Bergantin.

38 Endemics: *Polybotrya crassa, Dicliptera porphyrocoma, Dicliptera pyrrantha, Mendoncia albiflora, Mendoncia leucantha, Ruellia pterocaulon, Stenostephanus lasiostachyus*, Anthurium caripensis, Anthurium engelerianum, Oreopanax venezuelense*, Quararibea steyermarkii, Cinnamodendron venezuelense, Eupatorium bergantinense, Mikania monagensis, Senecio nigellus, Dioscorea lasseriana*, Croizatia neotropica,*

Besleria concinna, *Pterocarpus magnicarpa*, *Banisteria alternifolia*, *Miconia paupercula*, *Oncidium klotzschianum*, *Pleurothallis exilis*, *Peperomia steyermarkii*, *Piper elparamoense*, *Piper monagense*, *Monnina venezuelensis*, *Alseis microcarpa**, *Elaeagia laxiflora**, *Palicourea pustulata*, *Chrysophyllum steyermarkii**, *Pouteria simulans*, *Freziera steyermarkii*, *Pilea venosa*, *Duranta steyermarkii**.

Relicts: *Asplenium tinctum*, *Asplenium dentatum*, *Ctenitis ampla*, *Ctenitis nemophila*, *Ctenitis nigrovenia*, *Dryopteris sagittifolia**, *Dryopteris carrii**, *Hymenophyllum karstenianum**, *Meniscium macrophyllum*, *Macropharynx strigillosa**, *Prestonia brittonii*, *Tabernaemontana affinis*, *Anthurium palmatum*, *Cordia lineata*, *Cordia ucayaliensis**, *Tillandsia globosa*, *Vriesea scalaris*, *Capparis baducca*, *Vaccinium latifolium**, *Codonanthe uleana* var. *integrifolia**, *Besleria longipes**, *Clusia eugenioides*, *Tovomita cephalostigma*, *Hyptis colombiana*, *Machaerium angustifolium*, *Ormosia monosperma*, *Chlorophora brasiliensis*, *Bulbophyllum aristatum*, *Epidendrum sulphurum*, *Oncidium bicolor*, *Pleurothallis rubroviridis*, *Stanhopea oculata*, *Psychotria ctenophora*, *Psychotria grandis*.

References

Bermudez, P. J. (1969). Cuaternario y reciente en Venezuela. *Mem. Soc. Cienc. Nat. La Salle* **24**, 43–59.

Brown, K. (1975). Geographical patterns of neotropical Lepidoptera. Systematics and derivation of known and new Heliconiini (Nymphalidae: Nymphalinae). *J. Ent.* **44**, 201–242.

Camp, W. H. (1947). Distribution patterns in modern plants and the problems of ancient dispersals. *Ecol. Monogr.* **17**, 159–183.

Gannser, A. (1955). Ein Beitrag zur Geologie und Petrographie der Sierra Nevada de Santa Marta (Kolumbien, Südamerika). *Schweiz. Min. Pétrog. Mitt.* **35**, 209–279.

Garay, L. and G. C. K. Dunsterville (1976). *Venezuelan Orchids Illust.* **6**, 276–277.

Geel, B. van and T. van der Hammen (1973). Upper Quaternary vegetation and climatic sequence of the Fuquene area (Eastern Cordillera, Colombia). *Palaeogeog. Palaeoclimat. Palaeoecol.* **14**, 9–92.

Haffer, J. (1974). "Avian Speciation in Tropical South America." Nuttall Ornithology Club, No. 14, 1–390.

Hammen, T. van der (1961). Late Cretaceous and Tertiary stratigraphy and the tectogenesis of the Colombian Andes. *Geol. Mijnb.* **40**, 181–188.

Hammen, T. van der (1972a). Historia de la vegetación y el medio ambiente del norte sudamericano. Mem. Symp. 1st Cong. Lat. Am. Mex. Bot. 119–134.

Hammen, T. van der (1972b). Changes in vegetation and climate in the Amazon Basin and surrounding areas during the Pleistocene. *Geol. Mijnbouwkd.* **51**, 641–643.

Hammen, T. van der and E. González (1960). Upper Pleistocene and

Holocene climate and vegetation of the "Sabana de Bogotá" (Colombia, South America). *Leidse Geol. Meded.* **25**, 261–315.

Jablonski, E. (1967). In Botany of the Guayana Highland: Euphorbiaceae. *Mem. N. Y. Bot. Gard.* **17**, 122.

Leeuwenberg, A. J. M. (1969). Notes on American Gesneriaceae V. A new species of *Episcia* Mart. *Acta Bot. Neérl.* **18**, 585–588.

Maguire, B. (1970). On the Flora of the Guayana Highland. *Biotropica* **2**, 85–100.

Marrero, L. (1964). "Venezuela y sus Recursos." Cultural Venezolana, Caracas.

Mayr, E. and W. H. Phelps, Jr. (1967). The origin of the bird fauna of the South Venezuelan highlands. *Bull. Am. Mus. Nat. Hist.* **136**, 273–327.

Müller, P. (1972). Centres of dispersal and evolution in the neotropical region. *Stud. Neotrop. Fauna* **7**, 173–185.

Müller, P. (1973). The dispersal centres of terrestrial vertebrates in the neotropical realm. *Biogeographica* **2**, 10, 52–56, 62–79, 192–198.

Phelps, W. H., Jr. (1966). Contribución al analisis de los elementos que componen la avifauna subtropical de las cordilleras de la costa norte de Venezuela. *Bol. Acad. Cienc. Fis. Mat. Nat* **26**, 22–25.

Phelps, W. H. and W. H. Phelps, Jr. (1953). Eight new birds and thirty-three extensions of ranges to Venezuela. *Proc. Biol. Soc. Wash.* **66**, 125–146.

Phelps, W. H. and W. H. Phelps, Jr. (1955). Five new Venezuelan birds and nine extensions of ranges to Colombia. *Proc. Biol. Soc. Wash.* **68**, 47–58.

Pittier, H. *et al.* (1945). "Catalogo de la Flora Venezolana" Vol. 1. Tercera Conferencia Interamericana de Agricultura, Caracas.

Pittier, H. *et al.* (1947). "Catalogo de la Flora Venezolana" Vol 2. Tercera Conferencia Interamericana de Agricultura, Caracas.

Prance, G. (1973). Phytogeographic support for the theory of Pleistocene forest refuges in the Amazon Basin, based on evidence from distribution patterns in Caryocaraceae, Chrysobalanaceae, Dichapetalaceae, and Lecythidaceae. *Acta Amazon.* **3**, 5–28.

Royo, J. y Gomez (1956). "Stratigraphic Lexicon of Venezuela." Special Publication No. 1. Ministerio de Minas e Hidrobarbures, Dirección de Geología, Caracas.

Simpson, B. B. (1971). Pleistocene changes in the fauna and flora of South America. *Science* **173**, 771–780.

Simpson, B. B. (1975). Pleistocene changes in the flora of the high tropical Andes. *Paleobiology* **1**, 277–278.

Steyermark, J. (1957). Contribuciones to the Flora of Venezuela, part 4. *Fieldiana* **28**, 679–1190.

Steyermark, J. (1966). Contribuciones à la Flora de Venezuela, part 4. *Acta Bot. Ven.* **1**, 129–168.

Steyermark, J. (1967). Flora del Auyan-tepui. *Acta Bot. Ven.* **2**, 44–47, 64.

Steyermark, J. (1968). Contribuciones a la flora de la Sierra de Imataca, Altiplanicie de Nuria y region adyacente del Territorio Federal Amacuro al sur del río Orinoco. *Acta Bot. Ven.* **3**, 49–175.

Steyermark, J. (1973). Preservemos las cumbres de la Peninsula de Paria. *Def. Nat. Año* **2**, 33–35.

Steyermark, J. (1974a). Relación floristica entre la cordillera de la costa y la zona de Guayana y Amazonas. *Acta Bot. Ven.* **9**, 248–249.

Steyermark, J. (1974b). The summit vegetation of Cerro Autana. *Biotropica* **6**, 7–13.

Steyermark, J. (1975a). La región del Tamá debe ser conservada. *Natura* **57**, 5–8.

Steyermark, J. (1975b). Flora de la Sierra de San Luis (Estado Falcon, Venezuela) y sus afinidades fitogeograficas. *Acta Bot. Ven.* **10**, 131–218.

Steyermark, J. (1975c). Informe sobre la flora del Cerro Autana. *Acta Bot. Ven.* **10**, 219–233.

Steyermark, J. and G. Agostini (1966). Exploración botanica del Cerro Patao y zonas adyacentes a Puerto Hierro, en la Peninsula de Paria, Edo. Sucre. *Acta Bot. Ven.* **1**, 7–80.

Steyermark, J. and C. Brewer-Carias (1976). La vegetación de la cima del Macizo de Jaua. *Bol. Soc. Ven. Cienc. Nat.* **32**, 218–236, 241–242.

Steyermark, J. and G. Bunting (1975). Revision of the genus *Froesia* (Quiinaceae). *Brittonia* **27**, 172–178.

Steyermark, J. and L. Marcano-Berti (1966). Una especie nueva de *Sloanea*. *Bol. Soc. Ven. Cienc. Nat.* **26**, 467.

Vanzolini, P. E. (1970). Zoologia sistematica, geografica e a origem das especies. Univ. São Paulo, Inst. Geog., Teses e Mongr. No. 3.

Vila, P. (1960). "Geografiá de Venezuela." Vol. 1. Dirección de Cultura y Bellas Artes, Departamento de Publicaciones, Caracas.

Guayana, Region of the Roraima Sandstone Formation

B. MAGUIRE

New York Botanical Garden, USA

Introduction

Chiefly because of not yet completed background studies, these observations are presented in a preliminary outline of geologic, geographic and floristic considerations which are essential to an understanding and interpretation of the culminating contemporary physical and floristic structure and content of Guayana.

The following summary statement was presented by me at the Congreso Latinoamericano held in Mexico City from 3–9 December 1972. It may serve in part to set the tone and to introduce the subject of the present discussion.

The Guayana Highland of northern South America east of the Andes is the seat of a large ancient provincial flora containing a high incidence of relictual elements. Physiographically the Highland is characterized by spectacular disjunct flat-topped mesas, remnants of massive sandstone and related sediments which were laid down, toward the western part of the early continent of Gondwanaland, in a vast inland sea or seas during the Proterozoic some 1500–1800 million years ago, with accumulations exceeding three thousand meters and extending westward from near the limits of the present Atlantic coastal region beyond the present eastern flanks of the Andes. North and south this enormous lens of sedimentary outwash extended from the Amazon Valley to the Orinoco Valley, probably also with great accumulations south of the Amazon.

It is postulated that the sources of these vast sediments were the regions of the present basins of the Congo and Niger rivers, carried by them or their progenitor rivers by the way of the Amazon and Orinoco, to debouch before the elevation of the Andes from the western perimeter of Gondwanaland into the then unitary sea.

It is further postulated that ancient angiosperm floras, as commonly recognized, arose within evident proper evolutionary sequence in the equable climate of Gondwanaland, were well sorted out into the major ordinal and higher grouping before the breakup of the vast protocontinent of Gondwanaland, and reached the major patterns of distribution which now obtain in the disparate post-drift offspring continents. Such an orderly unfolding of correlated events would eliminate an inacceptable requirement of massive coordinated long distance intercontinental systematic floristic dispersal.

The putative relationships of the flora of Guayana to the floras of (1) the Brazilian Highland, (2) Africa, (3) the West Indies, (4) the Andes, and (5) the Malay–Indonesian region, are considered as a logical consequence in the context of the geophysical history of the tropical land masses of the earth.

As indicated above, Guayana, the region of the Roraima Sandstone Formation, characterized as a floristic province by Maguire (1970, 1972) and as the Dominio Guayano by Cabrera and Willink (1973), has evolved a spectacularly distinctive flora. The region's early botanical history commenced with the visits of Humboldt and Bonpland to the Alto Orinoco in 1800, of Martius in 1820 to the savannas of \raracoara on the Río Apaporis, then Brazilian territory; that of the redoubtable Robert Schomburgk, who during the years 1835–1844 carried out field work in Guiana and made the historic journey from Roraima on the Venezuelan–British Guianan–Brazilian frontier to Duida on the Alto Orinoco, the Casiquiare and the Río Negro; and the enormously productive journey of Spruce during the years 1851–1855 to the upper Rio Negro, the Casiquiare and the Alto Río Orinoco. Since, a host of biologists, geographers and geologists have given their special attentions to the region of the Roraima Formation.

Until recently, biological exploration had outrun geologic and geographic inquiry. Now, in the past decade, greatly important geologic findings relative to the Roraima sediments have been published by Snelling and McConnell (1969). The work of the mixed Brazilian–Venezuelan Boundary Commission, which was concluded in 1965, has established firmly and accurately the watershed demarcation between the two countries, the international Venezuelan–Brazilian boundaries, thus fulfilling early treaty arrangements. Both the Venezuelan and Brazilian governments have completed radar imagery of their respective areas of Guayana, thus providing accurate detail of geography, topography, and drainage systems. Further, the extensive helicopter-supported programmes of geographical and botanical exploration carried out by Brewer–Carías in 1972 and 1973 to Cerro Autana, the Cerro Jáua–Complex, and more recently to Cerros Ptari–tepuí and Ilu–tepuí; the numerous explorations of the staff of the Instituto Botánico, Caracas, of Lasser, Steyermark, Dunsterville, and their colleagues; and the helicopter visit of Tillett to Cerro Marahuaca, have added much critical detail to the knowledge of interior Guayana.

I and my associates have, since 1944, carried out 35 separate botanical explorations into Guayana, three of them to the critical boundary massif of Neblina and one to the frontier area of Territorio Roraima, Brazil. The botanical results of these explorations have been published as Parts I–X in the Memoirs of The New York Botanical Garden, 1953–1978. Preparation of Part XI is now in progress. Part XII is intended to carry the summation of all Guayana work.

It is now timely, therefore, to initiate a correlation and coordination of the variously accumulated botanic, pertinent geologic, and essential geographic information presently available. Here we attempt to present sufficient historic, geographic and topographic background to serve as a basis and framework for the subsequent consideration of the present distribution and composition of the indigenous Guayana flora, and the agencies leading to its formation.

Geographical Setting

As a general historic and geographic term, "Guayana" is usefully applied to the area of the large northern crystalline shield or plate of primal origin extending continentally northward from the contact beneath the course of the Amazon River with the southern abutting Brazilian Shield.

For the purposes of this presentation, the term Guayana refers in a restricted sense to that region of the Guayana Shield dominated by the superposed Roraima Sandstone Formation. Guayana so noted occupies a sigmoid configuration, lying between the 59th and 74th meridians and between the equator and the 7th parallel north, a region of some 1 000 000 square kilometers (somewhat less than 400 000 square miles), and is believed to support a flora in excess of 10 000 species. It is this region, both the spectacularly dissected Guayana Highland and related richly vegetated outwash lowland, that provides the essential base on which the Guayana flora rests, and the geographic framework within which the Guayana flora has developed.

Geological Setting

The Roraima sediments, an accumulation reaching some 3000 m in depth, consisting of sandstones, often quartzitized, conglomerates, graywacks, jasper, ash, etc., were laid down assumably in shallow inland Proterozoic seas overlying great portions of the more ancient matured surface of the Guayana crystalline shield. The Roraima sediments thus rest unconformably on the granites and metamorphosed original rock of the Guayana Shield.

Various evaluations of the geologic history of Guayana have been made over the past several decades. The resumption of debate and the presentation of massive geological and geophysical evidence through the past 15 years have established a formidable consensus requiring the actuality of a progenitor existence of the unitary land mass, the "super-continent" of Pangaea, being Gondwanaland in the south and Laurasia in the north. Our primary concern here lies with the history of Gondwanaland, embracing in time probably largely that of the Triassic and subsequent early Mesozoic, involving the separation of present Africa and South America and the consequent fates of the occupying vascular floras then so epocally asundered.

The physical and biological consequences of these sequences of great continental movements have determined in large part the content and evolutionary direction of the contemporary neotropical floras, and, within that dramatic series of events, of Guayana. Attempts are widely being made (van Steenis, 1962; Smith, 1967; Axelrod, 1970; Maguire, 1970; Raven and Axelrod, 1974; Schuster, 1976) to evalue the sequence of events as they would affect the initial postdrift floral content, and influence plant composition and evolution into contemporary time.

Until comparatively recently the vast geological literature, reviewed in part by Gansser (1954), considered the Roraima series to derive from Cretaceous sedimentation. However, McConnell (1968), McConnell and Williams (1970) and Snelling and McConnell (1969), by careful stratigraphic studies and radiometric dating, have determined that the diabasic intrusions into the Roraima series, at least in the east, are of 1700–1800 million years of age and, therefore, that the intruded sediments must necessarily be of greater antiquity and actually be assignable to the Proterozoic. Gansser (1974), 20 years later, as a consequence in a second consideration, used this dating for the beginning of the Roraima Formation in the east, but extended the estimate of duration of sedimentation especially in the west, and suggested that there might be even more recent sediments capping the Roraima in the east.

Further evidence developed by Gansser (1974) and Paba Silva and van der Hammen (1960), show the block-faulted Cordillera Macarena on the eastern flank of the Andes in Colombia to be of complicated orogeny,* but to be composed in part by sediments of which: "The lithological similarity of this section with the Roraimas, . . ., is surprising". (But should it be surprising?) Upper jasper horizons of the Cordillera are said to bear pollen indicators that are referable to the Paleocene. Superposed at the summit Gansser reports a grey shale of upper Eocene to lower Oligocene. Some 100 k to the east of Macarena he states "Roraima type sandstones" on which "the upper horizon is transgressed by shales of the upper Eocene type similar to the occurrences of the Macarena".

Thus, there is strong evidence that Roraima sedimentation or its comparable sequela experienced a continuing history of formation even into the mid-Tertiary. Gansser (1974) further states:

> No basement outcrops further towards the Orinoco in the table mountains of Yambi, Mapiripan and Inirida. The upper quartzitic sandstone is well developed and not unlike the Upper Roraima Member. With the Orinoco begin the outcrops of the Guiana Shield, and here again we find Roraima-type table mountains of the Sipapo block, which is still geologically poorly known. Towards the main bulk of the Roraima deposits in the Pacaraima and the Roraima range the table mountains increase in number until they merge into the large masses of the Roraima Formation with its associated basic sills and dykes. These igneous rocks coincide with the main and central mass of the Roraima sediments. They are not yet known from the marginal areas, in particular the western Shield region, where, however, the information is only scanty.

But basement outcrops strongly at Mitú on the Vaupés (70°W–0°30′N), and is westerly indicated on the 1942 geologic map of the American Geological Society and the American Geographical Society five millionth. Basement rock is shown there to involve much of Amazonian Colombia and, fragmentally, the eastern Andes.

In any event and by whatever nicety of geological nomenclature

---

* These statements on the complex structure of Macarena involving, as is reported, Andean elements and sandstones attributed to the Roraima Formation, lend support to information given to me by the late Eugene Jablonski (pers. comm.), earlier Chief Geologist for Socony–Mobil Company, that evidence deriving from exploration boring carried out in oil prospecting indicates there was Andean overthrust upon the western perimeter of the Guayana Shield (hence the Roraima Formation) in the consequent areas during one or more of the Andean uplifts. Dr Jablonski, who was also a botanist of stature, told me that drilling across the upper Amazon Basin had detected the contact of the Brazilian and Guayana Shields lying under the bed of the river itself.

employed, there is abundant outcrop of granitoid volcanic basement prominently exposed south and west of the Orinoco River, where one dramatic granite monolith, reaching nearly 1500 m, is the most spectacular feature between Cerro Duida on the Orinoco and Cerro de la Neblina on the Brazilian–Venezuelan frontier some 300 km to the south. Prominent rounded granitics lie unconformably at the base of Neblina, Duida, Huachamacari, Sipapo and Guanay, etc. Massive beds of jasper are exposed at river level in the Alto Paragua and at the base of Guaiquinima. At the base of the upper cliffs of Huachamacari are massive horizontally bedded dense pinkish tuff. These casual observations could be multiplied manifold, and any traveler on the Atabapo, the Casiquiare and the Guainía views these "lajas" constantly. Indeed, the twin dome of the granitoid Cocuy on the Rio Negro has conspicuously marked the frontier locus for many years.

Amazonian Venzuela and Amazonian Colombia have not yet received adequate systematic geological investigation, but the instructive observations such as those of Hitchcock (1931, 1947), Schultes (1945), Galavis (1947) and Reynolds (1955) have been neglected by contemporary writers.

The presumption remains that the deposition and subsequent elevation of equivalent sandstone sediments of the Guyana Pakaraimas, Venezuelan Bolívar and Amazonas, Colombian Amazonas and those of contiguous Brazil, however extended over great time intervals, provide sequential development of a long continued process undisturbed by cataclysmic events, and provide the physical and environmental homeland for the evolution of the extensive coherent contemporary flora of Guayana.

Physiography and Geography

Since the Roraima sediments in general incline from the southern Brazilian periphery northward, the greater part of the Highland drainage is northward into the Orinoco River, eastwardly and southeastwardly into the Essequibo system, or westwardly into the Rio Negro–Amazon systems. Short run-off streams from the high southern rim are gathered into major affluents of the Amazon.

The Highland is, thus, a huge eroded, dissected, sedimentary lens which feeds the embracing Rio Negro–Rio Branco–Amazon, the Orinoco, and the Essequibo, all waters finally debouching into the Atlantic via these three great river systems.

These three systems demark three distinctive areas, of which the easternmost, consisting of the Pakaraima Mountains and the Gran Sabana, is elevated into a plateau of 400–1500 m, upon which are superimposed numerous mountains or mesetas (one of which is Roraima, giving name to the great sandstone formation); the middle sector, that strongly eroded region west of the Río Caroní) studded with discrete mesetas whose vertical ringing cliffs and talus descend to Shield level, but the southern rim of which is more or less continuous; and the third, most westerly, that beyond the Río Negro–Guainía, also weathered to shield level but characterized by more disparate, smaller, less high mesetas and some gentle terraces. This last sector terminates with the Colombia Cordillera de Macarena.

Floristic composition and transition from east to west evidently reflect the historic sequence of elevation. Each sector is floristically characterized. The trans-Río Negro region, although botanically least known, perhaps because of more restricted topographic and ecologic variability, seems to support a less diversified flora. Macarena, because of its historic and orogenic involvement with the Andes, evidently partakes to a large extent of the Andean flora.

Each of the three major sectors is further compartmentalized physiographically and floristically by the enormously long history of erosion and establishment of drainage pattern. Such floristic compartmentalization has been noted by other botanists. Limited comment will be made below on this geographic–hydrographic floristic correlation.

It will be noted from Fig. 1 (originally prepared by the American Geographical Society and revised by us in the light of details deriving from the currently determined Brazilian–Venezuelan frontier, of certain portions of the interior provided by radar imagery and Brewer–Carías (1973, 1976), and data acquired by us on our last visit in 1965 to the Serra da Neblina via the Brazilian Rio Cauaburí) that the intimate groupings of principal relictual mesas and mesetas are delimited within the three major river systems by secondary river drainages. From west to east these are: I—the Río Guaviare in Colombia; II—the Ríos Inírida, Içana, Vaupés, Apaporis and Caquetá, all in Colombia; III—the Ríos Siapa and Pacimoni of Amazonian Venezuela, and the Rio Cauaburí of Amazonian Brazil; IV—the left-hand affluents of the Río Ventuari, and the Ríos Cunucunuma and Padamo, affluents of the Río Orinoco, Amazonas, Venzuela; V—the Río Sipapo and the right-hand affluents of the Río Ventuari, branches of the Orinoco, Amazonas, Venezuela; VI—the upper Ríos Caura and Paragua of Amazonas and Bolívar, Venezuela; VII—the central Río Rio Paragua; VIII—the middle Río Caroní) Bolívar, Venezuela; IX—the headwater affluents of the Caroní) Bolívar, Venezuela; X—the

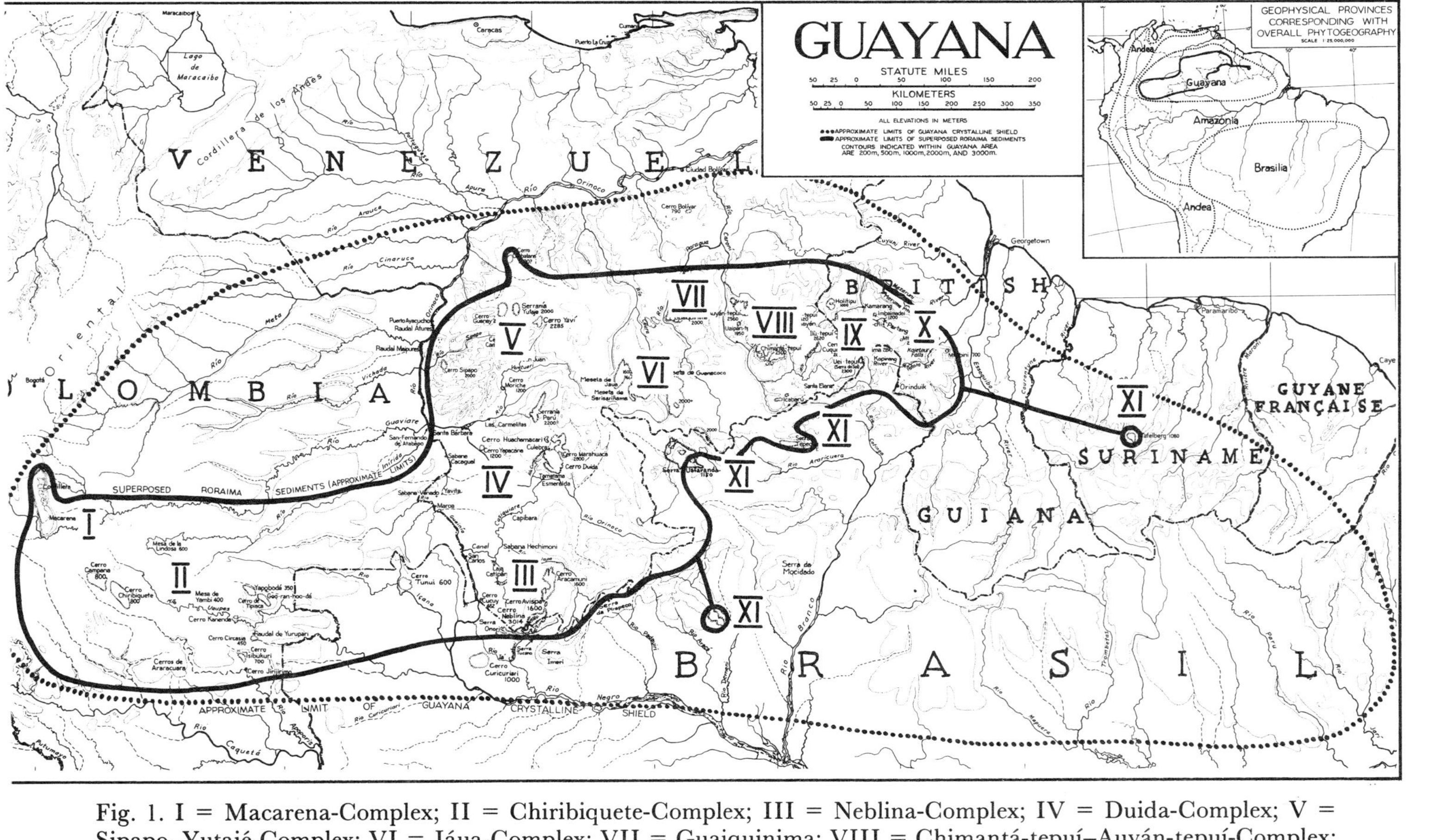

Fig. 1. I = Macarena-Complex; II = Chiribiquete-Complex; III = Neblina-Complex; IV = Duida-Complex; V = Sipapo–Yutajé-Complex; VI = Jáua-Complex; VII = Guaiquinima; VIII = Chimantá-tepuí–Auyán-tepuí-Complex; IX = Roraima–Ilu-tepuí-Complex; X = Merumé–Ayanganna-Complex. Satellite areas, southerly Roraima Formation disjuncts, Uafaranda, Tepequem, Aracá, and Tafelberg, are designated by XI.

Cuyuni, Mazaruni and Potaro rivers of the Essequibo system of Guyana; and XI—the disjunct satellite mesas of Tepequem and Uafaranda of the Rio Araricuera, of Aracá of the Rio Aracá—Demeni drainage in Brazilian Amazonas, and Tafelberg at the heads of the Saramacca and Coppename rivers in Surinam. The Rios Araricuera and Cotinga, headwaters of the Rio Branco, Roraima, Brazil, gather the southeastern short run-off of the Guayana Highland, and the Rios Cauaburí) Padauiri, Aracá and Demeni carry the short southwestern run-off of the Guayana Highland into the Rio Negro—Amazon system.

There are 10 such major meseta groupings which are hydrographically defined and contained, and correspond to the numbered drainage systems designated above. They are enumerated in Fig. 1.

The Trans-Río Negro, Colombian Guayana Subprovince

Macarena-Complex (I)

Defined by the Alto Río Guaviare, tributary to the Ríos Atabapo—Orinoco, Colombian Guayana. Includes, in addition to Macarena, the Mesa de la Lindosa of the Alto Río Inírida, ultimate affluent of the Río Guaviare. Little is known of the complex Andean—Roraima floras of Macarena (or Lindosa).

Chiribiquete-Complex (II)

Defined by the Colombian Alto Ríos Vaupés and Apaporis, tributaries to the Río Negro. Our knowledge of the Roraima floras of this complex derives largely from the explorations of Richard E. Schultes, José Cuatrecasas, and the contemporary staff of the Instituto de Ciencias Naturales, Bogotá.

The Río Caroní—Río Negro Subprovince

Neblina-Complex (III)

Defined to the north by the Ríos Pacimoni—Yatua, Baría, affluents to the Casiquiare—Río Negro; in the south the Rios Cauaburí—Negro, the Neblina-Complex consists of the Cerros Neblina (at 3015 m), the less high Avispa, Aracamuni, and southern Brazilian outliers. The floras of the latter two massifs, while 1000 m lower than Neblina, are little known but are expected to have a Neblina affinity. Avispa has been visited briefly via helicopter by Steyermark and Dunsterville.

The name Neblina derives from the Maguire (1955) report of the first New York Botanical Garden exploration of the mountain made in 1953–1954. During this initial trip we were accompanied by Charles D. Reynolds, Chief Geologist of the US Steel Corporation, who appended to the general report a brief but most informative geological assessment of the deeply eroded, block-faulted mountain, the sediments of which have been strongly folded, faulted and often overturned, showing the results of violent orogenic disturbance.

The flora of Neblina reflects isolation, extreme dissection, and great altitude, thus affecting the environmental diversity of the massif. By the record now established, Neblina has produced a greater degree of generic and specific endemism than any other of the Roraima mesetas, "tepuis" in the usage of Mayr and Phelps (1967). The mountain supports the only endemic family so far recorded for Guayana, Saccifoliaceae.

Duida-Complex (IV)

The Duida-Complex consists of the formidable Cerro Duida, at 2375 m altitude, the adjacent somewhat higher Cerro Marahuaca, and the nearby Huachamacari. Initially Duida was ascended by Tate and Hitchcock (1930), subsequently by Steyermark (1944) and Maguire (1949, 50, 51). Marahuaca has been visited by Maguire (1949), and by Tillett via helicopter (1975). Huachamacari was visited by Maguire and colleagues (1949).

Hitchcock (1931) describes Duida as a block-faulted meseta, much folded, deeply dissected, and the summit ". . . not a plateau, but an old mature surface partially broken by tectonic movements accompanying upfaulting, partly dissected by process of weathering and stream erosion".

Its flora, consistent with that of adjacent Marahuaca and Huachamacari, uniquely has developed the only ramiform species of the Guayana "pitcher plant", *Heliamphora*.

Sipapo–Yutajé-Complex (V)

This complex consists of Sipapo, Autana, Guanay, Yutajé, Yaví) and Camani, Moriche and lesser mesetas, and may be characterized by the endemic monotypic *Kunhardtia rhodantha* (Rapateaceae). These mesetas are conspicuously underlain by elevated granitic intrusions. Sipapo is strongly cut by diabase dikes. Principal mountains of the complex have been visited by Phelps and Maguire expeditions. Brewer-Carías carried out a spectacular helicopter and mountain-climbing trip to Autana, where large now elevated, once subterranean, stream channels were observed.

Jáua-Complex (VI)

The Jáua-Complex consists of the three mesetas of Jáua, Guanacoco and Sarisariñama. A fourth huge massif, name unknown to me, lying immediately to the south-east of Guanacoco, should be associated with the other three. The flora of the latter is totally unknown.

Early in 1932 Cardona visited this area, Brewer-Cariás in 1965, in 1967 Phelps and Steyermark conducted a helicopter-supported visit to the western portion of Sarisariñama, and in 1974 Brewer-Cariás made the extraordinary helicopter-supported expedition on behalf of geologic and floristic study (Steyermark and Brewer-Cariás 1976). One of the primary missions of the expedition was the exploration of the vast simas, sink holes, of Sarisariñama, and the equally vast interconnecting subterranean drainage channels associated with the simas. These greatly significant geologic findings are to be associated with similar features of Autana (Brewer-Cariás, 1976), adjacent Sipapo, Arrowhead Basin of Tafelberg in Surinam (Maguire, 1945), and probably interior Marahuaca.

Steyermark and Brewer-Cariás (1976) have listed in detail the plants collected in the 1974 and previous explorations and they use the term "nucleo" to designate the floristic–historic relationship of the plants of the three mesetas of the Jáua-Complex.

Guaiquinima (VII)

Guaiquinima stands by itself in the Río Paragua drainage, unless the unnamed massif to the southward, referred to above, may be associated with it. In this connection it is appropriate to refer to an astute statement made by Gansser (1974, p. 92) in which, as seen on Venezuelan government radar imagery:

> In the Auyantepui area and particularly in a table mountain much further to the southeast, the detailed fracture system is well exposed. Most surprising is here a saucer shaped marginal rim around the large table mountains, where, apparently, layers belonging to the Lower Roraima dip basin-like towards the central part of the block. The picture convincingly gives the impression as if the large block mountain was gently depressed (isostatically?) or that some kind of weak undulation affected the sandstone plateau and that only the sediments in the basin-like depressions are now preserved.

234 *B. Maguire*

I have been given the privilege of examining these Mosaicos de Radar, and see there directly west of Auyán-tepuí) Hoja NB-20-10, Cerro Guaiquinima on the Río Paragua with configuration similar to that described by Gansser (1974), and directly south, Hoja NB-20-14, along the 63°30′ meridian, the enormous feature (possibly as indicated above to be associated with the Jáua-Complex) of concentric "saucer shaped marginal rim". The two mountains do indeed seem in part to be subsidence products.

Cardona visited Guaiquinima at an earlier time, Maguire in 1951–1952, and recently Dunsterville and Steyermark by helicopter in 1977. The remarkable flora of the meseta is still little known, but is the locus of a second species of the genus *Pakaraimaea* of the subfamily Pakaraimoideae, referred to the Dipterocarpaceae. Further the flora of Guaiquinima is notable for the concentrated development of combretaceous *Terminalia* (Maguire and Exell, 1958).

The Guyana Pakaraima–Venezuelan Gran Sabana Subprovince

Auyán-tepuí–Chimantá-Complex (VIII)

Consisting also of Ilu-tepuí) Ptari-tepuí and outliers, this spectacular and productive complex is defined by the Alto Río Caroní and its principal tributary, the Aponguao.

Floristically, second to Neblina, the Auyán-tepuí–Chimantá-Complex has produced the most diversified flora. The closely endemic *Chimantaea*, certainly primitive among composits, is a physiognomy-dominating, dramatic pachycaul known nowhere else in Guayana. The unitypic *Ayensua*, the bromel that looks like a *Vellozia*, is narrowly distributed on Uaipán-tepuí)

The most popularly known feature of the complex is Angel Falls of Auyán-tepuí) which stands as one of the spectacular ornaments of the American tropics. An early expedition of the many mounted by the noted ornithologists, William H. Phelps, father and son, was directed to Auyán-tepuí) The plants collected were studied by Gleason and Killip. Steyermark (1967) has written statistically of plants of Auyántepuí)

Roraima–Ilu-tepuí-Complex (IX)

Defined by the headwaters of the Caroní and the western affluents of the upper Mazaruni. With the exception of Araracoara on the Apaporis in Colombia, as indicated above visited by Martius in 1820, Roraima has the oldest botanical history of any of the mesetas of Guayana. First seen by the elder Schomburgk brother, Robert (1841), and soon thereafter by Richard, Roraima, on the tricorners of Guyana, Venezuela and Brazil, has been a mecca for biologists and mountain climbers, having in the intervening century and a half been visited by more travellers than any of the other Guayana mesetas. Despite this long history, it appears that only the Venezuelan portion of the summit has been studied biologically. This "chain" of mountains, forming much of the boundary between Guyana and Venezuela, is the particular home of the handsome bromeliaceous genus *Connellia* of four species, one of which reaches Auyán-tepuí) The most extraordinary of the eastern *Heliamphora*, *H. ionasi*, festooning the walls of Ilu-tepuí) is conspicuous on Ilu-tepuí) Steyermark (1966) has written statistically of the floras of the mountains of this complex.

Merumé–Ayanganna-Complex (X)

Consisting of the upper Mazaruni Basin to the north and the Potaro drainage southward, both affluents of the Essequibo, the two commanding features are Mount Ayanganna, being the highest point in Guyana, and the eastern marginal and terminal Merumé Mountains. In the future these two river drainages and the mountains thereon may have to be treated as distinct complexes.

The upper Mazaruni Basin is conspicuously accented floristically by the recently discovered *Pakaraimaea dipterocarpacea* and the eastern-most genus *Epidryos* (Rapateaceae), containing three greatly disparate species, one in the Merumé Mountains, one in coastal south Colombia, and one in central Panama.

Southerly the Kaieteur Savanna and Gorge is the home of two monotypic genera of the Rapateaceae, *Windsorina* and *Potarophytum*, while *Brocchinia micrantha*, the largest of all bromeliads except *Puya raimondii*, dominates the open landscape of the Kaieteur Savanna and Kopinang Savannas. Known also from the northern periphery of the Gran Sabana above El Dorado.

Disparate satellite mesetas: Tafelberg, Tepequem, Uafaranda and Aracá (XI)

There are others to be accounted for to the southward of Neblina. Tafelberg, central Surinam, was visited by me in 1944. Its summit flora is typically Roraima. Particularly conspicuous is the development of species of the section *Clusiastrum* of the genus *Clusia*, and *Oedematopus quadratus* of the Clusiaceae.

Tepequem, Roraima, Brazil, has long been the locus of placer diamond mining. Its surface now contains only a degraded vegetation. However, an undescribed *Oedematopus*, which represents a connecting link with the genus *Clusia*, is one of the conspicuous escarpment shrubs. Tepequem was visited by us in 1954.

Aracá, a large, *c*. 1400 m high, prominent meseta, occupies an area between the Rios Aracá and Demini, Amazonas, Brazil. Recently discovered as a result of RADAM reconnaissance and radar imagery, Aracá has been visited by Murça Pires and associates. The collections have not at this time been evalued. An interesting collection is an undescribed species of *Stegolepis* (Rapateaceae), the first instance in this genus of extracontiguous Roraima occurrence.

Uafaranda, *c*. 1500 m high, lying athwart the Alto Rio Araricuera, is clearly defined by radar imagery (Projeto Radambrasil, 1975). Uafaranda is probably much the largest of the discontinuous satellite Roraima mesetas. Its flora is totally unknown.

Generalizations

As indicated earlier in this paper, the floristic province of the Roraima Formation is divisible into three subprovinces: Pakaraim–Gran Sabana subprovince; the Caroní-western Orinoco, Río Negro subprovince, and the trans Río Negro–Colombian Guayana subprovince, the three having strong but related floristic individuality. Each of the three major subprovinces is divisible into meseta groupings which are sharply hydrographically and floristically defined. They are so designated and delimited here.

I expect to elaborate, in Part XII of "The Botany of the Guayana Highland", on the phytogeographic organization as it is developed by the geographically and hydrographically defined mountain complexes, and as it reflects the floristic development and organization of Guayana over extended geologic time.

References

Axelrod, D. L. (1970). Mesozoic paleogeography and early angiosperm history. *Bot. Rev.* **36**, 277–319.

Brewer-Carías, (1973). Mesetas de Jáua, Guanacoco y Sarisariñama. *Bol. Soc. Venez. Cienc. Nat.* **XXX**, Suppl. 127.

Brewer-Carías, C. (1976). Las simas de Sarisariñama. *Bol. Soc. Venez. Cienc. Nat.* **XXII**, 349–623.

Cabrera, A. L. and A. Willink (1973). Biogeografia de America Latina. Monogr. 13, Ser. Biol. 67–68. Organización de los Estados Americanos.

Galavis, F. A. (1947). Petrography of rock specimens. *In* Hitchcock (1947). *Geog. Rev.* **37**, 562–564.

Gansser, A. (1954). Observations on the Guiana Shield (S. America). Eclog. *Geol. Helv.* **47**, 77–112.

Gansser, A. (1974). The Roraima Problem (South America). *Verhandl. Naturf. Ges. Basel* **84**, 80–100.

Hitchcock, C. B. (1931). Cerro Duida and the Guayana Highland. *Bull. Torrey Club* **58**, 284–287.

Hitchcock, C. B. (1947). The Orinoco–Ventuari Region, Venezuela. *Geog. Rev.* **37**, 525–566.

Humboldt, A. and A. Bonpland (1822). "Travels to the Equinoctial Region of the New Continent" (3rd edn). London.

Ijzerman, R. (1931). "Outline of the Geology and Petrology of Surinam." Martinus Nijhoff, Den Haag.

McConnell, R. B. (1968). Planation Surfaces in Guyana. *Geog. J.* **134**, 507–520.

McConnell, R. B. and E. Williams (1970). Distribution and provisional corelation of the Precambrian of the Guyana Shield. Proc. 8th Guiana Geol. Conf. I 1–20. Dep. Geol. Mines, Georgetown, Guyana.

Maguire, B. (1945). Notes on the geology and geography of Tafelberg, Suriname. *Geog. Rev.* **35**, 563–579.

Maguire, B. (1955). Cerro de la Neblina, Amazonas, Venezuela. *Geog. Rev.* **45**, 27–51.

Maguire, B. (1970). On the flora of the Guayana Highland. *Biotropica* **2**, 85–100.

Maguire, B. (1972). Guayana as a floristic province. Its relationship within the Neotropics and to the Paleotropics. *In* "Resumenes de los Trabajos I Congreso Latino Americano" 55–56. Soc. Bot. Mex.

Maguire, B. and A. W. Exell (1958). *Mem. N.Y. Bot. Gard.* **10**, p. 92–94.

Mayr, E. and Wm. H. Phelps, Jr. (1967). The origin of the bird fauna of the South Venezuelan Highlands. *Bul. Am. Mus. Nat. Hist.* **136**, 221–327, pl. 14–21.

Paba Silva, F. and T. van der Hammen, (1960). Sobre la geologia de la parte sur de la Macarena. *Bol. Geol.* **6**, 7–30.

Philipson, W. R., C. C. Doncaster and J. M. Idrobo (1951). An expedition to the Sierra de la Macarena, Colombia. *Geog. J.* 188–199.

Priem, H. N. A., N. A. I. M. Boelrijk, R. H. Verschure and E. H. Hebeda (1970). Isotopic geochronology in Surinam. Proc. 8th Guiana Geol. Conf. III 1–32. Dep. Geol. Mines, Georgetown, Guyana.

Raven, P. H. and D. I. Axelrod (1974). Angiosperm biogeography and past continental movements. Ann. Mo. Bot. Gard. **61**, 539–673.

Reynolds, C. D. (1955). Notes on the geology. *Geog. Rev.* **45**, 49–50.

Projeto Radambrasil (1975). "Levantamento de Recursos Naturais" Vol. 8,

Folhas NA 20 Boa Vista, NA 21 Tumucumaque, NB Roraima E NB 21, Rio de Janeiro.

Projeto Radambrasil (1976). "Levantamento de Recursos Naturais" Vol. 11, Folha NA 19 Pico da Neblina, Rio de Janeiro.

Schomburgk, R. H. (1841). "Travels in Guiana and on the Orinoco during the years 1835–1839" p. 1–202. Georg Wigand, Leipzig. English translation by W. E. Roth (1931). Argosy, Georgetown, Guyana.

Schultes, R. E. (1945). Glimpses of the little known Apaporis River in Colombia. *Chron. Bot.* **9**, 123–127.

Schuster, R. M. (1976). Plate tectonics and its bearing on the geographical origin and dispersal of angiosperms. *In* "Origin and Early Evolution of Angiosperms" (C. B. Beck, ed.) p. 48–138. Columbia University Press, Columbia.

Smith, A. C. (1967). The presence of primitive angiosperms in the Amazon Basin. *Atas Symp. Biota Amaz.* **4**, 37–59.

Smith, A. C. (1973). Angiosperm evolution and the relationship of the floras of Africa and America. *In* "Tropical Forest Ecosystems in Africa and South America" (B. J. Meggers *et al.*, eds) p. 49–61. Smithsonian Press, Washington.

Snelling, N. J. and R. B. McConnell (1969). The geochronology of Guyana. *Geol. Mijnb.* **48**, 201–213.

Spix, J. B. and C. F. P. Martius (1824). "Travels in Brazil, 1817–1820." London.

Spruce, R. (1908). "Notes of a Botanist on the Amazon and Andes." Macmillan, London.

Steenis, C. G. G. J. van (1962). The land-bridge theory in botany. Blumae **11**, 235–372.

Steyermark, J. A. (1966). Contribuciones a la flora de Venezuela. *Acta Bot. Venez.* **1**, 9–256.

Steyermark, J. A. (1967). Flora del Auyan-tepui. *Acta Bot. Venez.* **2**, 5–370.

Steyermark, J. A. and C. Brewer-Carías (1976). La vegetacion de la cima del Macizo de Jáua. *Bol. Soc. Venez. Cienc. Nat.* **22**, 179–405.

Tate, G. H. H. and C. B. Hitchcock (1930). The Cerro Duida region of Venezuela. *Geog. Rev.* **20**, 31–52.

Flora of the West Indies

R. A. HOWARD

Arnold Arboretum, Harvard University, USA

Introduction

One of the geographic areas covered in this symposium on Tropical Botany owing most to the botanists, collectors and writers of Scandinavian origin is the area of the West Indies. Five persons should be mentioned, and the first is Carl Linnaeus, whose "Species Plantarum" (1753) is the agreed starting point of the modern nomenclature of plants. The plants from the West Indies included in "Species Plantarum" were those described or illustrated by Charles Plumier, collected by Sir Hans Sloane and Patrick Browne, or assembled by Leonard Plukenet or James Petiver. However, Linnaeus' impact on the Caribbean area is significant—52% of the plants recorded from Barbados today were originally described by Linnaeus, a figure which contrasts with only 39% of those recorded in Vol. 3 of "Flora Europaea".

Olaf Swartz was a protégé of Linnaeus the younger, and spent three years in Jamaica and seven months in Haiti before publishing his "Prodromus" (1788) and the "Flora Indiae Occidentalis" (1797–1806). Of the 1005 species in these works, Swartz himself collected 782 in Jamaica and 155 in Hispaniola, of which 723 were described as new.

Martin Vahl, lector and director of the Copenhagen Botanical Garden, published "Symbolae Botanicae" (1790–1794) and the "Eclogae Americanae" (1796–1807) based on plants largely collected in the Danish Virgin Islands by Ryan, von Rohr and West. So close was the collaboration of Vahl and his collectors that West's own description of his travels (1793) on St Croix contains valid descriptions of plants mentioned the following year in Vahl's "Symbolae" Vol. 3.

Perhaps the most unusual person, whose biography needs to be compiled, was Baron von Eggers, who served in the Danish military force on St. Croix and St. Thomas from 1868 until 1885. Eggers published a study of St Croix and its vegetation in Swedish in 1876. Three years later (1879) the Smithsonian Institution in Washington, DC, published a similar treatment in English. The genesis of the latter work was revealed in correspondence between Eggers and Asa Gray, and it was Gray who edited the proofs and compiled the index, encouraged and cooperated with Eggers' desire to prepare a complete flora of the West Indies. Following Eggers' retirement from military service he apparently had freedom to travel and collect. He obtained successive support from the Berlin Academy of Sciences, the Danish Carlsberg Fund, the British Association of London, and Consul Leopold Krug, who is better known for his association with Ignatio Urban. Between 1887 and 1899 Eggers collected in Hispaniola, the Bahamas, Cuba, the British Virgin Islands, St Vincent, Barbados, Grenada, the Grenadines, Trinidad, Tobago, Venezuela and Ecuador, and his bibliography contains articles on many of these areas or the plants he collected.

Swedish-born Erik Ekman, perhaps the most cantankerous student of the West Indian flora, visited Cuba from 1914 until 1916, and Hispaniola from 1917 until his death in the Dominican Republic in 1931. He was buried in an unmarked free tomb for impoverished teachers. The tomb was located in 1950 in the town of Santiago de Las Vegas, and a marker was attached to the crypt. The Dominican government honoured him belatedly with a plaque on a monument in the central park of that town (Howard, 1952). Ekman's contribution to the knowledge of the Caribbean flora is the greatest of any single collector. His work resulted in approximately 35 000 numbers and contained over 2000 new taxa.

The significance of these Scandinavian botanists is seen in a count of the names published in the current floras of Barbados, Jamaica and Hispaniola. For the Barbados flora, 312 names are attributed to Linnaeus, 40 to Swartz, and five to Vahl. For Jamaica, where Swartz collected, 690 taxa were described by Linnaeus, 400 by Swartz, and 42 by Vahl. For Hispaniola, where Ekman collected, 669 species were first described by Linnaeus, 252 by Swartz, 60 by Vahl, 55 by Etman himself, and 1316 by Urban based on Ekman's collections.

The islands of the West Indies, the Antilles, form an archipelago of over 1000 land masses, varying in size, altitude, soil types, and the number of environmental niches. They extend 2720 km from Barbados on the east to the western tip of Cuba, while the distance from Grenada in the south to the northern point in the Bahamas is 1920 km. On the south Grenada is only 128 km from Tobago. On the north Cuba is 160 km from Key West, Florida, and 192 km from the Yucatan Peninsula of Mexico. The distance from the Bahamas to Florida is 96 km. Cuba is the largest island, comprising 70750 sq km with a flora of 6000 species, about 50% of them endemic. Hispaniola has the greatest diversity in altitude, ranging from the Enriquillo Basin, 34 m below sea level, to the highest peak of 3233 m. The major part of the land surface within the archipelago is below 300 m. Although there are vegetational changes with altitude, there is no timber-line within the Antilles, and no high altitude tundra or páramo. Volcanic activity was important in the formation of the Lesser Antilles, and significant eruptive events have occurred on two islands in the past five years, while fumerole activity is found on eight. Limestone formations dominate the islands of the Bahamas and some of the Lesser Antilles. In the Greater Antilles successions of uplifts of limestone reefs are evident, with overthrusts and graben faults common. Truncated synclines show outcrops of gypsum and of salt. The karst formations of Cuba, Jamaica, Hispaniola and Puerto Rico, are areas of botanical interest. Selective accumulations of bauxite soils in Jamaica and Hispaniola occur in limestone sink-holes, and often have characteristic plants tolerant of high aluminium content. Commercial nickel deposits occur in Cuba and Hispaniola, but the tolerant vegetation has not been studied for accumulator species. Serpentine soils run the length of Cuba, differing widely in constituents, but a recent study (Berazaín, 1976) reveals that while 47% of the plants of Cuba are endemic, 27% of these are found on serpentine soils. Siliceous savannas are found in western Cuba, with a disjunct area in Puerto Rico. A saline lake, below sea level and seemingly without phanerogamic vegetation, is Enriquillo in Hispaniola. The largest freshwater swamp is Cienaga Zapata of Cuba. Freshwater lakes in the Lesser Antilles are usually volcanic craters. Freshwater streams are few and short.

Rainfall records for the islands are few, incomplete, and scarcely comparable. A mountain forest station on Martinique recorded 5·3 m one year, and a mossy forest research station on Puerto Rico, 4·5 m. The airport in Havana received 1·1 m in the same year. An oil company working in the Enriquillo Basin experienced no rainfall during a three-year period. Many areas of the Antilles show periods of six or seven months of reduced rainfall occurring as two dry periods during which some vegetation may lose its leaves and show mass flowering at the onset of rain. According to Beard, the critical rainfall for the West Indies is 100 mm each month. The vegetation of the major part of the land surface in the West Indies has been selected for its tolerance of low rainfall conditions. Prevailing winds are from the south-east, and annual and frequent hurricanes have decimated the montaine forests in many areas.

Beard (1955) and Stehlé (1945) have suggested methods of classification of the vegetation types, utilizing physiognomic, floristic, or habitat considerations. A simpler method, offering the same ideas in readily recognized form, is to consider the coastal plant formations, those in lowlands, and those which are montane (Howard, 1974).

Coastal Formations

The shore lines of the Antilles may be of black (volcanic) sand, white sand, smooth boulders, or coral outcrops. Most are unstable, and coastal waters are often dangerous to bathers. *Canavalia*, *Cenchrus*, *Distichilis*, *Euphorbia*, *Ipomoea*, and *Sesuvium* are the common genera present. *Coccoloba* spp. dominate the woody vegetation of strand areas, but the unusual *Bontia* (Myoporaceae), *Strumpfia* (Rubiaceae), *Scaevola* (Goodeniaceae), *Suriana* (Simarubaceae), and *Mallatonia* (Boraginaceae) are encountered along with the poisonous *Metopium* and *Comocladia* (Anacardiaceae), and *Hippomane* (Euphorbiaceae).

Rock pavement substrata, in the form of uplifted coral reefs, may have such unusual plants as *Melocactus* and *Mammillaria* (Cactaceae), *Jacquinia* (Theophrastaceae), *Catesbaea* and *Scolosanthus* (Rubiaceae), *Phyllostylon* (Ulmaceae), tree forms of *Cnidoscolus* and *Victorina* (Euphorbiaceae), and cladophyllous species of *Phyllanthus* (Euphorbiaceae) and *Rhacoma* (Celastraceae). Reduced leaf size and extremes of coriaceousness and succulence will be encountered.

Mangrove formations are small, and the four genera, *Avicennia* (Verbenaceae), *Conocarpus* and *Laguncularia* (Combretaceae), and *Rhizophora* (Rhizophoraceae) are characteristic. *Pterocarpus* (Leguminosae) is occasional in riparian locations.

Lowland Formations

The thorn scrub formation dominates many areas of the Antillean land surface. The "thorns" occur as vascularized or unvascularized stem protuberances, modified stipules, rachises of compound leaves, modified axillary or terminal branches. Leaf spines may be terminal and straight, or recurved as hooks, or marginal, while prickles may be on one or both surfaces. Inflorescence bracts and floral envelopes may also have sharp, rigid projections. The abundant genera are, in the Leguminosae: *Acacia*, *Brya*, *Calliandra*, *Haematoxylon*, *Mimosa*, *Piscida*, *Pictetia*; in the Cactaceae: *Cereus* and *Opuntia* and their segregates; in the Rubiaceae: *Guettarda*, *Psychotria*, and *Rondeletia*; in the Euphorbiaceae: *Acidoton*, *Croton*, *Jatropha*, and *Phyllanthus*, as well as *Bursera* (Burseraceae), *Cassine* and *Rhacoma* (Celastraceae), *Rauvolfia* (Apocynaceae), *Samyda* (Flacourtiaceae), and *Tabebuia* (Bignoniaceae).

The serpentine soils of Cuba may be dominated by *Copernicia* (Palmae), *Byrsonima* (Malpighiaceae), *Buxus* (Buxaceae), *Leucocroton* and *Moacroton* (Euphorbiaceae), *Jacaranda* (Bignoniaceae), *Rondeletia* (Rubiaceae), *Erythroxylon* (Erythroxylaceae), and *Gochnatia* (Compositae). Reduction in leaf size is common, and unusual leaf forms occur in species of *Aristolochia* (Aristolochiaceae), and *Passiflora* (Passifloraceae).

The flatlands of Pinar del Rio, Cuba, support a vegetation with species of genera of distinct distribution such as *Befria*, *Kalmia*, *Lyonia*, and *Pieris* (Ericaceae), *Lachnocaulon* (Eriocaulaceae), *Lechea* (Cistaceae), and *Quercus* (Fagaceae). Two genera of the Cycadaceae, the endemic *Microcycas* and localized species of *Zamia*, are found in this area.

Montane Formations

The *Pinus* forests of Cuba and Hispaniola represent one unusual distributional pattern for the Gymnospermae of the Antilles. *Juniperus* occurs only in the Bahamas, Cuba, and Hispaniola of the Greater Antilles. *Podocarpus* has five species in Cuba, three in Jamaica, one in Hispaniola, and one in Puerto Rico and the Lesser Antilles. Pine forests on limestone soils in Hispaniola, and serpentine soils in Cuba are associated with many unusual endemic phanerogams.

The montane, broad-leafed, sclerophyllous forests of the Antilles are best seen in the Sierra Maestra range in Cuba, the Blue Mountains and the John Crow Mountains in Jamaica, and the middle islands of the Lesser Antilles. Genera are numerous within this vegetation type, and Stehlé (1945) lists 175 genera with 385 species for the islands of Guadeloupe and Martinique. These are represented as 14% Burseraceae, 10% Lauraceae, 9% Elaeocarpaceae, and 7% Myrtaceae. Asprey and Robbins (1953) indicate that the dominants of this forest in Jamaica are *Alchornea* (Euphorbiaceae), *Calophyllum* (Guttiferae), and *Ficus* (Moraceae). In the Lesser Antilles the dominants are a few species of *Dacryodes* (Burseraceae), *Sloanea* (Elaeocarpaceae), *Licania* (Chrysobalanaceae), *Marila* (Guttiferae), *Pouteria* (Sapotaceae), *Talauma* (Magnoliaceae), and *Ormosia* and *Swartzia* (Leguminosae).

The vegetation is reduced in stature and increases in density at the summits of the volcanic peaks of the Antilles. The peaks are cloaked in clouds much of the time, light is reduced, and humidity is high. The woody plants are commonly laden with vascular and nonvascular epiphytes (Howard, 1968). The principal genera found are *Clusia* (Guttiferae), *Didymopanax* and *Oreopanax* (Araliaceae), *Inga* (Leguminosae), *Hibiscus* (Malvaceae), *Ilex* (Aquifoliaceae), *Freziera* (Theaceae), *Richeria* (Euphorbiaceae), *Weinmannia* (Cunoniaceae), and *Hedyosmum* (Chloranthaceae).

Geologic interpretations of the tectonic Caribbean plate are conflicting and not revealing. Paleopalynological records are few. For the extant flora, no reliable figure is available for the total number of species in the archipelago. Specific endemism is estimated at 50% for Cuba, 33% for Hispaniola, 20% for Jamaica, and 4% for Puerto Rico. Estimates of 13% and 12% specific endemism have been suggested for the Bahamas and the Lesser Antilles, respectively. The genus *Picrodendron* of the monotypic family Picrodendraceae, once considered the only endemic family in the West Indies, has now been assigned to the Euphorbiaceae (Hayden, 1977). However, there are 187 endemic genera in 49 families in the Antillean flora. The largest numbers are found in the Compositae (22), Euphorbiaceae (16), and Rubiaceae (28). The phytogeographic relationships in the area are complex. A group of species, considered a western continental unit, extended from South America north through the Yucatan peninsula and eastward in the Greater Antilles, diminishing abruptly in Puerto Rico and ending in Guadeloupe (Howard, 1975). Another group, a southern continental unit from Central America and northern South America, proceeds northward in the Lesser Antilles, ending equally abruptly in the Leeward Islands north of Guadeloupe.

Current Work on the Antilles

The following summarizes the floristic work extant or in progress.

Cuba

León (1946) and León and Alain (1951, 1953, 1956) published the first volumes of a "Flora de Cuba", and the work was finished by Alain

(1962), with a supplement published in 1969. Currently a committee of European and Cuban botanists have indicated they plan to prepare several works as new floras of Cuba.

Hispaniola

Moscoso (1943) published a "Catalogus Florae Domingensis". Jiménez (1966) added a supplement, and smaller papers more recently. Liogier is working toward a new flora of the island, having issued a "Diccionario Botanico de Nobres Vulgares de la Española" in 1974.

Jamaica

Adams (1972) "Flowering Plants of Jamaica" was badly needed. Stearn was working to complete the missing volumes of Fawcett and Rendle's "Flora of Jamaica", but the status of this work is uncertain.

Cayman Islands

Proctor has prepared a flora of these islands, and the manuscript is in press.

Puerto Rico

Britton and Wilson (1923–1924, 1925–1930) published "Flora of Puerto Rico and the Virgin Islands". The nomenclature was updated in a series of papers by Liogier (1965, 1967). D'Arcy (1967, 1975), Fosberg (1976), Little (1969) and Little *et al.* (1976) have published lists, additions, and descriptions of several smaller islands of the Virgin Island area.

The Bahamas

Britton and Millspaugh (1920) published "The Bahama Flora". Gillis *et al.* (1973) published additions to this, and Gillis (1974) later corrected the nomenclature. Currently Correll (1974, 1977) and Gillis† are working independently toward new floras of the Bahamas, and each frequently describes new taxa or adds nomenclatural notes.

The Lesser Antilles

Howard (1974a, b, 1977b, 1979) has published a comprehensive flora of the Lesser Antilles. Additional volumes and completion can be expected.

Martinique and Guadeloupe

The French islands were represented by Duss (1897) in "Flore Phanérogamique des Antilles Françaises" which was reprinted in 1972. Fournet (1978) in his "Flore Illustrée des Phanérogames de Guadeloupe et de Martinique" represents a significant new contribution.

Dominica

Hodge (1954) began a "Flora of Dominica", publishing as part one the ferns and the monocotyledons. More recently, following considerable field work, staff members of the Smithsonian Institution, Washington, DC, now especially Nicolson, have almost completed manuscripts for the dicotyledons.

Barbados

Gooding, Loveless and Proctor (1965) published a "Flora of Barbados". The senior author is revising the publication for a second edition.

Trinidad and Tobago

Although not considered a part of the Antilles, publications about this flora can be applied usefully to the plants of the Lesser Antilles. A flora of Trinidad and Tobago was begun in 1928, with initial parts by Williams (1928), Cheesman (1947), and Williams and Cheesman (1929), followed at intervals by contributions by additional authors for various families. The most recent publication in 1977 represents a new cooperative effort with the staff at Kew.

Although there is and has been collaboration in recent years between the few people working on the flora of the Antilles, there are conspicuous examples of competition and differences of opinion, some political and some scientific. The preparators of floras, although more than compilers, still cannot be monographers of all genera. Published monographs are taken into consideration when such treatments are available. The work comprising "Flora Neotropica" has in each instance been useful, but still such families as the Lauraceae, Melastomataceae, Piperaceae, and Urticaceae defy the efforts of a single individual.

A consideration of the plant life of Cuba illustrates some of the difficulties facing a current worker (Howard, 1977a). Under the Castro administration, work proceeds toward a new flora of Cuba. Papers on the subject have been written by nearly two dozen workers, and have appeared in almost as many scattered periodicals. In some, the major contribution is a change of varietal status to subspecific level. However, records are on hand of over 241 new taxa described, with the majority based on collections not distributed and not available on loan.

When a new species description may include references to collections of León or Alain, or even my own, there appears to be a difference of opinion on the definition of a species. When Britton and Rose (1922) treated the genus *Melocactus* (as *Cactus*) they recognized five species from the Antilles, including one from Cuba. In 1934 León added three more for Cuba. In 1976 two workers on the Cuban flora seem to conflict. Areces Mallea (1976a, b) added two species in papers published in Cuba. Mészáros (1976), in a paper published in Hungary, added five species, and two varieties to another species, with the suggestion that there are other taxa to be described. Surely *Melocactus holguinensis* Areces is the same as *Melocactus jakusii* Mészáros from the same isolated location, but the dates of the two publications as to day and month cannot be determined. Interestingly, the types of the two species are located in herbaria less than 10 miles apart, in Havana and in Santiago de Las Vegas.

Britton and Rose (1922) list 17 synonyms for their *Cactus intortus*, each name representing the same plant on a different island. Trelease and Stehlé, who felt that plants from Martinique and Guadeloupe were different species from those on the intervening island of Dominica, or on the British islands to the north or south, described many species of *Piper* and *Peperomia*. In a recent study (Howard, 1973), the 50 species recognized in the Lesser Antilles were reduced to 29, with 103 specific and 52 varietal synonyms. Even this did not give the consideration required in a comprehensive monograph to the 164 species currently in the floras of the islands of the Greater Antilles. Trelease (1913) recognized 37 native species of *Agave* in the Antilles, describing as new 14 species, each from a single island in the Lesser Antilles. The agaves of the area are large, spiny plants that flower infrequently and are monocarpic, and are often bulbiferous rather than producing fruit; these are difficult to collect, and especially to represent adequately in herbarium material. Less than 30 herbarium specimens of all species have been prepared since Trelease (1913). For the modern monographer, the plants in Cuba are not accessible, and adequate material of any species is difficult to obtain. It would be most unusual if each island had different species and only one species.

There is no doubt that the flora of the Antilles is overdescribed, but the single flora of the Antilles, described by Eggers to Asa Gray as "without regard to political division, which in my opinion is the worst possible distinction where science is concerned," remains unplanned for the future.

References

Adams, C. D. (1972). "Flowering Plants of Jamaica" 848 pp. University of the West Indies, Mona, Jamaica.

Alain, H. (1962). "Flora de Cuba" Vol. 5, 362 pp. University of Puerto Rico, Rio Padres.

Areces Mallea, A. E. (1976a). Una nueva especie de *Melocactus* Link & Otto de Cuba. *Cienc. Bot.* **9**, 3–11.

Areces Mallea, A. E. (1976b). *Melocactus holguinensis*: Una nueva especie de Cuba oriental. *Cienc. Bot.* **10**, 3–12.

Asprey, G. F. and R. G. Robbins (1953). The vegetation of Jamaica. *Ecol. Monogr.* **23**, 359–412.

Beard, J. S. (1955). The classification of tropical American vegetation-types. *Ecology* **36**, 89–100.

Berazaín, R. (1977). Estudio preliminar de la flora serpentinicola de Cuba. *Cienc. Bot.* **12**, 11–26.

Britton, N. L. and J. N. Rose (1922). "The Cactaceae" Vol. 3, p. 220–238. Carnegie Institute, Washington, DC.

Britton, N. L. and C. F. Millspaugh (1920). "The Bahama Flora" 695 pp. New York Academy of Sciences, New York.

Britton, N. L. and P. Wilson (1923–1924). "Scientific Survey of Porto Rico and the Virgin Islands" Vol. 5, 626 pp. New York Academy of Sciences, New York.

Britton, N. L. and P. Wilson (1925–1930). "Scientific Survey of Puerto Rico and the Virgin Islands" Vol. 6, 663 pp. New York Academy of Sciences, New York.

Cheesman, E. E. (1947). "Flora of Trinidad and Tobago" Vol. 1, Part 3, p. 465–531. Guardian Printery, Port-of-Spain, Trinidad.

Correll, D. S. (1974). Flora of the Bahama Islands—new additions. *Fairchild Trop. Gard. Bull.* **29**, 11–12, 15.

Correll, D. S. (1977). New species and a new combination from the Bahamas, Caicos and Turks Islands. *J. Arnold Arb.* **58**, 40–51.

D'Arcy, W. G. (1967). Annotated checklist of the dicotyledons of Tortola, Virgin Islands. *Rhodora* **69**, 385–450.

D'Arcy, W. G. (1975). Anegada Island: vegetation and flora. *Atoll Res. Bull.* **188**, 1–40.

Duss, R. P. (1897). Flore phanérogamique des Antilles francaises. *Ann. Inst. Col. Mars.* **3**, 1–656.

Eggers, H. F. A. (1876). St Croix's flora. *Vid. Meddel. Naturh. Foren. Kjobenhaven.* 33–158.

Eggers, H. F. A. (1879). The flora of St Croix and the Virgin Islands. *Bull. U. S. Nat. Mus.* **13**, 1–133.

Fosberg, F. R. (1976). Revisions in the flora of St Croix, U.S. Virgin Islands. *Rhodora* **78**, 79–119.

Fournet, J. (1978). "Flore Illustrée des Phanérogames de Guadeloupe et de Martinique" 1654 pp. Institut National de la Recherche Agronomique, Paris.

Gillis, W. T. (1974). Name changes for the seed plants in the Bahama flora. *Rhodora* **76**, 67–138.

Gillis, W. T., R. A. Howard and G. R. Proctor (1973). Additions to the Bahama flora since Britton and Millspaugh. *Rhodora* **75**, 411–425.

Gooding, E. G. B., A. R. Loveless and G. R. Proctor (1965). "Flora of Barbados" 486 pp. HMSO, London.

Hayden, W. J. (1977). Comparative anatomy and systematics of *Picrodendron*, Genus incertae sedis. *J. Arnold Arb.* **58**, 257–279.

Hodge, W. H. (1954). Flora of Dominica, Part 1. *Lloydia* **17**, 1–238.

Howard, R. A. (1952). Society of Plant Taxonomists' plaque honoring Erik L. Ekman. *Bull. Torrey Club* **79**, 80–84.

Howard, R. A. (1968). The ecology of an elfin forest in Puerto Rico. 1. Introduction and composition studies. *J. Arnold Arb.* **49**, 381–418.

Howard, R. A. (1973). Notes on the Piperaceae of the Lesser Antilles. *J. Arnold Arb.* **54**, 377–411.

Howard, R. A. (1974a). The vegetation of the Antilles. *In* "Vegetation and Vegetation History of Northern Latin America" (A. Graham, ed.) p. 1–38. Elsevier, New York.

Howard, R. A. (1974b). "Flora of the Lesser Antilles, Leeward and Windward Islands. Vol. 1, Orchidaceae" (L. A. Garay and H. R. Sweet, eds.) 235 pp. Arnold Arboretum, Jamaica Plain, MA.

Howard, R. A. (1975). Modern problems of the years 1492–1800 in the Lesser Antilles. *Ann. Mo. Bot. Gard.* **62**, 368–379.

Howard, R. A. (1977a). Current work on the flora of Cuba—A commentary. *Taxon* **26**, 417–423.

Howard, R. A. (1977b). "Flora of the Lesser Antilles, Leeward and Windward Islands. Vol. 2, Pteridophyta" (G. R. Proctor, ed.) 414 pp. Arnold Arboretum, Jamaica Plain, MA.

Howard, R. A. (1979). "Flora of the Lesser Antilles, Leeward and Windward Islands. Vol. 3, Monocotyledoneae" (R. A. Howard *et al.*, eds) 586 pp. Arnold Arboretum, Jamaica Plain, MA.

Jiménez, J. J. (1966). "Catalogus Florae Domingensis" Suppl. 1, p. 1–278. University of Santo Domingo, Dominican Republic.

León, H. (1934). El genero *Melocactus* en Cuba. *Mem. Soc. Cubana Hist. Nat.* **8**, 1–9.

León, H. (1946). "Flora de Cuba" Vol. 1, 441 pp. Cultural SA, Havana, Cuba.

León, H. and H. Alain (1951). "Flora de Cuba" Vol. 2, 456 pp. P. Fernandez, Havana, Cuba.

León, H. and H. Alain (1953). "Flora de Cuba" Vol. 3, 502 pp. P. Fernandez, Havana, Cuba.

León, H. and H. Alain (1956). "Flora de Cuba" Vol. 4, 556 pp. P. Fernandez, Havana, Cuba.

Liogier, A. (1965). Nomenclatural changes and additions to Britton and Wilson's "Flora of Porto Rico and the Virgin Islands". *Rhodora* **67**, 315–261.

Liogier, A. (1967). Further changes and additions to the Flora of Porto Rico and the Virgin Islands. *Rhodora* **69**, 372–376.

Liogier, H. and H. Alain (1969). "Flora de Cuba Supplemento" 150 pp. Editorial Sucre, Caracas, Venezuela.

Liogier, H. and H. Alain (1974). "Diccionario Botanico de Nombres Vulgares de la Espanola" 813 pp. Jardin Botanico, Santo Domingo, Dominican Republic.

Little, E. L., Jr. (1969). Trees of Jost Van Dyke (British Virgin Islands). *U.S.D.A. For. Serv. Res. Pap.* **ITF–9**, 1–12.

Little. E. L., Jr., R. O. Woodbury and F. H. Wadsworth (1976). Flora of Virgin Gorda (British Virgin Islands). *U.S.D.A. For. Serv. Res. Pap.* **ITF–21**, 1–36.

Mészáros, Z. (1976). The *Melocactus* species of Cuba. *Act. Bot. Acad. Sci. Hung.* **22**, 127–147.

Moscoso, R. M. (1943). "Catalogus Florae Domingensis" 732 pp. L. and S. Printing, New York.

Stehlé, H. (1945). Forest types of the Caribbean islands. *Carib. For.* **6** (Suppl.), 273–408.

Trelease, W. T. (1913). Agave in the West Indies. *Mem. Nat. Acad. Sci.* **11**, 1–55.

West, H. (1793). "Bidrag til beskrivelse over St Croix." Friderik Wilhelm Thiele, Copenhagen.

Williams, R. O. (1928). "Flora of Trinidad and Tobago" Vol. 1, Part 1, p. 1–22. Government Printing Office, Port-of-Spain, Trinidad.

Williams, R. O. and E. E. Cheesman (1929). "Flora of Trinidad and Tobago" Vol. 1, Part 2, p. 23–164. Government Printing Office, Port-of-Spain, Trinidad.

Ecological Aspects of the Vegetation of Curaçao, Netherlands Antilles. 1. Water Relations

A. L. STOFFERS

University of Utrecht, Netherlands

Introduction

The island of Curaçao forms part of a row of small islands stretching from west to east along the north coast of Venezuela. Geologically and geomorphologically it has much in common with the nearby islands of Aruba and Bonaire. As a result of the geological development the interior of the western and eastern parts of the island consists mainly of diabase, which is usually deeply weathered and much denuded. Several low gently sloping hills are to be found in this area. The higher summits occur in the western part (Seroe Christoffel and surroundings), and are formed from cherts and tuffs, which here compose what is known as the Knip formation. Several steep slopes occur in this area. The western portion of the central part of the island is occupied by the Midden Curaçao beds, which consist of sandstone, shales and conglomerates. Coral limestone partly encircles these older formations and becomes prominent in the central part of the island and along the north and north-east coast. It forms conspicuous table-mountains and plateaus. Several hand-shaped, landlocked bays occur along the north-east and south-western coasts. Alluvial deposits are found along these bays and along the edges of the salínas.

In Curaçao, as well as in Aruba and Bonaire, three groups of vegetation types may be distinguished: (i) edaphic communities, e.g. mangrove woodland, beach and dune vegetations and vegetations of the saltflats; (ii) a series of deciduous seasonal formations on the diabase and the Knip formation (lacking in the limestone area); (iii) a series of dry evergreen formations in the limestone area. The seasonal and dry evergreen formation series are used in the sense as proposed by Beard (1944, 1949, 1955).

The seasonal and dry evergreen series differ not only in floristic composition but especially in structure. It is generally assumed that the seasonal formations indicate physical drought of the environment and consequently deciduous trees are more common than hard-leaved evergreens. As the habitat becomes more adverse the height of the dominant trees is reduced and evergreens become rarer. Under the driest conditions thorniness and microphylly becomes characteristic.

The dry evergreen formations on the other hand are the expression of physiological rather than physical drought of the habitat. Species typical for these formations are hard-leaved evergreens of which the leaves may be either thickened and fleshy or thin and brittle, whereas the upper surface is often very shiny and thickly cutinized. Under drier conditions microphylly is characteristic but never thorniness.

On account of the marked differences between the vegetation of the limestone plateaus and that in the non-calcareous, particularly the diabase areas, the island seemed to be very suitable for an ecological investigation as limestone and diabase are found side by side in the same macroclimate.

Since the island is located in an area of low rainfall, which extends along the north coast of South America, rainfall is—generally speaking—scanty particularly from February until September; the total amount of precipitation in this period is about 200 mm. In the rainy season from October until January the average precipitation amounts to 300–350 mm.

Meteorological observations do not indicate that there are differences in amounts of precipitation between limestone and diabase areas through which the differences in vegetation could be explained. This is, however, questionable in the case of the Seroe Christoffel region which supposedly has a higher precipitation, but meteorological records are not available. Humidity also seems to be much higher because large numbers of epiphytic lichens are present.

Since the dry-evergreen vegetations on limestone are more luxuriant than the seasonal vegetations on diabase, it is hard to understand how in a region of low rainfall, characterized by a large number of dry months and giving raise to seasonal vegetation types, an additional physiological drought results in the existence of a more luxuriant vegetation.

Considering the definite correlation between seasonal formations and diabase on the one hand and between evergreen formations and limestone on the other, one could characterize them as calcifuge resp.

calcicole vegetations. However, no woody species is known to occur in the diabase region only. Species that could be taken into consideration as calcicole are e.g. *Antirrhoea acutata* (Rubiac.), *Erithalis fruticosa* (Rubiac.), *Condalia henriquezii* (Rhamnac.), *Rhacoma crossopetalum* (Celastrac.), and *Bulbostylis curassavica* (Cyperac.). Some species, e.g. *Coccoloba swartzii* (Polygonac.), *Bumelia obovata* (Sapotac.) and *Metopium brownei* (Anacardiac.), may be very common in the limestone area, yet cannot be considered calcicole because they are also frequently found in areas of the Knip formation and on the slopes of Seroe Christoffel.

The purpose of the investigation was to gain insight in the ecological differences between the two formation series. Five primary ecological factors can be discerned: water, heat, light, chemical and mechanical factors. The present paper focuses on the water factor only. It was decided to start the investigation at the end of the rainy season and study the effect of increasing drought on a number of species characteristic for the two vegetation series.

Results and Discussion

Transpiration

The rate of transpiration in the course of the day was measured by Stocker's "Momentanmethode" (Steubing, 1965) giving transpiration as loss of water in mg / min / g of leaf (fresh weight). Evaporation was measured with the aid of the Piche evaporimeter.

Figure 1 shows transpiration curves for a number of species growing on limestone. For most species the curve is characterized by two maxima in the course of the day: the first early in the morning and the second in the afternoon. In some species transpiration decreases around noon to values too low to be measured with the applied method.

The sharp fall in transpiration as well as the low values attained indicate that water loss is extremely well controlled in plants which are common on limestone. Of these *Antirrhoea*, *Coccoloba* and *Rhacoma* have xeromorphic, strongly cutinized leaves whereas *Erithalis* has somewhat succulent leaves with a thick cuticle. *Phyllanthus* (Euphorbiac.) and *Casearia* (Flacourtiac.), like *Coccoloba* also frequent in the non-calcareous area, can curl the leaves, inducing additional decrease in transpiration. An entirely different type of curve is seen in *Croton flavens* (Euphorbiac.), where transpiration is proportional to the evaporation curve with its slow increase in the morning hours and a gradual decrease in the afternoon. In Fig. 2 a third type of transpiration curve is shown.

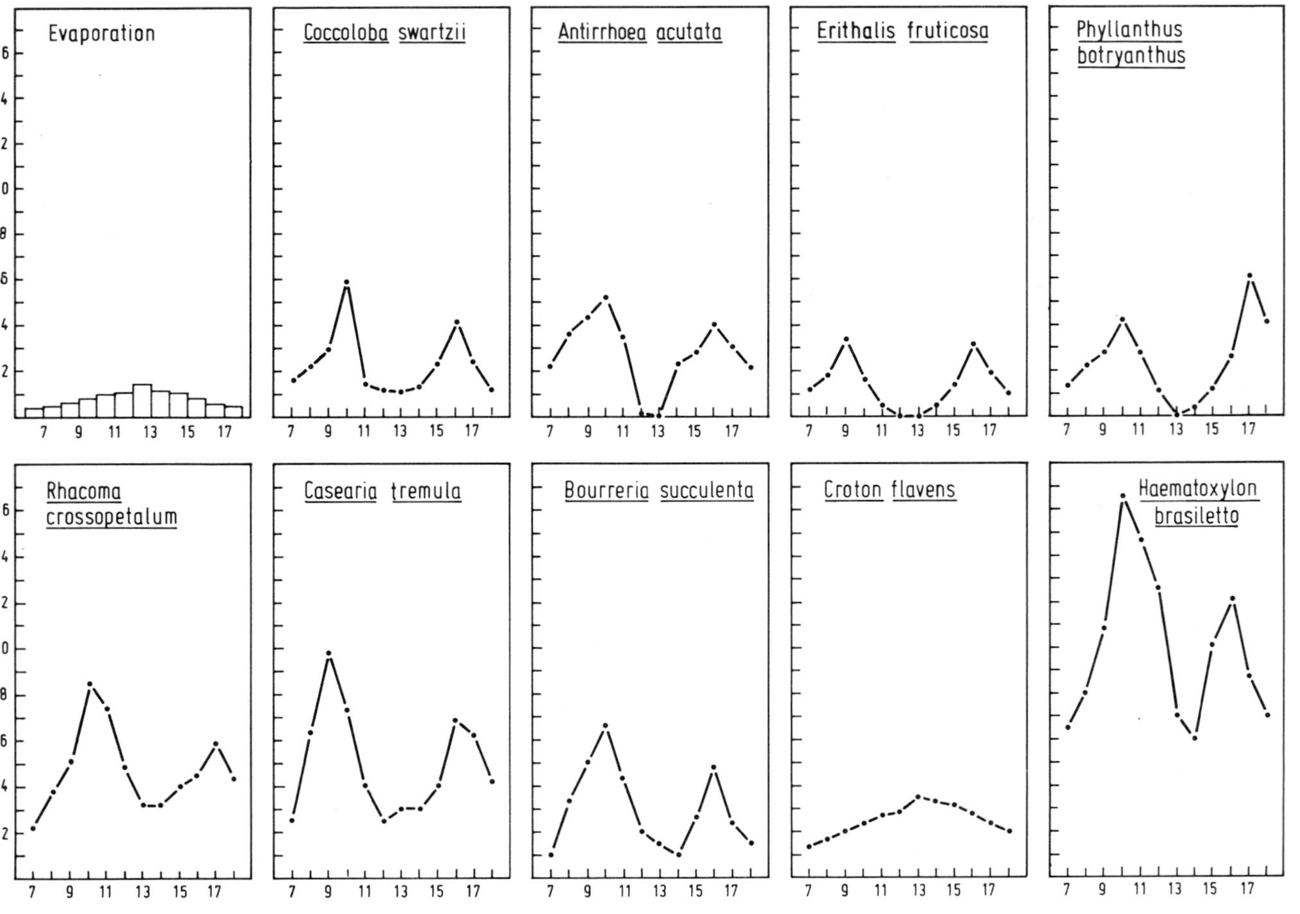

Fig. 1. Evaporation and transpiration curve of some species growing on limestone.

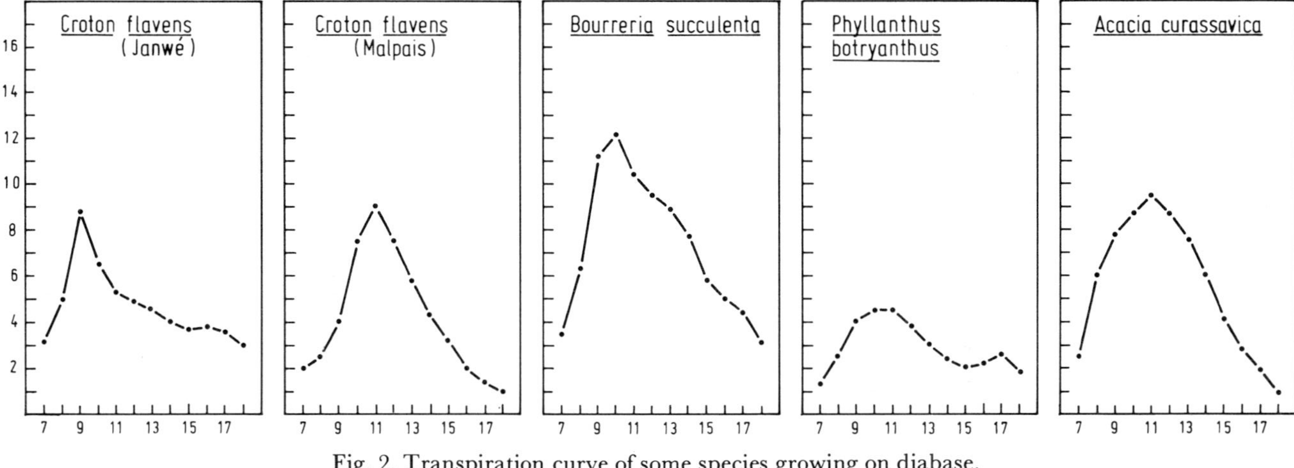

Fig. 2. Transpiration curve of some species growing on diabase.

With the exception of *Phyllanthus botryanthus* all investigated species have a more or less similar transpiration curve. The curves are characterized by a maximum of transpiration in the morning hours followed by a rather sharp fall in transpiration. A second maximum, as on limestone, is not observed, although a weakly two-topped curve was found in *Phyllanthus*. This is probably connected to the leaf-curl as is mentioned before. The time at which the maximum is reached varies not only from species to species but also depends on the habitat of the plant. In *Croton flavens* growing near Janwé, a site under strong influence of the trade-wind, the maximum was attained early in the morning, whereas in plants growing at Malpais, a site protected against the trade-winds, the maximum was shiften to around noon. This phenomenon has been observed in several species growing on wind-exposed as well as sheltered sites.

The three described types of transpiration curves clearly reflect the soil water relation (see e.g. Stocker, 1956). If the amount of water in the soil is adequate the transpiration curve runs proportionate to the evaporation curve: an increase during the morning hours reaching a maximum at noon, followed by a gradual decrease in the afternoon (type I). This is clearly the case with *Croton* growing on limestone. In the single peaked curve transpiration decreases disproportionate to the decrease in evaporation (type II). This implies that the soil water supply is impeded and the water relation becomes severe. The decrease of transpiration does not improve the water relation and consequently recovery of transpiration in the afternoon fails. In contrast the two-topped curve is found in plants which limit their transpiration by closing the stomata as soon as a slight loss of water occurs (type III). The strong decrease in transpiration improves the water relation and the stomata are reopened again in the afternoon.

In addition to the species listed in Fig. 1 and Fig. 2, transpiration curve type I was also found in some other species growing on limestone, e.g. *Melochia tomentosa* (Sterculiac.), *Croton ovalifolius* and *Waltheria americana* (Sterculiac.); type II in *Acacia tortuosa*, *Caesalpinia coriaria*, *Cordia curassavica* (Boraginac.), and *Haematoxylon brasiletto* (Caesalpiniac.) growing on diabase whereas type III was found in a number of species growing on limestone such as *Cordia curassavica*, *Guaiacum officinale* (Zygophyllac.), *Jacquinia barbasco* (Theophrastac.), and *Bumelia obovata*, i.e. species having a thick cuticle or strongly cutinized cell walls. From the transpiration relation as found some weeks after the rainy season it is clear that at this time of the year the limestone areas are more favourable than the diabase soils with regards to the water relation.

Water Saturation Deficit

The loss of water by transpiration leads to a water saturation deficit in the cells of the leaves and consequently the cells start loosing their turgidity. The saturation deficit will increase faster and reach higher values in plants with lower drought resistance. This means that in plants unable to decrease transpiration adequately in response to an impeded water supply, a high water deficit will be found. The water saturation deficit was determined by a slightly modified method of Čatský (Čatský, 1960; Steubing, 1965; Stoffers and Elassaiss, 1967). In Table 1 the water saturation deficit is given as percentages of the fresh weight at maximal saturation. The results show a clear-cut ecological difference between plants growing on limestone and those growing on diabase. In the species on limestone a water saturation deficit was found up to a maximum of about 12%. This value is far below the maximum deficit attained in the diabase plants. The differences are particularly significant in the cases where the deficit was measured in plants of those species growing both on limestone and diabase. These results support the conclusion that water relations are less severe in the limestone area than in the diabase region at this time of the year.

Suction Pressure

It will be clear that water deficit is connected with suction pressure. The suction pressure is dependent on momentary meterological and

Table 1. Water saturation deficit in some plants growing on diabase and limestone (in %).

	Diabase		Limestone	
Haematoxylon brasiletto	24	30	10	7
Croton flavens	28	17	6	9
Bourreria succulenta	27	20	10	9
Phyllanthus botryanthus	30	37		
Balanites aegyptica	21	17	5	7
Coccoloba swartzii			9	7
Cordia curassavica			6	7
Metopium brownei			4	10
Erithalis fruticosa			11	8
Antirrhoea acutata			3	9
Lantana camara			8	12
Rhacoma crossopetalum			4	6
Casearia tremula			6	8

soil water conditions. As a result the suction pressure often shows considerable daily fluctuations. From data obtained with the aid of the Shardakov method (Table 2) it appears that in species growing on

Table 2. Suction pressure in some plants growing on diabase and limestone, at 9.30, 13.00 and 17.00 hr.

	Diabase			Limestone		
Haematoxylon brasiletto	21	45	48	20	20	15
Croton flavens	12	32	24	9	16	13
Caesalpinia coriaria	10	31	28	20	20	22
Bourreria succulenta	8	30	26	8	20	19
Randia aculeata	12	29	29	16	16	21
Phyllanthus botryanthus	16	33	37	14	17	17
Antirrhoea acutata				6	10	10
Rhacoma crossopetalum				14	15	15

limestone daily fluctuations in suction pressure are much less than in the plants growing on diabase. This once more indicates that the decrease in transpiration is due to shortness of soil water available to the plant.

Soil Water Content

An impression of differences in soil water available to the plant was obtained by means of modified plasterblock method (Bouyoucos and Mick, 1940). Small-mesh wire nettings are separated by a 3 mm thick plastic ring and to both sides of these electrodes plastic rings (3 mm) are attached. The entire structure is filled with plaster of Paris. The electric resistance between the electrodes, (measured with the aid of a Wheatstone bridge) was used as an indication for the soil humidity. Results shown in Fig. 3. It appears that at the beginning of the dry season the soil at a depth of 25 cm and lower is moister in the limestone than in the diabase area. The difference increases strongly in relatively short time. The greater retention capacity of the limestone soil is attributed to a better developed clay complex since granulometric composition of soil samples showed an appreciable higher fraction smaller than 2 μm than in diabase samples. This is in good agreement with the granulometric analysis presented by Hamilton and Sesseler (1945) and Westermann and Zonneveld (1956).

Whereas in humid climates fine-textured clay soils provide moister habitats than soils with a coarser texture, in arid regions, however,

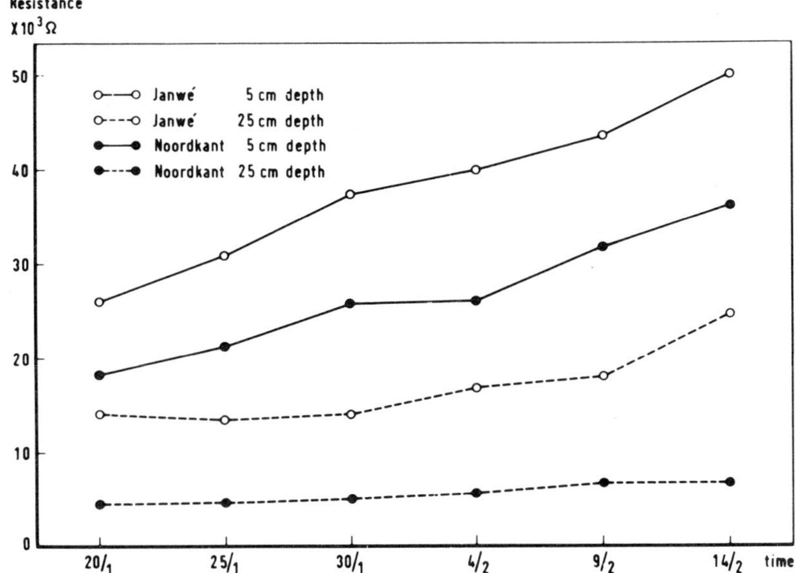

Fig. 3. Soil moisture; data obtained by the modified Bouyoucos-method.

quite contrary situations exist (Walter, 1964). It must be emphasized that on the limestone plateaus of Curacao soil is only found in fissures and rock crevices, which at some distance below the surface is entirely protected against evaporation. Although the island has an arid climate, clay soils under conditions as described above provide a wet habitat as compared with the diabase soils.

Conclusions

In the diabase areas a shallow soil is found. Owing to the hilly character of these areas much of the rain water flows to the sea during and shortly after a shower. Only a small portion of the water soaks into the soil, part of which adds to the ground water. Due to its low water retention capacity the soil becomes rapidly saturated. The upper layers dry out quickly after the rainy season through evaporation and water uptake by herbs, while shrubs and trees drain the deeper layers. As a result, shortly after the last shower the soil water is exhausted. Since the coarse soil texture causes a slow water succession and greatly restricted root growth, the drought will be experienced by the plant very soon. This results in temporary or permanent wilting and leaf abscission.

On the limestone plateaus the situation is entirely different. Here all the water flows into the cracks and holes of the rocks in which a soil with a high water retention capacity is found. This soil also becomes saturated, though more gradually than the diabase soil. After the last showers the water content in the upper layers decreases due to evaporation and uptake by the herb layer. In the deeper soil layers the greater part of the rain water is at the disposal of the plants, and provides sufficient water during the majority of the year for deep rooting plants and plants able to develop a high suction pressure. Moreover this water is used very economically as most plants limit their evapotranspiration. Therefore the presence of a more luxuriant vegetation on limestone plateaus in the arid climate of Curaçao is seen as another example of what Walter (1964) indicates as "Das Gesetz der relativen Standortskonstanz und des Biotopwechsels". This might be an explanation for the phenomenon that species restricted to the limestone plateaus in this area are found on non-calcareous soils in other parts of the West Indies where the amount of precipitation is appreciably larger.

The question remains why plants as e.g. *Cordia curassavica*, *Casearia tremula* and *Haematoxylon brasiletto* show a double peaked transpiration curve when growing on limestone and single peaked curve when growing on diabase. Unless the specimens of the diabase area belong to another infraspecific taxon than the specimens of the limestone, one would suppose that they would show a similar type of transpiration curve. However in transpiration two components have to be discerned: stomatal and cuticular transpiration. The latter is strongly dependent on the structure of the cuticle. It has been proved that cuticles may be permeable to water and to polar as well as non-polar substances. The permeability depends on the specific structure of the cuticle, which is influenced not only by the genetic background of the plant but also by climate, humidity and other environmental circumstances. A complete absence of cuticular transpiration found in e.g. *Antirrhoea acutata* and *Erithalis fruticosa*, is due to the process of "incipient drying" in which the presence of epicuticular wax and the impregnation of the cuticle with wax lamellae is of more importance than the thickness of the cuticle. (For general surveys see Linskens *et al.*, 1965; Martin and Juniper, 1970.) Unfortunately no generalization can be made on the role of the cuticle in preventing the loss of water from plants; each case needs to be examined on its own merits (Martin and Juniper, 1970, p. 10).

There are indications that low temperatures during the growing season or the occurrence of short lasting growing seasons hinder the synthesis of a well-developed cuticle. Consequently a strong cuticular transpiration can take place. Since the growing season in the limestone area is actually longer than in the diabase area as a result of the better water relation in the soil, this might be important in explaining the occurrence of hard evergreens in the limestone area and the nearly complete absence of this lifeform in the diabase area. In order to get more insight into the structure of the cuticle and leaf for various vegetation types, light microscopic investigations are in progress. Scanning electron and transmission electron microscopic investigations are in preparation.

Acknowledgements

The investigation was made possible by a grant of the Netherlands Foundation for the Advancement of Tropical Research (WOTRO). The author is indebted to Dr I. Kristensen, Director of the Caribbean Marine Biological Institute (CARMABI) in Curaçao for providing facilities, to Dr S. R. Gradstein (Utrecht) for correcting the English text, to Mr T. Schipper for the drawings and to Mrs E. G. van Bemmel-Henken for typing the manuscript.

References

Beard, J. S. (1944). Climax vegetation in tropical America. *Ecology* **25**, 127–158.

Beard, J. S. (1949). The natural vegetation of the Windward and Leeward Islands. Oxford For. Mem. No. 21.

Beard, J. S. (1955). The classification of tropical American vegetation-types. *Ecology* **36**, 89–100.

Bouyoucos, G. J. and A. J. Mick (1940). An electrical resistance method for the continuous measurement of soil moisture under field conditions. Mi. State Coll. Agric. Ex. Stn. Tech. Bull. No. 172.

Čatský, J. (1960). Determination of water deficit in disks cut out from leaf blades. *Biol. Plant (Praha)* **2**, 76–78.

Hamilton, R. and W. M. Sesseler (1945). Bijdrage tot de bodemkundige kennis van (Nederlandsch) West Indië. Med. Kon. Ver. Ind. Inst. Amsterdam No. 63.

Köhn, M. and H. Person (1950). Über die Bestimmung der Bodenfeuchtigkeit auf elektrischem Wege. 2. Mitteilung Jahresber. mit Abhand. des Bad. Landeswetterdienstes, 31–34.

Linskens, H. F., W. Heinen and A. L. Stoffers (1965). Cuticula of leaves and the residue problem. *Res. Rev.* **8**, 135–178.

Martin, J. T. and B. E. Juniper (1970). "The Cuticles of Plants." Edward Arnold, London.

Shardakow, W. (1956). Die Bestimmung der Bewässerungstermine der Baumwollpflanzen mit Hilfe der Saugkraft der Blätter. Art. Akad. Wiss. Usbek SSR Taschkent.

Steubing, L. (1965). "Pflanzenökologisches Praktikum." Verlag Paul Parey, Berlin and Hamburg.

Stocker, O. (1956). Die Abhängigkeit der Transpiration von den Umweltsfaktoren. *In* "Handbuch der Pflanzenphysiologie" (W. Ruhland, ed.) Vol. III, p. 436–488. Springer Verlag, Berlin and New York.

Stoffers, A. L. (1956). The vegetation of the Netherlands Antilles. Thesis, Utrecht.

Stoffers, A. L. and C. J. A. M. Elassaiss (1967). On the water-relation in limestone and diabase vegetations in the Leeward Islands of the Netherlands Antilles. *Acta Bot. Neerl.* **15**, 539–556.

Walter, H. (1964). "Die Vegetation der Erde 2. Aufl. Gustav. I." Fischer Verlag, Jena.

Westermann, J. H. and J. I. S. Zonneveld (1956). Photo-geological observations and land capability and land use survey of the island of Bonaire (Netherlands Antilles). Med. Kon. Inst. Tropen, Amsterdam, No. 73.

IV. Taxonomic and Biological Examples

The Subtribe Hypochoeridinae (Asteraceae, Lactuceae) in the Tropics and the Southern Hemisphere

H. WALTER LACK

Botanischer Garten und Botanisches Museum Berlin–Dahlem

Introduction

Within the large family Asteraceae the tribe Lactuceae is the most easily recognized group; a recent summary (Tomb, 1978) lists 70 genera with a total of approximately 2300 species for Lactuceae. The tribe can be regarded as the best known group within Asteraceae: nearly all genera are known cytologically, a number of genera have been admirably revised (Stebbins, 1940a; Babcock, 1947; Chambers, 1955) and, in addition, there is general agreement on the subtribal arrangement first proposed by Stebbins (1953) and later modified in details only (Jeffrey 1967).

Stebbins (1953) recognized eight subtribes of Lactuceae, of which three, i.e. Stephanomeriinae, Microseridinae and Dendroseridinae, are confined to the New World, all of them small groups comprising less than 10% of the total number of species of the tribe. The remaining five subtribes cover the bulk of the Lactuceae and show a predominantly extratropical, Old World distribution, with the centre of diversity lying in temperate Eurasia and the Mediterranean area. These include the subtribes Scolyminae, Cichoriinae, Scorzonerinae, Hypochoeridinae and, by far the biggest subtribe, the Crepidinae with such widespread and well-known genera as *Crepis* L., *Hieracium* L., *Lactuca* L. and *Sonchus* L.

The Subtribe Hypochoeridinae

The Hypochoeridinae are characterized as: perennial cr annual herbs mostly with coarse, spreading hirsute pubescence, the hairs often forked; involucres various in size and appearance; receptacle paleaceous or naked; flowers nearly always yellow; achenes mostly fusiform or beaked; pappus of coarse, plumose setae, or occasionally paleaceous or coroniform; pollen grains echinolophate; stigma branches elongate. Basic chromosome number is x = 7,6,5,4, and 3′ (Stebbins, 1953).

The subtribe comprises three widespread genera of moderate size: *Hypochoeris* L., *Picris* L. and *Leontodon* L., and six small satellite genera: *Helminthotheca* Zinn, *Rhagadiolus* Scop., *Garhadiolus* Jaub. Spach, *Urospermum* Scop., *Hedypnois* Mill. and *Microderis* DC.

With the exception of the monotypic genus *Microderis* endemic to the Azores, all genera of Hypochoeridinae can be found in the Mediterranean area and the Near East (Fig. 1). Only *Hypochoeris*, *Picris* and

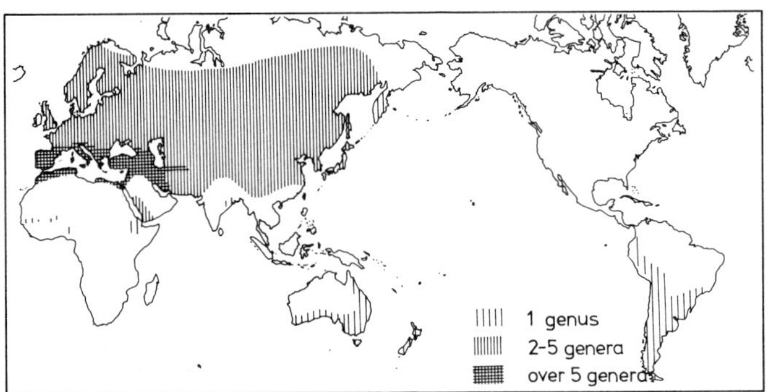

Fig. 1. Distribution of the subtribe Hypochoeridinae.

Leontodon extend considerably beyond this area; in the extratropical part of Eurasia *Hypochoeris* reaches East Siberia (Vasil'ev, 1964) and *Picris* extends even further eastwards to Kamchatka and Attu Island in the Aleutian archipelago. However, I am not aware of a single record for a native species of the subtribe from continental North America. A number of Mediterranean Hypochoeridinae, notably *Hypochoeris radicata* L., *H. glabra* L. and *Helminthotheca echioides* (L.) Holub, have been introduced as weeds into many parts of the world and have become firmly established in the vegetation. Only two genera of the subtribe Hypochoeridinae have native representatives in the tropics: *Hypochoeris* in the neotropics and *Picris* in the paleotropics. Both genera extend into the southern hemisphere, *Hypochoeris* to Tierra del Fuego and *Picris* as far as Tasmania and New Zealand.

Hypochoeris in South America

Hypochoeris, a clearly circumscribed genus unique within the subtribe in having conspicuous receptacular paleae, comprises about 50 species (Stebbins, 1971; *c.*100 according to Tomb, 1978), the greater part of which are native to South America (Fig. 2) and already described in the

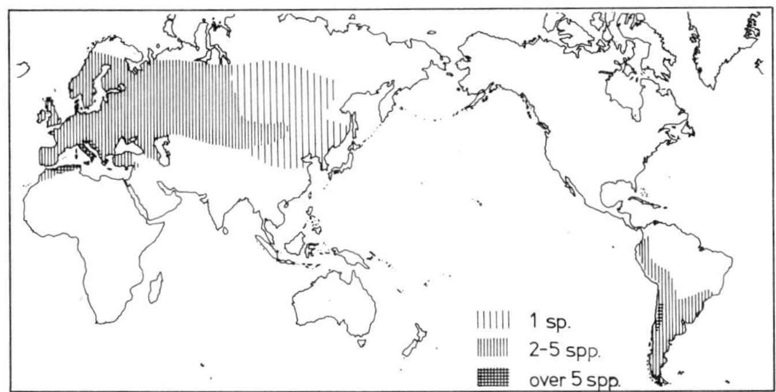

Fig. 2. Distribution of the genus *Hypochoeris* L. (based in part on Stebbins, 1940b, Fig. 5).

pioneer papers on the genus (Schultz–Bipontinus, 1845, 1859). Apart from a study of the species growing in the southern parts of South America (Cabrera, 1963) unfortunately no revision of the whole genus, which also includes some Eurasian and Mediterranean species, has been published in recent years.

The South American species of *Hypochoeris* are very diverse in appearance, ranging from tall, branched perennials of humid regions to high-alpine cushion plants like *Hypochoeris sonchoides* Kunth (Fig. 3). They also occupy diverse habitats: the regions high above the timber-line in Colombia, the tropical areas of South Brazil and the arid plains of Patagonia (Stebbins, 1971). In addition some South American species have white or pink ligules, an extremely rare colour within Lactuceae, and show a number of unusual pollen characteristics. (Heusser, 1971; Nordenstam, pers. comm.). On the basis of the highly distinctive, strongly asymmetrical and bimodal karyotype in the South American species known so far—a character not found in any of the Eurasian and Mediterranean taxa of *Hypochoeris*—it was argued that the large and most diverse centre of variation in South America is probably secondary and that the genus originated in temperate Eurasia (Stebbins, 1971). In view of the absence of native *Hypochoeris* species from North America, the main difficulty with this assumption is the question of how the genus reached South America, a problem well

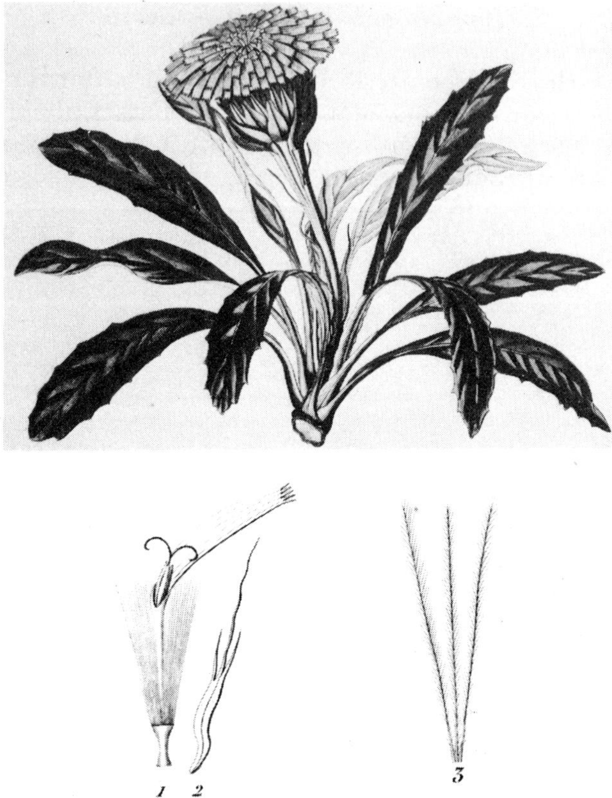

Fig. 3. *Hypochoeris sonchoides* Kunth: 1 = flosculus; 2 = palea; 3 = pappi pili plumosi (from Humboldt *et al.*, 1820, Table 301).

known for a number of genera with similar distribution e.g. *Alchemilla* L. The hypothesis has been put forward that *Hypochoeris* migrated southward via North America like many other groups and that, due to fairly recent climatic changes and new competition with more aggressive plants *Hypochoeris* has become extinct in North America (Stebbins, 1971). Independently a similar interpretation maintains that the southward extension of *Hypochoeris* may well have happened by means of long-distance dispersal in the Pleistocene and more recent times (Raven and Axelrod, 1974). However, it should be noted that the data available at present (in particular of the chromosome numbers) are far from extensive and that a full revision of the whole genus will be necessary in order to find additional arguments for an interpretation of the distribution pattern in *Hypochoeris*.

The more common situation of an amphiboreal distribution in the New World is exemplified by the genus *Microseris* D. Don of the predominantly North American subtribe Microseridinae. All the species of *Microseris* subgenus *Microseris* grow in western North America with the exception of *Microseris pygmaea* D. Don, an autogamous annual endemic to central Chile, unique in having 10 pappus elements per achene rather than the five typical for all other annual species of the subgenus (Chambers, 1955). Astonishingly, hybrids between the Chilean *M. pygmaea* and the Californian *M. bigelovii* (Gray) Sch. Bip.—two species distinct in a number of characters—could be obtained producing apparently healthy fruits which gave rise to relatively fertile F_2 generations (Chambers, 1963; Bachmann and Chambers, 1978). The two species are therefore interfertile and directly related to each other. In view of the complete absence of any relatives of the autogamous *M. pygmaea* in South America and the presence of all related species in western North America, a southward migration of some ancestor of modern *M. pygmaea*, possibly in the pre-Pleistocene, was assumed (Chambers, 1963). A similar explanation was advanced to explain the distribution of *Agoseris* Rafin., the second genus of Microseridinae showing a bipolar distribution (Chambers, 1963). The only genus of Lactuceae with a holarctic distribution and a continuous range in the New World from Alaska to Tierra del Fuego and the Falkland Islands is the enormous genus *Hieracium* L. of the subtribe Crepidinae (Zahn, 1921–1923).

Picris in Africa, Asia and Australia

Turning back to Hypochoeridinae, the only other genus of the subtribe extending into the tropics is *Picris*, currently being revised by me. The generic limits of *Picris* are unusually clearcut—all species have a branched habit, all possess striking anchor-shaped glochids on the vegetative parts, the receptacle is always naked and all species studied so far exhibit the chromosome number 2n = 10 (Lack, 1975). This widespread Eurasian genus has one centre of diversity in the Mediterranean region; it extends into the paleotropics and was gathered in tropical Africa, tropical Arabia, South-East Asia and tropical Australia; its second centre of diversity lies in temperate Australia with outposts in Tasmania and New Zealand (Fig. 4).

At present only the species from the eastern Mediterranean and western Asiatic region, as well as the representatives from tropical Africa and tropical Arabia have been fully revised (Lack, 1975, 1977,

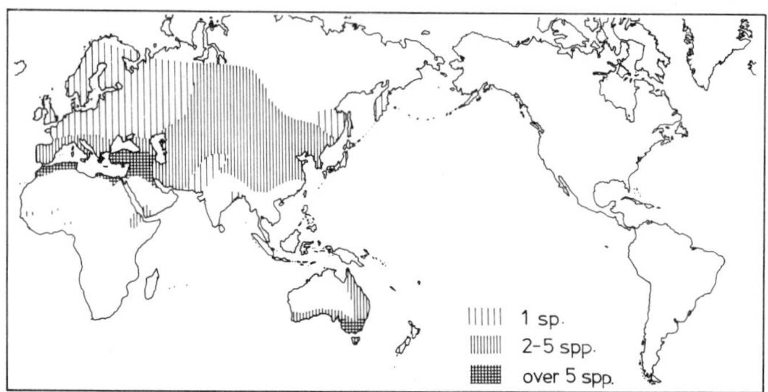

Fig. 4. Distribution of *Picris* L.

1979a, b), whereas work is in progress on the other groups (Lack, 1979c).

The total number of *Picris* species growing in the tropics is small compared to the agglomeration of species in the Mediterranean region. In the vast area of tropical Africa only three native species have been found, all of them growing in clearly circumscribed areas north of the equator. One of them is *P. humilis* DC., a tiny ephemeral annual restricted in distribution to the regularly flooded banks of the Niger and Sénégal rivers, which probably represents a specialized off-shoot of some western Mediterranean species (Lack, 1979a). The two other species are perennial montane plants developing massive woody root-stocks; both are native in the Ethiopian Highlands, but one of them, *P. xylopoda* Lack, has also been found on the Bauchi Plateau, the largest highland of Nigeria, well known for its peculiar flora comprising many East African or Zambesian elements not found anywhere else in tropical West Africa (Morton, 1972). The present disjunct distribution of many Afromontane elements has been interpreted either as a result of long-distance dispersal or as a result of past climatic changes, which may have led to a situation in which Afromontane habits were not more than *c*.200 km apart thus enabling migration (Morton, 1972). Generally it is assumed that migration of Afromontane elements took place in a westward direction and this may well have also been the case with *P. xylopoda* (Lack, 1979a).

Tropical Arabia supports only four native species of *Picris*, two of them widespread desert annuals of the Arabian Peninsula (Lack, 1979b). The other species, one of them *P. scabra* Forssk. (Fig. 5), are perennial rosette plants endemic to the high mountains of Yemen, which are closely related to the montane species of the Ethiopian Highlands. The perennials from Ethiopia and Yemen have no relatives

Fig. 5. *Picris scabra* Forssk. (A = habit; B, C = capitula; D = bracts; E = corolla; F = fruit; G = glochids on rosette leaf). Material cultivated in the Berlin Botanic Garden from the Yemen Arab Republic, Jebel Sabir above Taiz, Hepper 5926, K.

in North Africa or the Near East and probably represent a distinct evolutionary line starting from some central Asiatic ancestor, which extended its range southwestward (Lack, 1979a,b).

In the eastern part of Asia south of the tropic of Cancer a single and so far undescribed species of *Picris* was found; however, the southern provinces of China and also Taiwan support a few perennial species, all of them restricted in distribution to the alpine or high-alpine zone, e.g. *P. divaricata* Vaniot from the Yunnan province and *P. ohwiana* Kitam. from the alpine meadows of Taiwan.

As far as I am informed not a single species of the genus *Picris* was ever collected in the vast area of Malesia; it is particularly remarkable that the genus has not been found in the montane to alpine regions of New Guinea well known for supporting a considerable number of Eurasian elements, like *Epilobium* L., *Rubus* L. or *Viola* L. It is also in this area that the genera *Lactuca* L. and *Ixeris* Cass. of the subtribe Crepidinae reach their southernmost distribution in Australasia (Koster, 1976).

The presence of indigenous species of *Picris* in Australia, Tasmania and New Zealand came somewhat as a surprise since most standard floras for this region (e.g. Bentham, 1867; Cheeseman, 1925; Curtis, 1963) list for the genus *Picris hieracioides* L. only, an Eurasian perennial, and regard it as an introduced weed. This was, however, the result of a misidentification of native species. Copious material presently under study indicated at least a dozen species new to science, a few of which have been recently recognized (Lack, 1979c). From preliminary studies of the Australian gatherings it is evident that *Picris*—like *Hypochoeris* in South America—has spread into the most diverse habitats including the tropical lowlands of Queensland, the alpine meadows of the Snowy Mountains, the coastal sand dunes on the islands off the coast of South Australia and the Mediterranean regions in West Australia. There is no doubt that these novelties actually belong to *Picris* as all species show the characters typical for the genus and even the scanty cytological data currently available from Australian material corroborate this assumption. In contrast to the situation in Australia both Tasmania and New Zealand support very few species, some of them already known from the Australian mainland.

In view of the complete absence of any other native Hypochoeridinae in Australasia, it can be assumed that the genus spread southward from Central Asia, the hypothetic centre of origin not only of this subtribe, but of the whole Lactuceae (Babcock, 1947; Stebbins, 1953). The migration of the mountain species from Eurasia into Australia probably took place only several million years ago; before this time migration would have had to take place across a wide gap of lowland tropical forest between China and Australia (Raven and Raven, 1976). Within the Lactuceae only the genus *Sonchus* L. of the subtribe Crepidinae shows a distribution pattern in Australia similar to *Picris*. However, in *Sonchus* two tetraploid aberrant taxa have evolved which were recently regarded even as belonging to a separate genus (Roux and Boulos, 1973; Boulos, 1973, 1974).

A comparison of the thoroughly studied genus *Crepis* L. of the Crepidinae with *Picris* is of particular interest: the centre of origin for *Crepis* is assumed to lie in Central Asia from where a number of evolutionary lines or migration routes seem to have started (Fig. 6; Babcock, 1947). A preliminary summary of the situation in *Picris* reveals striking parallels—three possible evolutionary lines westward, one eastward (but

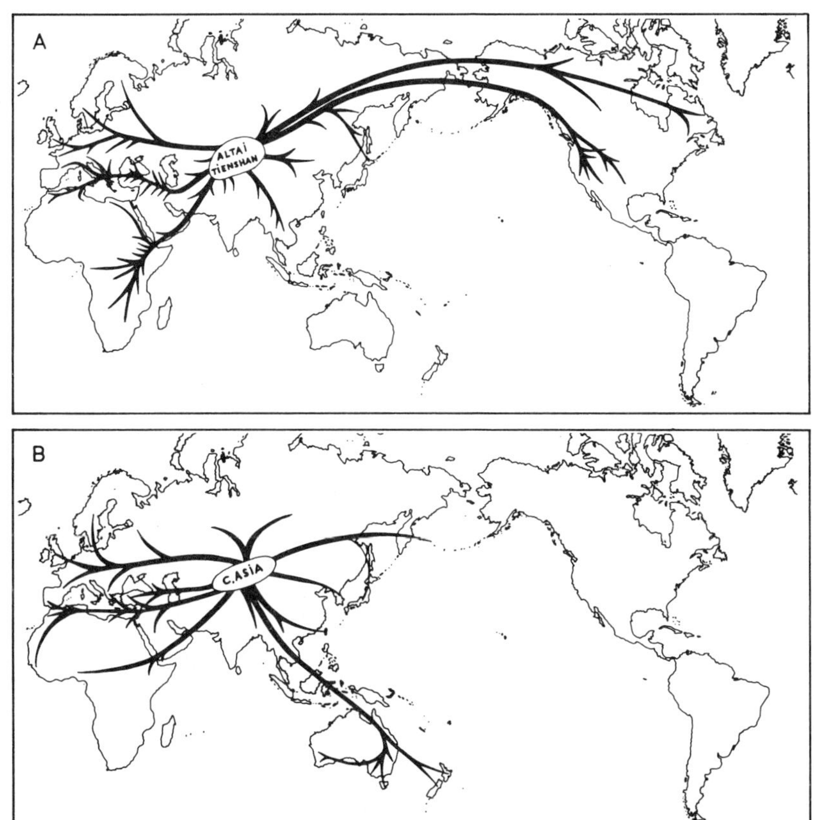

Fig. 6. Diagrammatic representation of hypothetical migration routes in *Crepis* L. (A) and *Picris* L. (B). (A) is based on Babcock (1947, Fig. 11).

not reaching continental North America), one southwestward (but compared to *Crepis* with very few species in tropical Africa) and one evolutionary line southeastward, much more prominent than in *Crepis* resulting in a second and in all probability secondary centre of diversity in temperate Australia (Fig. 6). In contrast not a single indigenous species of *Crepis* was reported from Australia, Tasmania and New Zealand (Babcock, 1947).

Conclusion

Comparing the data available on the systematics and the distribution of Crepidinae (in particular *Crepis*, *Hieracium* and *Sonchus*) and Hypochoeridinae (*Hypochoeris* and *Picris*) it may be assumed that both subtribes, though primarily having an extratropical Old World distribution, have produced independently in the Old World as well as in the New World a relatively small number of tropical and southern hemisphere taxa, which seem to have evolved along parallel routes in a parallel manner.

Acknowledgements

The drawings have been made by E. Dieckmann (Figs 1,2,4 and 6) and H. Lünser (Fig.5). Thanks are due to J. R. Edmondson, who revised the style of the text, and to H. Merxmüller and G. Wagenite, who kindly looked through the manuscript and made valuable suggestions.

References

Bachmann, K. and K. L. Chambers (1978). Pappus part number in annual species of *Microseris* (Compositae, Cichoriaceae). *Pl. Syst. Evol.* **129**, 119–134.

Babcock, E. B. (1947). The genus *Crepis*. *Univ. Cal. Publ. Bot.* **21–22.**

Bentham, G. (1867). "Flora Australiensis" Vol. 3. Lovell Reeve, London.

Boulos, L. (1973). Révision systématique du genre *Sonchus* L.s.1. IV. Sousgenre 1. *Sonchus*. *Bot. Notiser* **126**, 155–196.

Boulos, L. (1974). Révision systématique du genre *Sonchus* L.s.1.VI. Sousgenre 3. *Origosonchus*. Genres *Embergeria*, *Babcockia* et *Taeckholmia*. Species exclusae et dubiae. Index. *Bot. Notiser* **127**, 402–451.

Cabrera, A. L. (1963). Estudios sobre el genero *Hypochoeris*. *Bol. Soc. Arg. Bot.* **10**, 166–195.

Chambers, K. L. (1955). A biosystematic study of the annual species of *Microseris*. *Contr. Dudley Herb.* **4**, 207–312.

Chambers, K. L. (1963). Amphitropical species pairs in *Microseris* and *Agoseris* (Compositae: Cichorieae). *Q. Rev. Biol.* **38**, 124–140.

Cheeseman, T. F. (1925). "Manual of the New Zealand Flora" (2nd edn). W. A. G. Skinner, Wellington.

Curtis, W. M. (1963). "The Student's Flora of Tasmania" Vol. 2.G. Shea, Hobart.

Handel-Mazzetti, H., ed. (1936). Compositae. *In* "Symbolae Sinicae" Vol. 7 p. 1084–1186. Springer, Vienna.

Heusser, C. J. (1971). "Pollen and Spores of Chile." University of Arizona Press, Tucson.

Humboldt, A., A. Bonpland and C. S. Kunth (1820). "Nova Genera et Species Plantarum" Vol. 4. Staatsbibliothek Preussischer Kulturbesitz, Berlin.

Jeffrey, C. (1966). Notes on Compositae: I The Cichorieae in East Tropical Africa. *Kew Bull.* **18**, 427–486.

Koster, J. Th. (1976). The Compositae of New Guinea V. *Blumea* **23**, 163–175.

Lack, H. W. (1975). Die Gattung *Picris* L., *sensu lato*, im ostmediterran-westasiatischen Raum. Dissertation 116, University of Vienna.

Lack, H. W. (1977). *Picris sinuata* (Lam.) Lack (Asteraceae, Lactuceae), eine verkannte Art aus Nordafrika. *Willdenowia* **8**, 49–65.

Lack. H. W. (1979a). The genus *Picris* L. (Asteraceae, Lactuceae) in Tropical Africa. *Pl. Syst. Evol.* **131**, 35–52.

Lack, H. W. (1979b). The genus *Picris* L. (Asteraceae, Lactuceae) in Tropical Arabia. (In preparation).

Lack, H. W. (1979c). New species of *Picris* L. (Asteraceae, Lactuceae) from Australia. *Phylologia* **42**, 209–214.

Li, H.-L. (1978). Compositae. *In* "Flora of Taiwan" (H.-L. Li *et al.*, eds) Vol. 4, p. 768–965. Epoch Publishing, Taipeh.

Morton, J. K. (1972). Phytogeography of the West African mountains. *In* "Taxonomy Phytogeography and Evolution" (D. H. Valentine, ed.). p. 221–239. Academic Press, London and New York.

Raven, P. H. and D. I. Axelrod (1974). Angiosperm biogeography and past continental movements. *Ann. Mo. Bot. Gard.* **61**, 539–673.

Raven, P. H. and T. E. Raven (1976). The genus *Epilobium* (Onagraceae) in Australasia: a systematic and evolutionary study. *N.Z. Dep. Sci. Ind. Res. Bull.* **216**.

Roux, J. and L. Boulos (1973). Révision systématique du genre *Sonchus* L. s.1.II. Etude caryologique. *Bot. Notiser* **125**, 306–309.

Schultz–Bipontinus, C. H. (1845). Hypochoerideae. *Nov. Act. Acad. Caes. Leop.–Carol. Nat. Cur.* **21**, 87–172.

Schultz–Bipontinus, C. H. (1859). Revisio critica generis *Achyrophori. Jahresber. Pollichia* **16–17**, 45–73.

Stebbins, G. L. (1940a). Studies in the Cichorieae: *Dubyaea* and *Sorseris*, endemics of the Sino–Himalayan region. *Mem. Torrey Bot. Club* **19**.

Stebbins, G. L. (1940b). Additional evidence for a holarctic dispersal of flowering plants in the mesozoic era. Proc. 6th Pac. Sci. Cong. Vol.3, 649–660.

Stebbins, G. L. (1953). A new classification of the tribe Cichorieae, family Compositae. *Madroño* **12**, 65–81.

Stebbins, G. L. (1971). "Chromosomal Evolution in Higher Plants." Edward Arnold, London.

Tomb, A. S. (1978). Lactuceae—systematic review. *In* "The Biology and Chemistry of the Compositae" (V. H. Heywood, J. B. Harborne and B. L. Turner, eds) Vol. 2, p. 1067–1079. Academic Press, London and New York.

Vasil'ev, V. N. (1964). *Achyrophorus* Scop. *Flora SSSR* **29**, 201–204.

Zahn, K. H. (1921–1923). *Hieracium* in "Pflanzenreich" (A. Engler, ed.) Series IV, Vol. 280, Parts 75–77 (1921), 79 (1922), 82 (1923).

Comments on the Distribution and Evolution of the Genus *Phyllanthus* (Euphorbiaceae)

L. B. HOLM-NIELSEN

Botanical Institute, University of Aarhus, Denmark

This Linnean genus typified by *Phyllanthus niruri* L. was based on only six species, mainly of tropical distribution. Of these *P. niruri* and *P. urinaria* L. are weeds and *P. emblica* L. is a well known palaeotropic plant of economic importance. Today about 700 species of this genus are known. They exhibit pronounced variation in morphology, palynology, cytology and distribution patterns.

The genus belongs to the relatively unspecialized and primitive subfamily Phyllanthoideae of Euphorbiaceae, which, during its long history has caught the interest of many botanists both in the field and especially in the herbarium and laboratory. It is not possible to give a complete history of the investigation of these interesting plants. However, it is necessary to comment on a few of the milestones in the progression from six species described to the present day 700, placed in 10 subgenera and a number of sections.

In the first monograph of Euphorbiaceae Jussieu (1824a) listed the genera formerly described (*Cicca* L., *Phyllanthus* L., *Xyllophylla* L., *Agyneia* L., *Emblica* Gaertn., *Kirganelia* A. L. Juss, Epistylium Sw., and *Glochidion* Forst.) In addition he described *Gynoon*, *Anisonema* and *Leptonema*. Jussieu (1824b) drew attention to the narrow generic concept and pointed out that most of these genera could just as well be placed in *Phyllanthus*. Baillon (1858) maintained this narrow generic concept for the complex. Mueller Argovensis (1863, 1866) based his monumental work on a huge amount of precise observation. He lumped the previously recognized genera into a system of 44 sections. These were, however, highly artificial due to his pre-evolutionary approach.

Later authors, Bentham (1878), Bentham and Hoooker (1880),

Hooker (1887), and Pax (1890) modified Mueller's system. Hooker (1887) excluded *Glochidion* from the genus, which has been accepted by later workers. The group has been revised regionally by several authors: Africa, Hutchinson (1913, 1925) and Brunel (1975); New Caledonia, Moore (1921); Indo–China, Gagnepain and Beille (1927); Suriname, Lanjouw (1931, 1932); Argentina, Lourteig and O'Donell (1942) and Lourteig (1952); Peru, Macbride (1951); Cuba, Alain (1954); The West-Indies, Webster (1956–58); Madagascar, Leandri (1958); Java, Meuse and Adelbert (1962); and Thailand Airy-Shaw (1971); New Guinea, Webster and Airy-Shaw (1972).

Present Stage of Infrageneric Taxonomy

There is no recent world monograph, but several authors have dealt with the infrageneric taxonomy. The system of Pax and Hoffman (1931) was artificial, as pointed by Croizat (1943 a,b). Webster (1956–58) proposed a phylogentic system based mainly on the West-Indian representatives of the genus. Webster (1967 a,b), Webster and Airy-Shaw (1972) and Brunel (1975) expanded the original system of Webster (1956–58). Bancilhon (1971), did not make any major changes to Webster's system, but gave more evidence in support of this phylogenetic system.

Excluding *Glochidion* and other palaeotropic genera, *Breynia*, *Sauropus*, and *Synostemon*, the genus *Phyllanthus* consists of 10 subgenera, most of which have been treated as genera by previous authors. There is, however, no evidence for splitting the genus into several genera because at the subgeneric level the boundaries between the groups are somewhat undefined. Some of the infrageneric groups are probably not of monophyletic origin, reticulate evolution may have played a role, as the evolutionary tendencies recognized today are more like a network than straight lines.

Subgenus *Isocladus* Webst.

Herbs or shrubs, branching unspecialized, androecium trimerous, phyllotaxy spiral (sect. *Isocladus*) or distichous (sect. *Loxopodium* Webst.). The sections *Anisolobium* Muell. Arg. and *Macraea* (Wight.) Baillon also distichous differ from *Loxopodium* in their pollen morphology. This subgenus is pantropical (Fig. 1) and is, together with subgenus *Botryanthus* Webst., the only one with unspecialized branching.

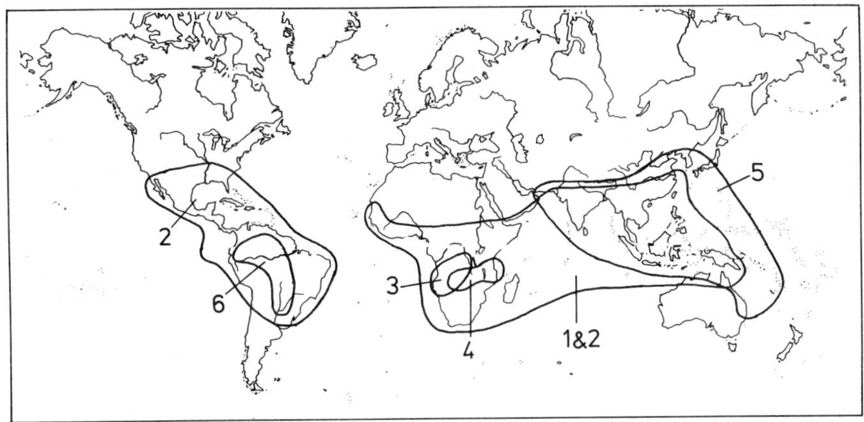

Fig. 1. Subgenus *Isocladus:* 1 = *Isocladus*, 2 = *Loxopodium*, 3 = *Pseudomenarda*, 4 = *Anisolobium*, 5 = *Macraea*, 6 = *Salviniopsis*.

Subgenus *Kirganelia* (Juss.) Webst.

Herbs, shrubs or trees, branching phyllanthoid (Webster, 1956–58, p. 104), androecium pentamerous. The sections *Pentandra* Webst. and *Floribundi* Pax and Hoffm. are herbaceous. Section *Kirganelia* is woody, and is morphologically rather similar to the excluded genus *Breynia*. Subgenus *Kirganelia* is a palaeotropic group (Fig. 2).

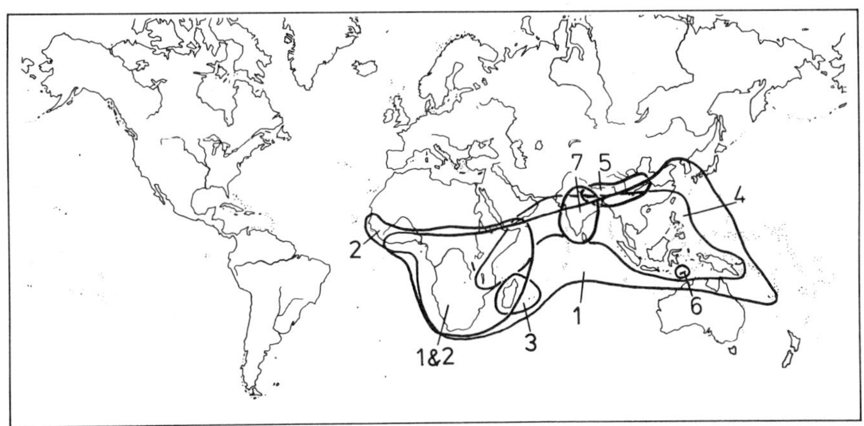

Fig. 2. Subgenus *Kirganelia*: 1 = *Kirganelia*, 2 = *Floribundi*, 3 = *Menarda*, 4 = *Chorizandra*, 5 = *Flueggeopsis*, 6 = *Neoscepasma*, 7 = *Peltrandra*.

Subgenus *Gomphidium* (Baill.) Webst.

Trees or shrubs, branching phyllanthoid, disc reduced, style entire. This subgenus shows remarkable evolutionary tendencies, the reduction of the disc, which is present in all other subgenera and the development of the gynoecium shows the relationship to the genus *Glochidion*. *Gomphidium* is Austral–Asian with relatives on Madagascar and Central America (Fig. 3).

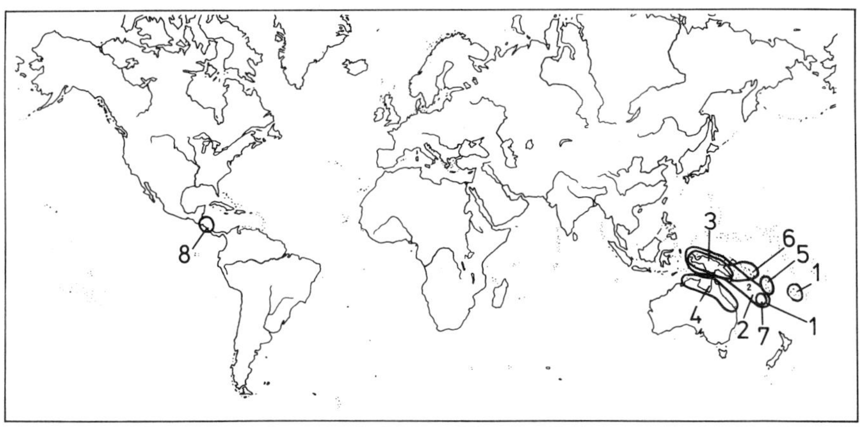

Fig. 3. Subgenus *Gomphidium*: 1 = *Gomphidium*, 2 = *Adenoglochidion*, 3 = *Tetraglochidion*, 4 = *Synostemon*, 5 = *Schleroglochidion*, 6 = *Nymania*, 7 = *Physoglochidion* A.O., 8 = *Calodictyon*.

Subgenus *Cicca* (L.) Webst.

Trees or shrubs, branching phyllanthoid, androecium normally tri- or tetramerous, fruit indehiscent. Section *Ciccopsis* Webst. and the cauliflorous sect. *Cicca* are monoecious with more or less drupaceous fruits. Section *Aporosella* (Chodat) Webst., which is also cauliferous, has a woody indehiscent fruit and no disc. Subgenus *Cicca* has exclusively neotropical distribution (Fig. 4).

Subgenus *Emblica* (Gaertn.) Webst.

Trees, branching phyllanthoid, androecium trimerous, fruits capsular with fleshy exocarps. The two sections *Emblica* and *Microglochidion* Muell. Arg. are connected by their pollen morphology. Section *Emblica* is of Indo–Malaysian distribution, sect. *Microglochidion* neotropical (Fig. 5).

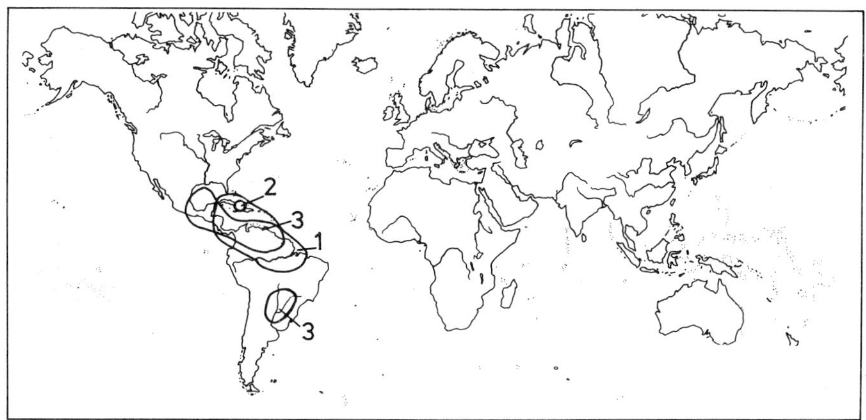

Fig. 4. Subgenus *Cicca*: 1 = *Cicca*, 2 = *Ciccopsis*, 3 = *Aporosella*.

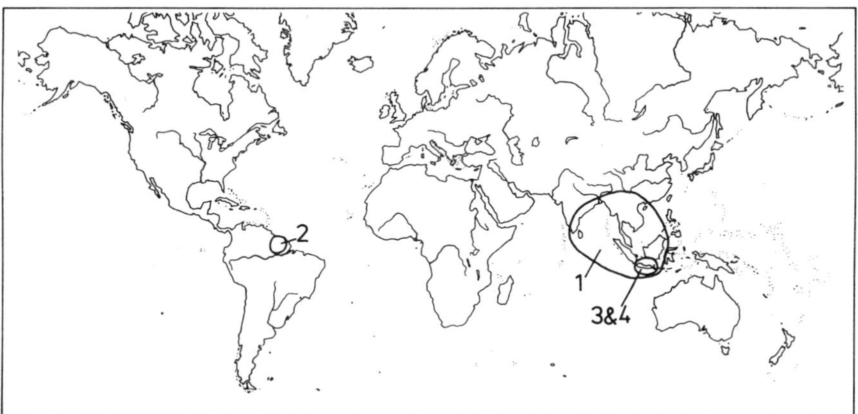

Fig. 5. Subgenus *Emblica*: 1 = *Emblica*, 2 = *Microglochidion*, 3 = *Hedycarpidium*, 4 = *Ceramanthus*.

Subgenus *Phyllanthus*

Herbs, branching phyllanthoid, androecium di- or trimerous. Section *Callictrichoides* Webst. and sect. Cyclanthera Webst. with their colum- nar androecium also differ from sect. *Phyllanthus* in pollen morphology and from sect. *Urinaria* (L.) Webst. in seed and stipular morphology. This subgenus is pantropical (Fig. 6).

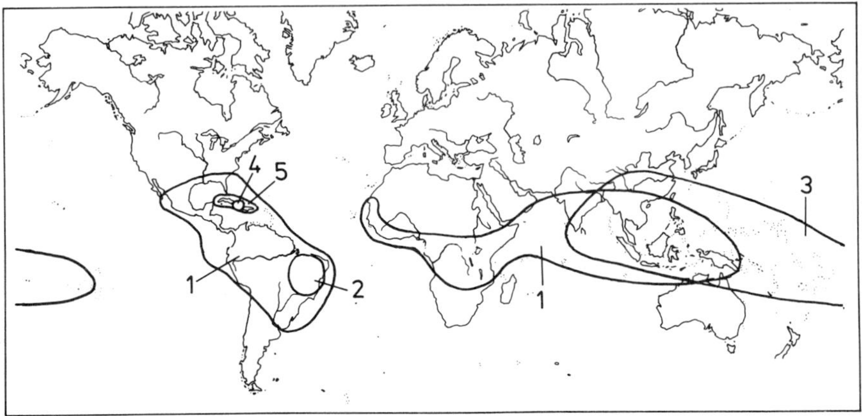

Fig. 6. Subgenus *Phyllanthus*: 1 = *Phyllanthus*, 2 = *Choretropsis*, 3 = *Urinaria*, 4 = *Callitrichoides*, 5 = *Cyclanthera*.

Subgenus *Eriococcus* (Hassk.) Croiz. and Metc.

Shrubs, branching phyllanthoid, tepals often lacerate, androecium di- or trimerous. Subgenus *Eriococcus* has an Indo–Malaysian–New Guinean distribution (Fig. 7).

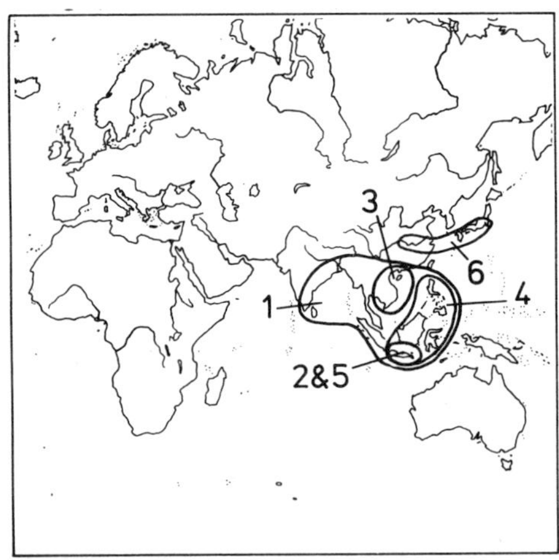

Fig. 7. Subgenus *Eriococcus*: 1 = *Eriococcus*, 2 = *Eriococcodes*, 3 = *Nymphanthus*, 4 = *Scepasma*, 5 = *Emblicastrum*, 6 = *Hemicicca*.

Subgenus *Conami* (Aubl.) Webst.

Shrubs or trees, branching phyllanthoid, androecium tri- or pentamerous. Differs palynologically from other subgenera. Sections *Apolepis* Webst. (trimerous) and *Brazzaeni* Brunel (pentamerous) have free stamens. Section *Conami* has filaments united into a central column. This subgenus is neotropical and West-African (Fig. 8).

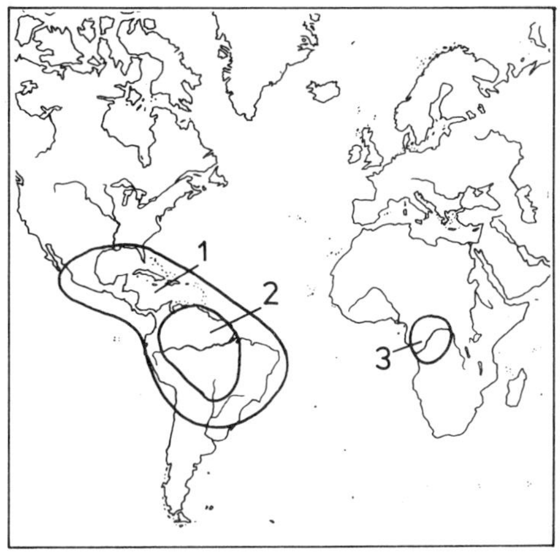

Fig. 8. Subgenus *Conami*: 1 = *Conami*, 2 = *Apolepis*, 3 = *Brazzaeani*.

Subgenus *Botryanthus* Webst.

Trees or shrubs, branching unspecialized, androecium trimerous, filaments connate. Palynologically related to subgenus *Xyllophylla*. This subgenus is exclusively neotropical (Fig. 9).

Subgenus *Xyllophylla* (L.) Pers.

Trees or shrubs, branching phyllanthoid, androecium of several differentiated types. This subgenus has a distinct pollen morphology and exhibits several evolutionary lines. Section *Xyllophylla* has phylloclades. This subgenus is exclusively neotropical (Fig. 10).

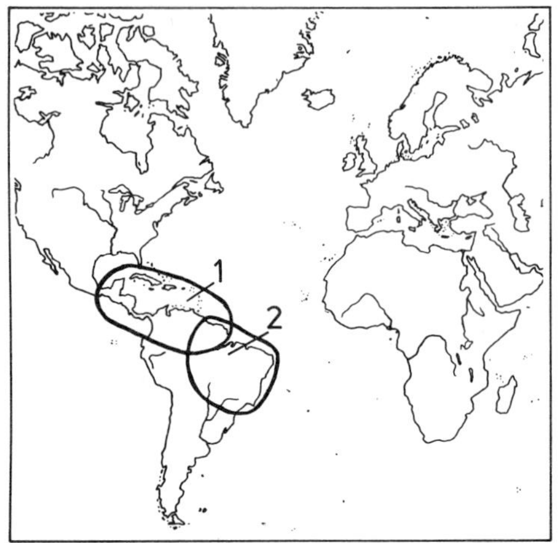

Fig. 9. Subgenus *Botryanthus*: 1 = *Botryanthus*, 2 = *Diplocicca*.

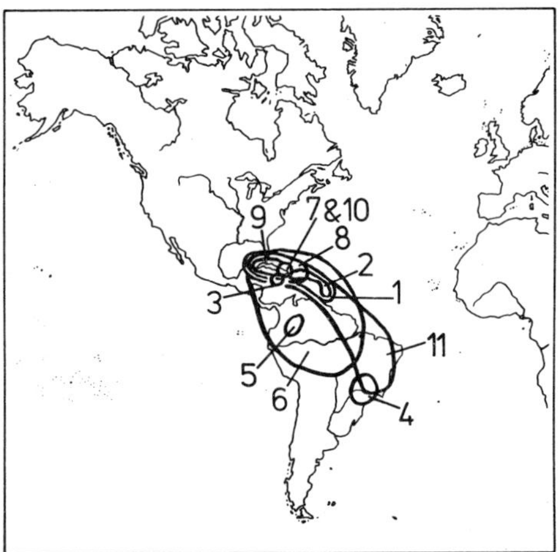

Fig. 10. Subgenus *Xyllophylla*: 1 = *Xyllophylla*, 2 = *Hemiphyllanthus*, 3 = *Epistylium*, 4 = *Ciccastrum*, 5 = *Oxalistylis*, 6 = *Asterandra*, 7 = *Glyptothamnus*, 8 = *Omphacodes*, 9 = *Williamia*, 10 = *Thamnocharis*, 11 = *Orbicularia*.

Review of Experimental Results

In addition to the morphological investigations during recent decades there have been much embryological, cytological, and palynological data produced which have helped to show the course of evolution and have also indicated the large amount of variation within the genus. A review and a general discussion is found in Bancilhon (1971).

The embryology was studied by Maheshwari and Chowdry (1937) and Banerji and Dutt (1945). They compared the development of the female gametophyte with other Euphorbiaceae. Different developmental types were demonstrated, but only a few species were studied. Webster and Rupert (1972) demonstrated correlation between the nuclear number of the male gametophyte and the development of ovules in Euphorbiaceae. These embryological data, however, are too sporadic to be of help in reconstructing the subgeneric evolutionary lines.

The cytology has been examined of species from most of the subgenera and sections. However, the cytology of *Phyllanthus* is still far from well known. A huge variety of chromosome numbers has been reported. In *Phyllanthus* basic numbers between 6 and 15 have been proposed and polyploidy and aneuploidy have been demonstrated. The number 13 was proposed by Webster and Ellis, 1962. This number is well known within the Phyllantheae. Further results will undoubtedly help to clarify the situation. A number of theories concerning chromosomal evolution have been proposed, for example Bancilhon (1971) suggested that apomixis occurs in some taxa.

The palynology together with the revaluation of the branching pattern and other data were used by Webster (1956–58) for his phylogenetic taxonomy of the genus. Punt (1962) in his treatment of the pollen morphology of the Euphorbiaceae described a number of pollen types and concluded that pollen morphology is useful in the natural classification of Phyllanthoideae. Köhler (1967) demonstrated and discussed the possible correlation between polyploidy and the evolution of pollen types. These results begin to indicate the possible evolutionary tendencies of the group.

Anatomy and biochemistry have scarcely been used so far in the investigation of this genus, but they will surely provide us with more information about its evolution.

The Patterns of Distribution

Some difficulties must be taken into consideration for the interpretation of distribution maps of the infrageneric taxa. The intensity of collections is markedly unequal, the taxonomy of some groups is still uncertain, the species concept is not uniform throughout the genus, perhaps too many monotypic sections have been described, and finally some of the distribution areas have a complicated phytogeography.

The distributional patterns of the taxa studied vary as much as the morphology. Three major distribution types are important in this study:

(i) Narrow endemic species—species known from a single stand or part of an island for example *P. pseudocicca* Griseb. (subgenus *Cicca*), which is endemic to the eastern mountains of Cuba, and *P. acacioides* Urb. (subgenus *Xyllophylla*) endemic to Tobago.

(ii) Disjunct species—*P. chacoensis* Morong, endemic to the Chaco, is probably conspecific with *P. elsiae* Urb. from Central and northern South America. These are the only two species of sect. *Aporosella* (subgenus *Cicca*). Likewise a number of island species show remarkable disjunctions. For example *P. bourgeoisii* (Baill.) Muell. Arg. bridges the gap between New Caledonia and New Guinea.

(iii) Widespread species—*P. reticulatus* Poir. (subgenus *Kirganelia*) which is found throughout the palaeotropics, and has been introduced to the New World. *P. carolinensis* Walt. (subgenus *Isocladus*) extends from temperate North to temperate South America. No pantropical natives are known, but a number of pantropical weeds are found, for example the Old World *P. urinaria* L. (subgenus *Phyllanthus*).

A similar variety of distributional patterns occurs in the sections. A number of sections have evolved as the result of island isolation, for example, the section of subgenus *Comphidium* in the New Caledonian–New Guinea area. Some of these sections are narrow possibly monotypic endemics while others have become more widespread. The sect. *Orbicularia* (Baill.) Griseb., subgenus *Xyllophylla*, is a complex of Cuban plants, of which only *P. nummularioides* Muell. Arg. is found in a mountainous area in central Hispaniola. The group is endemic to these two islands which possibly were connected in the past.

An example of a remarkable disjunction occurs in the Cuban sect. *Williamia* (Baill.), Muell. Arg., subgenus *Xyllophylla* which is closely related to the monotypic sect. *Oxalistylis* (*P. salviaefolius* H.B.K.) of the Ecuadorean and Colombian Andes. A similar disjunction occurs in the monotypic sect. *Asterandra* (Kl.) Muell. Arg., subgenus *Xyllophylla* where *P. juglandifolius* Willd. spp. *juglandifolius* is found on the Greater Antilles and spp. *cornifolius* (H.B.K.) Webst. in northern South America and Trinidad. In continental areas the sections tend to have a larger number of species and to be more widespread, for example sect. *Phyllanthus*.

The distribution of sections does not appear unusual when considered alone (for example see Figs 4 and 7 for sections of subgenera *Cicca* and *Eriococeus*). However, when they are considered together some have rather usual and expected distribution patterns, but others have more unusual distributions, for example the widespread South-American plants of sect. *Conami* and *Apolepis* Webst. are related to sect. *Brazzaeni* Brunel an exclusively West-African group (see Fig. 8 for subgenus *Conami*). The peculiar distribution of subgenus *Comphidium* (see Fig. 3) is noteworthy because the Central-American sect. *Calodictyon* Webst. is the only representative outside the New Guinean–New Caledonean region. Further phylogenetic evidence is needed for a full understanding of some of the apparently illogical distributional patterns.

Centres of Diversity

The Euphorbiaceae is generally considered to be a young group of plants which evolved after the large southern continent was subdivided in the plates known today Raven (p. 3); Hawkes and Smith (1965), Raven and Axelrod (1972), Raven (1972, 1975).

The dispersal of the taxa of the genus *Phyllanthus* following the evolution from a common ancester would indicate that transoceamic long distance dispersal has been effective assuming that the group is as recent as indicated. The group must also have used intercontinental islands as a stepping stone system which also gave rise to secondary evolution. To the subfamily Phyllanthoideae belongs, however, a number of relatively unspecialized genera. These undoubtedly evolved early in the history of Euphorbiaceae and the lesser distance between the continental plates during earlier geological periods could have been of importance to the early distribution of these taxa. During time the separation of the populations would have been gradually enlarged by still larger barriers of deep and epicontinental seas due to the changes in the position of the continental plates. In addition another important factor was their distribution through archipelago systems which was effective in the separation of populations. The resulting different selectionary stress has stimulated the evolution of the known spectrum of Phyllanthoide types.

Figure 11 shows the number of sections per area. Considering the sections are those of the Webster system which is phyllogenetic, the concentration of sectional representation to some areas or centres of diversity, is remarkable. The less specialized groups in subgenus *Kirganelia* and the *Gomphidium* and *Glochidion* complexes are eastern palaeotropic, which could indicate Asian–Malesian origin of Phyllathoide plants followed by secondary evolution in other areas. The picture of possible subgeneric relationships based on systematic and phytogeographic criteria as known today may be the result of a reticu-

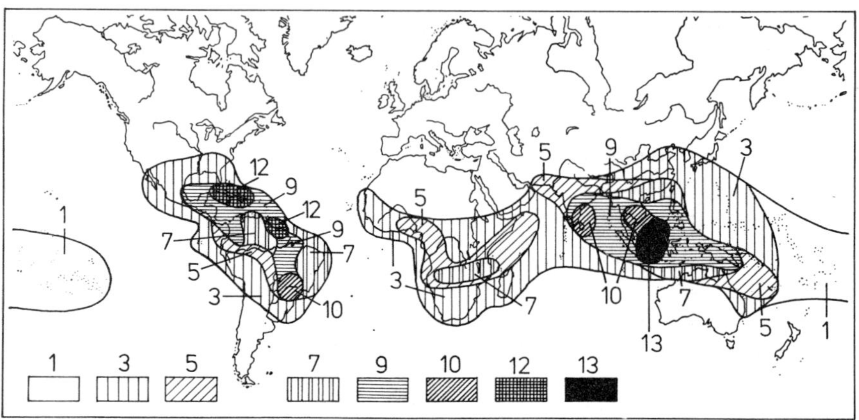

Fig. 11. Evolutionary centres.

late evolutionary system. If the infrageneric taxa are natural, differentiation has occurred in the neo- and paleotropics, however, several possible transoceanic relationships are also demonstrated.

Acknowledgements

This study has partly been carried out as a background study for my work with Ecuadorean plants. I wish to thank Benjamin Øllgaard for numerous valuable discussions. For valuable suggestions and criticism of the manuscript I am indebted to Dr G. T. Prance and H. Balslev.

References

Airy-Shaw, H. K. (1972). The Euphorbiaceae of Siam. *Kew Bull.* **26**, 191–363.

Alain, H. (1954). "Flora de Cuba" Vol. 3. Cont. Occ. Mus. Hist. Nat. Colegio de la Salle, 13, Habana.

Baillon, H. E. (1858). "Etude Générale du Groupe des Euphorbiacées." Masson, Paris.

Bancilhon, L. (1971). Contribution à l'étude taxonomique de genre *Phyllanthus* (Euphorbiacées). *Boissiera* **18**, 9–81.

Banerji, I. and M. K. Dutt (1945). The development of the female gametophyte in some members of the Euphorbiaceae. *Proc. Ind. Acad. Sci.* **XXB**, 51–60.

Bentham, G. (1878). Notes on Euphorbiaceae. *J. Linn. Soc. Bot.* **17**, 185–267.

Bentham, G. and J. D. Hooker (1880). "Genera Plantarum" p. 239–375. Reeve, London.

Brunel, J. F. (1975). Contribution a l'etude de quelques *Phyllanthus africans* et à la taxonomie de genre *Phyllanthus* L. (Euphorbiaceae). Thesis, Strasbourg.

Croizat, L. (1943a). Notes on Polynesian *Glochidion* and *Phyllanthus. Occ. Pap. Bishop-Museum* **XVII**, 207–214.

Croizat, L. (1943b). Notes on American Euphorbiaceae. *J. Wash. Acad. Sci.* **33**, 11–20.

Gagnepain, F. and L. Beille (1927). *In* "Flora Génerale de l'Indo-chine V" p. 230–673. Didot, Paris.

Hawkes, J. G. and P. Smith (1965). Continental drift and the age of angiosperm genera. *Nature, Lond.* **207**, 40–50.

Hooker, J. D. (1887). *In* "Flora of British India V" p. 239–477. Reeve, London.

Hutchinson, J. (1913). *In* "Flora of Tropical Africa VI, 1" p. 692–736. W. T. Thieselton Dyer, London.

Hutchinson, J. (1925). *In* "Flora Capensis V" p. 386–401. W. T. Thieselton Dyer, London.

Jussieu, A. de (1824a). "De Euphorbiacearum Generilens Medisque Earundium Viribus Tentamen." Didot, Paris.

Jussieu, A. de (1824b). Euphorbiaceae. *Ann. Sci. Nat.* **1**, 146–167.

Köhler, E. (1967). Über Beziehungen zwischen Pollenmorphologie und Polyploidiestufen im Verwandtschaftsbereich der Gattung Phyllanthus (Euphorbiaceae). *Fed. Rep.* **74**, 159–165.

Lanjouw, J. (1931). "The Euphorbiaceae of Surinam." Amsterdam.

Lanjouw, J. (1932). *In* "Flora of Surinam II" p. 13–24. Amsterdam.

Leandri, J. (1958). *In* "Flora du Madegascar et des Comores I" p. 21–104. Didot, Paris.

Lourteig. A. (1952). Euphorbiaceae Argentinae. Adenda II. *Ark. Bot.* **3**, 71–74.

Lourteig, A. and C. A. O'Donnell (1942). Euphorbiaceae Argentinae and Adenda I. *Lilloa* **9**, 77–177.

Macbride, J. F. (1951). *In* "Flora of Peru" Vol. 1, p. 34–47. Bot. Ser. Field Mus. Nat. Hist. XIII, III A.

Maheshwari, P. and O. R. Chowdry (1937). A note on the development of the embryo sac in *Phyllanthus miruri* L. *Curr. Sci.* **5**, 535–536.

Meuse, A. D. J. and A. G. L. Adelbert (1962). *In* "Flora of Java" p. 441–505. Groningen.

Moore, S. (1921). Plants from New Caledonia. *J. Linn. Soc.* **XLV**, 393–409.

Meuller Argovensis, J. (1863). Euphorbiaceae. Vorläufige Mitteilungen aus dem für De Candolles Prodromus bestimmten Manuscript über diese Familie. *Linnaea* **XXXII**, 1–126.

Mueller Argovensis, J. (1866). *In* "De Candolle Prodromus" p. 274–436. Masson, Paris.

Pax, F. (1890). *In* "Engler Nat. Pflanzenfamilien 3" p. 1–119. Engelmann, Leipzig.

Pax, F. and K. Hoffmann (1931). *In* "Engler Nat. Pflanzenfamilien 19 c". p. 11–233. Engelmann, Leipzig.

Punt, W. (1962). Pollen morphology of the Euphorbiaceae with special reference to taxonomy. *Ventia* **7**, 1–110.

Raven, P. (1972). Plant species disjunctions: A summary. *Ann. Mo. Bot. Gard.* **59**, 234–246.

Raven, P. (1975). Summary of the Biogeography Synposium. *Ann. Mo. Bot. Gard.* **62**, 380–385.

Raven, P. and D. I. Axelrod (1972). Plate tectonics and Australasian palaeobiogeography. *Science* **186**, 1379–1386.

Webster, G. L. (1956–58). A monographic study of the West Indian species of *Phyllanthus*. *J. Arnold Arbor.* **37**, 91–122, 217–268, 340–359; **38**, 51–80, 170–198, 295–373; **39**, 49–100, 111–212.

Webster, G. L. (1967a). The genera of Euphorbiaceae in the Southeastern United States. *J. Arnold Abor.* **48**, 303–430.

Webster, G. L. (1967b). A remarkable new *Phyllanthus* (Euphorbiaceae) from Central America. *Ann. Mo. Bot. Gard.* **54**, 194–198.

Webster, G. L. and J. R. Ellis (1962). Cytotaxonomic studies in the Euphorbiaceae subtribe Phyllanthinae. *Am. J. Bot.* **49**, 14–18.

Webster, G. L. and H. K. Airy-Shaw (1972). A provisional synopsis of the New Guinea taxa of *Phyllanthus* (Euphorbiaceae). *Kew Bull.* **26**, 85–109.

Webster, G. L. and E. A. Rupert (1974). Phylogenetic significance of pollen nuclear number in the Euphorbiaceae. *Evolution* **27**, 524–531.

The Distribution of Araceae

T. B. CROAT

Missouri Botanical Garden, Saint Louis, USA

The Araceae, a family of 110 genera and approximately 2500 species, is worldwide in distribution but has most species in tropical areas. There are 46 genera in the New World and 75 genera in the Old World (Tables 1, 2). Only 10 genera are found in both the New and the Old World. North America has 24 genera and South America has 38. The West Indies is poor in both numbers of genera and species with only nine genera (assuming *Caladium* is not native) which have rather few species. Asia (in the broad sense here to include Indonesia and the Philippines) has 50 genera, Africa 31, Europe nine and Australia eight genera. There are 11 genera with 20 species restricted to the Asian islands.

Centres of distribution include both Asia and America with 34 genera restricted to Asia and 35 genera restricted to America. Indonesia is again considered as part of Asia. In all there are approximately 600 species in the genera restricted to Asia (in the broad sense) and approximately 1400 species in the genera restricted to tropical America. About 1350 species, 55% of the total for the family, occur in the American tropics and subtropics. There are 16 genera with approximately 100 species restricted to tropical Africa and Madagascar. Genera of Araceae from Africa, Madagascar and the Seychelles Islands are as follows: *Ambrosina, Amorphophallus, *Anchomanes, *Anubias, Arisaema, Arisarum, *Arophyton, Arum, Biarum, *Callopsis, *Carlephyton, *Cercestis, *Colletogyne, *Culcasia, Cyrtosperma, Dracunculus, Eminium, *Gonatopus, *Nephthytis, Pistia, Pothos, *Protarum, *Pseudohydrosme, Remusatia, Rhaphidophora, *Rhektophyllum, Sauromatum, *Stylochiton, Typhonium* (naturalized), *Typhonodorum, *Zamioculcas* and *Zantedeschia.*

*Endemic.

Table 1. Araceae in major geographical areas (no. genera—no. species). Each genus is included in only one of the geographic categories.

	North Temperate	Tropical America	South America	Subtropical and temperate South America, Andean Regions	Old World	Tropical Africa and Madagascar	Asia	Miscellaneous
Pothoideae								
Pothoeae					1–50		2–10	Borneo 1–1
Heteropsideae		1–12						
Anthurieae		1–700						
Culcasieae						1–20		
Zamioculcaseae						2–6		
Acoreae	1–2							Australia 1–1
Monsteroideae								
Monstereae		3–57			1–60		3–48[4]	
Spathiphylleae		1–45[2]						New Guinea 1–3
Calloideae								
Symplocarpeae	3–4							
Calleae	1–1							
Lasiodeae								
Lasieae		2–33	2–3		1–19[1]		2–5	Borneo 1–1 India 1–2
Pythonieae					1–100	2–12	2–8	India 1–1
Nephthytideae						3–15		
Montrichardieae		1–2						

	North Temperate	Tropical America	South America	Subtropical and temperate South America, Andean Regions	Old World	Tropical Africa and Madagascar	Asia	Miscellaneous
Philodendroideae		1–350					2–240[1]	Borneo and western Malaysia 1–10 Borneo 5–10 New Guinea 1–1
Anubiadeae						1–14		
Aglaonemeae				-			1–21	Sumatra and Borneo 1–1
Dieffenbachieae		1–30						
Zantedeschieae						1–6		
Typhonodoreae						1–1		
Peltandreae	1–4							
Colocasioideae								
Colocasieae		1–45	2–19	2–2	1–2		6–95[3,4]	Venezuela 1–1
Syngonieae		2–28						
Ariopsideae							1–1	
Aroideae								
Stylochitoneae						1–21		
Arophyteae						3–11		
Asterostigmateae				8–23				
Protareae								Seychelles 1–1

	North Temperate	Tropical America	South America	Subtropical and temperate South America, Andean Regions	Old World	Tropical Africa and Madagascar	Asia	Miscellaneous
Callopsideae						1–2		
Zomicarpeae			3–3	2–4				
Areae					1–6		4–76	Worldwide 1–150 Europe 1–15 Mediterranean 5–12 Mediterranean to Asia 1–5 Asia and Australia 1–25
Pistoideae								Worldwide 1–1
Total	6–11	14–1302	7–25	12–29	6–237	16–108	23–504	26–241

[1] A few species in tropical America.
[2] A few species in Asia.
[3] A few species in Africa.
[4] Also partly in Australia.

Table 2. Distribution of Araceae.

Subfamily–tribe	Genus	Number of species	Distribution
Pothoideae			
Pothoeae	*Pothos* L.	50	India and Ceylon, Himalayas, South-East Asia, Malaysia to South China, Australia (1 sp. in Madagascar and Comoro Islands)
	Pothoidium Schott	1	Philippines, Java, Celebes, Molucca
	Pedicellarum M. Hotta	1	Borneo
	Anadendrum Schott	9	South-East Asia, Malaysia
Heteropsideae	*Heteropsis* Kunth	12	American tropics, mostly South America
Anthurieae	*Anthurium* Schott	700	American tropics (Tamaulipas, Mexico to Argentina and Paraguay, West Indies)
Culcasieae	*Culcasia* Beauv.	20	Tropical Africa
Zamioculcaseae	*Zamioculcas* Schott	1	Tropical East Africa
	Gonatopus Hook. f.	5	Tropical East Africa and South Africa (2 spp. in South Africa, eastern Transvaal and northern Natal)
Acoreae	*Acorus* L.	2	North temperate and subtropical (America, Europe and Asia)
	Gymnostachys R. Br.	1	Eastern Australia
Monsteroideae			
Monstereae	*Rhaphidophora* Hassk.	60	West Africa, India and Ceylon, Nepal, South-East Asia
	Epipremnum Schott	15	Burma and South-East Asia to Malaysia, Formosa, Japan, the Philippines and Marshall Islands; Australia

Subfamily–tribe	Genus	Number of species	Distribution
	Amydrium Schott	8	Malaysia
	Scindapsus Schott	25	India and Sikkim, South-East Asia, South-East China and Malaysia to Solomon Islands; 1 sp. in Brazil
	Stenospermation Schott	20	American tropics, Nicaragua to Bolivia
	Rhodospatha P. & E.	15	Mexico to Brazil
	Monstera Adans.	22	Mexico to Brazil; Lesser Antilles
Spathiphylleae	*Spathiphyllum* Schott	45	Mostly American tropics; a few in Asia and Pacific Islands
	Homochlamys Engler	3	New Guinea
Calloideae			
Symplocarpeae	*Lysichiton* Schott	2	Pacific North America, East Siberia, Japan
	Symplocarpus Nuttall	1	North-East Asia, Japan, Atlantic North America
	Orontium L.	1	Atlantic North America
Calleae	*Calla* L.	1	North temperate and subarctic
Lasioideae			
Lasieae	*Cyrtosperma* Griff.	19	Malaysia to Solomon Islands and Fiji Islands; 1 sp. in tropical West Africa; 2 spp. in American tropics
	Lasia Lour.	3	India and Ceylon, South-East Asia, Malaysia
	Anaphyllum Schott	2	Southern India
	Podolasia N. E. Br.	1	Borneo
	Urospatha Schott	20	American tropics, mostly South America
	Dracontioides Engl.	1	Brazil (Bahia coast)
	Echidnium Schott	2	Tropical South America (Brazil, Guianas)

Subfamily–tribe	Genus	Number of species	Distribution
	Dracontium L.	13	Tropical America
	Pycnospatha Gagnep.	2	Thailand, Laos
Pythonieae	*Pseudohydrosme* Engler	2	West Africa
(Amorphophalleae)	*Anchomanes* Schott	10	Tropical Africa
	Plesmonium Schott	1	Northern India
	Thomsonia Wall.	1	Himalayas, Assam, Thailand
	Pseudodracontium N. E. Br.	7	Thailand, Indochina
	Amorphophallus Dcne.	100	Tropical Africa and Asia (25 spp. in Africa); Australia
Nephthytideae	*Nephthytis* Schott	5	Tropical Africa
	Cercestis Schott	9	West Africa (Guinea to Gabon)
	Rhektophyllum N. E. Br.	1	Tropical West Africa
Montrichardieae	*Montrichardia* Crueg.	2	Tropical America
Philodendroideae			
Philodendreae	*Homalomena* Schott	140	Mostly tropical Asia (perhaps 10 spp. in South America)
	Diandriella Engler	1	New Guinea
	Schismatoglottis Zoll. & Mor.	100	Mostly Malaysia (a few species in South-East Asia and Burma; 3 spp. in northern tropical South America)
	Bucephalandra Schott	2	Borneo
	Phymatarum M. Hotta	1	Northern Borneo
	Aridarum Ridley	5	Northern Borneo
	Heteroaridarum M. Hotta	1	Borneo
	Hottarum Bogner & Nicolson	1	Borneo
	Piptospatha N. E. Br.	10	Borneo, western Malaysia

Subfamily—tribe	Genus	Number of species	Distribution
	Philodendron Schott	350	American tropics
Anubiadeae	*Anubias* Schott	14	West Africa
Aglaonemeae	*Aglaonema* Schott	21	Burma, South-East Asia, Malaysia
	Aglaodorum Schott	1	Sumatra, Borneo
Dieffenbachieae	*Dieffenbachia* Schott	30	American tropics
Zantedeschieae	*Zantedeschia* Spreng.	6	South Africa
Typhonodoreae	*Typhonodorum* Schott	1	Tropical East Africa, Mascarenes, Madagascar, Zanzibar, Pemba
Peltandreae	*Peltandra* Raf.	4	Southern United States, Atlantic North America
Colocasioideae			
Colocasieae	*Steudnera* C. Koch	8	Northern India, South-East Asia, southern China, Malay Peninsula
	Remusatia Schott	2	1 sp. Asia–Himalaya to Formosa; 1 sp. tropical Asia, tropical Africa, Australia
	Gonatanthus Kl.	2	Himalayas to Thailand and South-West China
	Hapaline Schott	5	Himalayas, South-East Asia, Malaysia
	Caladiopsis Engler	4	Northern tropical South America
	Caladium Vent.	15	Tropical South America
	Aphyllarum S. Moore	1	Central Brazil (Mato Grosso)
	Chlorospatha Engler	1	Colombia (Andean)
	Xanthosoma Schott	45	American tropics
	Colocasia Schott	8	Western India, Cochinchina, Polynesia
	Alocasia Neck.	70	India and Ceylon, South-East Asia, Malaysia, southern China and Japan to Caroline Islands; Australia
	Xenophya Schott	2	New Guinea, Thailand, eastern Malaysia
	Jasarum Bunting	1	Venezuela

Subfamily–tribe	Genus	Number of species	Distribution
Syngonieae	*Porphyrospatha* Engler	3	Central America (probably not distinct from *Syngonium*)
	Syngonium Schott	25	Tropical America, especially Central America
Ariopsideae	*Ariopsis* J. Graham	1	Eastern and northern India, western Malaysia
Aroideae			
Stylochitoneae	*Stylochiton* Lepr.	21	Central and South-East Africa (tropical)
Arophyteae	*Carlephyton* Jumelle	3	Madagascar
	Colletogyne S. Buchet	1	Madagascar
	Arophyton Jumelle	7	Madagascar
Asterostigmateae	*Mangonia* Schott	2	Tropical southern Brazil (Rio Grande do Sul); Uruguay
	Taccarum Schott	4	Southern Brazil, northern Argentina, possibly Paraguay
	Asterostigma Fisch. & Meyer	5	Southern Brazil (Minas Gerais and Bahia) south to northern Argentina
	Gorgonidium Schott	1	Bolivia
	Synandrospadix Engler	1	Northern Argentina, Bolivia
	Gearum N. E. Br.	1	Southern Brazil (Goias)
	Spathantheum Schott	2	Andean Bolivia, northern Argentina
	Spathicarpa Hook.	7	Southern Brazil, Paraguay, northern Argentina
Protareae	*Protarum* Engler	1	Seychelles Islands
Callopsidae	*Callopsis* Engler	2	East Africa (Kenya and Zanzibar)
Zomicarpeae	*Scaphispatha* Schott	1	Southern Peru and Brazil
	Zomicarpa Schott	3	Subtropical and warm temperate Brazil (Atlantic slope)

Subfamily–tribe	Genus	Number of species	Distribution
Areae	*Zomicarpella* N. E. Br.	1	Northern South America (Colombia)
	Filarum Nicolson	1	Peru (Loreto)
	Ulearum Engler	1	South America (Upper Amazon Valley)
	Arum L.	15	Europe and Mediterranean region (2 spp. in North Africa)
	Dracunculus Mill.	2	Mediterranean region (including North Africa)
	Helicodiceros Schott	1	Balearic Islands, Sardinia, Corsica
	Theriophonum Blume	7	Peninsular India, Ceylon
	Typhonium Schott	25	India and Ceylon, Himalayas, South-East Asia, China, Japan, Malaysia; Australia (1 sp. naturalized in Comoros, Ghana, Venezuela, Brazil)
	Sauromatum (Blume) Schott	6	India and Himalayas to western Malaysia (1 sp in tropical Africa)
	Eminium Schott	5	Eastern Mediterranean to central Asia
	Biarum Schott	5	Mediterranean region (3 spp. in North Africa)
	Arisarum Targ. Tozz.	3	Mediterranean region (2 spp. in North Africa)
	Arisaema Mart.	150	Tropical Asia, Atlantic North America to Mexico, East Africa (4 spp.)
	Pinellia Tenore	7	China, Japan
	Ambrosinia Bassi	1	Italy, Algeria
	Lagenandra Dalz.	12	Western India (Western Ghats, Assam, Ceylon)
	Cryptocoryne Wydl.	50	India and Ceylon, South-East Asia, Malaysia
Pistoideae	*Pistia* L.	1	Tropics and subtropics

Important local centres of diversity in the Araceae include subtropical and warm temperate South America with 11 endemic genera (Table 3), and Malaysia with 13 endemic genera. Genera of Araceae

Table 3. Genera of Araceae endemic to South America

Geographically and ecologically widespread	*Caladium*
Andean regions	*Caladiopsis, Chlorospatha, Zomicarpella*
Amazonia and Guiana Highlands	*Echidnium, Filarum, Ulearum, Jasarum*
Brazilian Atlantic coast (tropical)	*Dracontioides*
Subtropical and warm temperate	*Aphyllarum, Mangonia, Taccarum, Asterostigma, Gearum, Synandrospadix, Spathantheum, Spathicarpa, Scaphispatha, Zomicarpa, Gorgonidium*

from Malaysia include: *Aglaonema, *Aglaodorum, Alocasia, *Amydrium, Anadendrum, *Aridarum, Ariopsis, Arisaema, *Bucephalandra, Cryptocoryne, Cyrtosperma, *Diandriella, Epipremnum* (Burma to Samoa, Marshall Islands), *Hapaline, *Heteroaridarum, *Holochlamys, Homalomena, *Hottarum, Lasia, *Pedicellarum, *Piptospatha, *Phymatarum, Pista, *Podolasia, *Pothoidium, Pothos, Remusatia, Rhaphidophora, Sauromatum, Schismatoglottis, Scindapsus* (India to Solomon Islands), *Steudnera, Typhonium* and *Xenophya*.

Malaysia is used here as defined by H. K. Airy-Shaw (Willis, 1966) to include the Malay Peninsula, Indonesia, all of New Guinea and the Philippines. The South American continent is also rich in diversity of Araceae with 20 genera containing 50 species (assuming that *Caladium* has been introduced by man into Central America) restricted to that continent (Table 3). It is impossible to estimate the number of species of Araceae which exist in South America but it probably has at least two-thirds of the known species for the New World. Moreover, most of the species are known from tropical and subtropical areas (including Andean regions) of the continent.

Another local area of diversity is the Mediterranean region and Southern Europe with five endemic genera, *Dracunculus, Helicodiceros, Biarum, Arisarum* and *Ambrosina*. All of these are in the subfamily Aroideae. It is interesting to note that of the 11 genera which are endemic to subtropical or warm temperate South America, 10 are also in the subfamily Aroideae (one genus *Aphyllarum* is in the Colocasioideae).

*Endemic.

Unlike the northern hemisphere which has several genera from colder temperate and boreal regions, the colder temperate and boreal parts of South America contain no genera. Most warm temperate genera from South America range into subtropical parts of the continent and few range south of 30° south latitude. In contrast, temperate and subtropical genera from the northern hemisphere in the Americas mostly range north of 40° latitude north and some, such as *Calla* and *Lysichiton*, range well beyond 50° latitude north. The South American warm temperate species have much narrower ranges, for the most part, while north temperate genera tend to be circumboreal or nearly so.

With the exception of *Orontium*, a genus endemic to eastern North America, temperate genera of North America also occur in Europe and/or Asia. On the other hand, tropical genera, in the Americas especially, tend to be more restricted to particular continents. For example, of the 39 genera known from tropical parts of the Americas, only four genera are also found in the Old World. The genus *Spathiphyllum* has several species which do not occur in the American tropics (some of these on oceanic islands). The genus *Homalomena*, on the other hand, is chiefly Asiatic with only a few, poorly known species in the American tropics. The genus *Scindapsus*, with 25 species ranging from India to the Pacific islands, has one species in Brazil. The genus *Schismatoglottis* is mostly from Malaysia, with three species in South America. *Arisaema* is worldwide in distribution and reaches tropical areas of Mexico. The only genera shared between America and Africa are *Cyrtosperma*, *Arisaema* and *Pistia* (all three are also in Asia). Africa (including Madagascar) and Asia share seven genera, *Arisaema*, *Pothos*, *Rhaphidophora*, *Cyrtosperma*, *Amorphophallus*, *Remusatia* and *Sauromatum*.

Although Asia and America are about equally rich in genera, the number of species is much higher in the American tropics than in the Asian tropics. The average number of species per genus for the family is 22·3. The average number of species per genus in Asia is 21·2 and in America is 30·2. The American tropics has substantially more species than Asia (1400 v. 900). This is because two genera in the American tropics, *Anthurium* and *Philodendron*, are the largest genera in the family with about 700 and 350 species respectively. Both of these genera have been given higher estimates than those given in Willis (1966), based upon my own estimates of the numbers of new species believed to be present in the American tropics. It is possible that similar reevaulations will have to be made on certain tropical genera of the Old World tropics, such as *Homalomena* (c. 140 spp.), *Schismatoglottis* (100 spp.), *Alocasia* (70 spp.), and *Cryptocoryne* (c. 50 spp.). There may be species inflation in many of the smaller genera in both the Old and New World tropics but, owing to their smaller number of species, they are not likely to increase the total number of species substantially. Unlike the Asian tropics, the New World tropics still has vast areas which have been poorly explored and it is there that the largest number of new species is to be expected.

Although 282 names of *Anthurium* were reported for Ecuador by either Sodiro or Engler, many represent oversplitting. Despite this, there are believed to be many undescribed species in Ecuador. No doubt even more new species will be discovered in Colombia. Preliminary work in the Department of Choco shows many novelties. Also Panama, the richest country of North America for *Anthurium*, has many new species. The genus *Philodendron*, although smaller, also has numerous new species, based on studies in Costa Rica and Panama.

Some geographical regions, though moderately rich in numbers of genera, are not rich in species. For example, the Mediterranean region with five endemic genera averages only 2·4 species/genus. Temperate North America with six genera, averages only 2 species/genus. The nine genera inhabiting generally temperate regions of the world have an average of only 2·2 species/genus. These genera are *Acorus*, *Gymnostachys*, *Lysichiton*, *Symplocarpus*, *Orontium*, *Calla*, *Thomsonia*, *Peltandra* and *Pinellia*. *Arisaema*, with 150 species, is strongly temperate but with species occurring also in tropical areas. Nor is low species diversity restricted to temperate regions. The 10 genera restricted to New Guinea, Borneo and Sumatra have an average of fewer than 2 species// genus. Those endemic to India have an average of only slightly more than 1 species/genus. The 20 genera endemic to South America have an average of 2·7 species/genus.

The tribal distribution of the Araceae is variable from continent to continent. Although most subfamilies are represented at least by a few species on each continent, they may be much better represented on one continent (Table 4).

Pothoideae (11 genera/802 species)

The subfamily Pothoideae is tropical except for the tribe Acoreae with the widespread genus *Acorus* and the Australian endemic *Gymnostachys*. The tribes Culcasieae and Zamioculcaseae, with 26 species, are strictly African. These make up only about 3% of the subfamily. Four Asian genera, *Pothoidium*, *Anadendrum*, *Pothos* and *Pedicellarum* account for 8% of the species in the subfamily. One of the Asian genera is *Pothos*, a genus of about 50 species.

Table 4. Distribution of subfamilies of Araceae in percentages.

Subfamily no. genera/ no. species	Temperate	Tropical	American	African	Asian
Pothoideae 11/802	0·4	96·6	89	3	8
Monsteroideae 9/213		100	48	0·1	51
Calloideae 4/5	100		80		60
Lasioideae 19/201	1	99	20	27	53
Philodendroideae 17/688	1	99	58	3	39
Colocasioideae 16/193	1	99	49	0·5	51
Aroideae[1] 33/354	20	80	10	13	71
Pistoideae 1/1		100		Pantropical	

[1] Exact percentages are difficult to calculate since as much as 43% of the total is for *Arisaema* which has an uncertain ditribution.

The American tropics has only the small genus *Heteropsis*, which in many ways is an intermediate between the Monsteroideae and Pothideae, and the genus *Anthurium*, the largest genus in the family. The American genera of Pothoideae represent 89% of the subfamily.

Monsteroideae (nine genera/213 species)

The subfamily is 100% tropical and is almost equally divided between Asia and America. A total of 48% of the subfamily is represented in the New World by four medium sized genera. *Monstera* has a centre of species diversity in Costa Rica and Panama with secondary centres in

Mexico and the northern Andes of South America and has no endemic species in the West Indies, the Guiana Highlands or in southern Brazil. *Stenospermation* and *Rhodospatha* apparently have a similar distribution but with perhaps a greater degree of speciation in Andean South America. *Spathiphyllum*, though chiefly American, has two species on Pacific Islands and three species in Indonesia and the Philippines. In the American tropics it is rather evenly distributed in Central and South America.

The subfamily Monsteroideae is scarcely represented in Africa. There are only two species of *Rhaphidophora* in western tropical Africa. Asia, with 51% of the total species for the subfamily, has five genera, four of them in the Monstereae, namely *Scindapsus*, *Epipremnum*, *Amydrium* and *Rhaphidophora*. Only the latter with 60 species is very large. The small genus *Holochlamys*, a relative of *Spathiphyllum*, is endemic to New Guinea. With the exception of *Amydrium*, which is endemic to Malaysia, the other genera of the Monstereae are widespread in Asia from India to the Pacific Islands.

Calloideae (four genera/five species)

The subfamily Calloideae is the smallest in terms of both genera and species and is restricted to temperate and boreal regions of the northern hemisphere. The genus *Lysichiton* ranges from Pacific North America to Alaska, eastern Siberia and Japan. *Symplocarpus* has an Atlantic North American–eastern Asian distribution. *Orontium* is endemic to eastern North America, ranging from New York and Massachusetts to northern Florida. *Calla* is widespread in northern and middle Europe, northern United States and southern Canada, ranging as far north as Alaska and eastern Siberia.

Lasioideae (19 genera/201 species)

The subfamily Lasioideae is approximately 99% tropical and is the only subfamily well represented in Africa with 27% of its species in Africa. These include *Cyrtosperma*, *Anchomanes*, *Amorphophallus*, *Pseudohydrosme*, *Cercestis*, *Rhektophyllum*, and *Nephthytis*. The last three genera are in the tribe Nephthytideae which is restricted to Africa. The genus *Cyrtosperma*, chiefly Asian, has a single species in Africa and two species in South America.

The subfamily is much better represented in Asia than in America with 53% and 20% of the species found in each area respectively. It is represented in America by the smaller genera *Urospatha*, *Dracontium*, *Dracontioides*, *Echidnium* and *Montrichardia*. All but the latter are in the tribe Lasieae and most are relatively rare. *Montrichardia* in the tribe Montrichardieae is a widespread genus usually growing in standing water.

The Asian genera *Lasia*, *Anaphyllum*, *Pycnospatha* and *Pseudodracontium* are from India and Ceylon, South-East Asia or Malaysia. *Plesmonium* and *Thomsonia* are from northern India or the Himalayas and *Podolasia* is from Borneo. *Cyrtosperma* and *Amorphophallus* are more widespread, though the former genus is most well represented in New Guinea. All of the Asian genera except *Thomsonia* are in the tribe Lasieae.

Philodendroideae (17 genera/688 species)

This subfamily is nearly all tropical with 58% of its species in America. The tribe Philodendreae contains both of the largest genera, namely *Philodendron* with about 350 species and *Homalomena* with about 140 species. *Philodendron* is restricted to the American tropics and *Homalomena* is mostly Asian with perhaps 10 species in the American tropics, mostly in South America. *Schismatoglottis* with 100 species is chiefly Asian with only three species in South America. Other American genera include the widespread *Dieffenbachia* and the small genus *Peltandra* (tribe Peltandreae) from Atlantic North America, the only temperate member of the subfamily.

Only about 3% of the subfamily are from Africa (including Madagascar). These include *Anubias* and *Typhonodorum*, each of which is in its own tribe.

Asian genera represent 39% of the subfamily. In addition to *Homalomena* and *Schismatoglottis*, the two large genera discussed above, the tribe Philodendreae also has *Diandriella*, *Aridarum*, *Bucephalandra*, *Phymatarum*, *Hottarum* and *Heteroaridarum* from Borneo or New Guinea (each with a few species) and *Piptospatha* ranging from Borneo to western Malaysia. Also in Asia is the Aglaonemeae with *Aglaodorum* (one species in Sumatra and Borneo) and *Aglaonema* with 21 species in South-East Asia and Malaysia.

Colocasioideae (16 genera /193 species)

The subfamily is almost entirely tropical with only a few subtropical species known from the Himalayan region (*Ariopsis*, *Gonatanthus* and

Hapaline). It is nearly equally represented in the American tropics and Asian tropics with 49% and 51% of the species in each area respectively. Only the small genus *Remusatia* is represented in Africa. In the American tropics, the Colocasieae has two moderately large genera, *Caladium* and *Xanthosoma*, and four genera with one or a few species each, namely *Jasarum*, *Caladiopsis*, *Chlorospatha* and *Aphyllarum*. The closest relative of *Jasarum* is the Old World genus *Alocasia* according to Bogner (1977).

The Syngonieae with *Syngonium* and *Porphyrospatha* (doubtfully distinct from *Syngonium*) are strictly American and the Ariopsideae with *Ariopsis* is restricted to the Himalayas.

The Colocasieae in Asia contains seven genera, most of which are fairly widespread. These are *Steudnera*, *Gonatanthus*, *Alocasia*, *Xenophya*, *Remusatia*, *Colocasia* and *Hepaline*.

Aroideae (33 genera/354 species)

The subfamily Aroideae is the most diverse generically with 33 genera containing nearly twice as many genera as the next largest subfamily, the Lasioideae. Only two subfamilies, the Philodendroideae with 688 species and the Pothoideae with 802 species, are larger in terms of numbers of species. The group is less than 80% tropical and has a relatively large percentage (13%) of African species. Four of the seven tribes in the subfamily are endemic to Africa or Madagascar. The tribe Stylochitoneae, with 21 species of *Stylochiton*, is endemic to tropical Africa. The tribe Arophyteae with three genera and 11 species is endemic to Madagascar. These genera are *Carlephyton*, *Colletogyne* and *Arophyton*.

The Protareae with a single species of *Protarum* (perhaps better placed in the Colocasioideae fide D. H. Nicolson) is endemic to the Seychelles Islands and the Callopsidae with a single species of *Callopsis* is endemic to eastern Africa. In addition to these endemic tribes, a number of typically European and Mediterranean genera are in northern Africa. These include *Arum*, *Dracunculus*, *Biarum*, *Arisarum*, *Arisaema* and *Ambrosina*.

About 10% of the species in the subfamily are American, and here again, generic diversity is great. The tribe Asterostigmateae consist of eight subtropical or warm temperate genera. The tribe Zomicarpeae contains five small genera in South America. *Filarum* and *Ulearum* are from the Amazon region whereas *Zomicarpella* is from Colombia and *Scaphispatha* is from southern Peru and Brazil. The genus *Zomicarpa* is from eastern Brazil (Bahia, Ceara). All but the latter genus have a single species each.

Pistoideae (one genus/one species)

This monospecific subfamily is represented by *Pistia stratiotes*, a widespread aquatic occurring in fresh water in tropical and subtropical areas throughout the world. The subfamily is not very typical for the family and is in many ways intermediate with the Lemnaceae.

Summary

The family Araceae has major centres of diversity in the tropics of Asia and America with a lesser centre in tropical Africa. Local centres of diversity include Malaysia, subtropical South America and to a lesser extent the Mediterranean region. The Asian tropics exhibit high generic diversity but relatively low species diversity. The American tropics are characterized by relatively lower generic diversity but with correspondingly higher specific diversity.

Acknowledgements

This study was partly by the National Science Foundation, Grant DEB 77–14414. The author wishes to recognize Josef Bogner and D. H. Nicolson who reviewed the manuscript.

References

Bogner, J. (1977). *Jasarum steyermarkii* Bunting (Araceae). *Aqua Planta* **2/3**, 4–7.
Willis, J. C. (1966). "A Dictionary of the Flowering Plants and Ferns." Cambridge University Press, Cambridge.

Distribution of the Filmy Ferns in Palaeotropics

K. IWATSUKI

Kyoto University, Japan

Introduction

The filmy fern family Hymenophyllaceae is comprised of some 600 species, mostly in the tropics and the southern hemisphere of both the Old and New World. Due to its simple organization and specialized involucres this family is distinct among the ferns, and is generally considered as having evolved to adapt to the warm moist environment. Only fragmental records of fossils of this family are recorded and it is impossible to note any of its history based solely on fossil records. To discuss the origin, diversification and dispersal of the filmy ferns, therefore, we usually use a comparative study of the extant species.

Copeland (1933, 1937, 1938, 1947) made comprehensive contributions to the elucidation of interspecific relationships in this family, especially those of the Old World tropics. He revised *Trichomanes* s. lat. in the Old World and then *Hymenophyllum* s. lat. Based on these studies he arrived at the conclusion that the classical bigeneric system of this family was unnatural, and proposed a new system. Including some additional observations, he later recognized 34 genera in this family in his "Genera *Filicum*".

Morton (1942, 1947, 1968) criticized Copeland's system and failed to recognize 34 genera in this family. He discussed Copeland's system in 1942, revised *Hymenophyllum* sect. *Sphaerocionium* in 1947, and later proposed a system admitting only six genera, four of which are monotypic, including nine subgenera and 36 sections, four years before his unexpected death.

The Copeland and Morton systems are distinctly different if compared only in generic rank, although they are similar to each other in recognizing the relationships among the species of the filmy ferns.

Subdivisions of the Filmy Ferns and their Distribution

Copeland (1938) developed a short discussion on the distribution of filmy ferns. He suggested the origin of this family in Antarctica, and extended this idea to most of the fern families (Copeland, 1939). Along with the fact that the proportion of the filmy ferns is high in the south, one of the most important facts on which Copeland based his hypothesis of Antarctic origin was that many monotypic genera are known in the south.

Among the subgroups of Hymenophyllaceae, *Hymenoglossum*, *Serpyllopsis*, *Rosenstockia* and *Cardiomanes* were recognized both by Copeland (1947) and Morton (1968) as monotypic genera. Each of these genera or species is peculiar in some way and has no relationship with any other species of this family. All these genera are confined to the south in their distribution: *Hymenoglossum cruentum* is common in certain parts of the Chilean Andes but no western slopes; *Serpyllopsis caespitosa* is in Antarctic America, the Falkland Islands and Juan Fernandez; *Rosenstockia rolandi–principis* is known only from a few collections from New Caledonia; and *Cardiomanes reniforme* is endemic in New Zealand.

Copeland (1938) recorded eight others as monotypic genera, all of which were recognized by Morton (1938) in the ranks below genus: one in a distinct subgenus, six in different sections, and one in a rank of subsection. *Craspedophyllum marginatum* of Tasmania and New South Wales was treated by Morton as a distinct subgenus of *Hymenophyllum*, and *Leptocionium dicranotrichum* in southern Chile was included in sect. *Sphaerocionium* by Morton as to form a monotypic subsection. The other six are *Apteropteris malingii* in Tasmania and New Zealand, *Myriodon odontophyllum** from New Guinea, *Polyphlebium venosum* from Tasmania, Queensland and New Zealand, *Lecanolepis membranacea*† in the West Indies, Nicaragua to Bolivia and Venezuela, *Davalliopsis elegans* in tropical America, and *Abrodictyum cumingii*‡ known from Taiwan, the Philippines, Celebes and Moluccas. These species are isolated from each other as well as from the other members of the Hymenophyllaceae. Most of them are confined to the extreme south, and it is evident that the filmy ferns include more peculiar forms in the south than in the tropics where the number of the species is much higher.

---

* Copeland (1947) includes *M. brassii* as a second species of this genus, although he was doubtful about the specific distinction of these two; Morton (1968) considered them conspecific.

† Copeland (1938) listed this species under an invalid generic name *Lecanium*.

‡ I added a second species of *Abrodictyum* (Iwatsuki, 1958) and am inclined to doubt the distinction of this "genus" from *Macroglena*.

Of the subgenera of *Hymenophyllum* and *Trichomanes* enumerated by Morton (1968), *Hemicyatheon* was treated as monotypic, although the genus *Hemicyatheon* Copel. was established as including two species. *Hemicyatheon* is distinct and may be safely referred to *Meringium* or *Hymenophyllum* sect. *Ptychophyllum* (Iwatsuki, 1977). Contrary to this, I propose to treat *H. levingei* of the Himalayas as having a distinct genus.

As many monotypic groups are in the south we can enumerate various examples in the flowering plants, especially in some gymnospermous families. Of 16 genera of the Cupressaceae, monotypic ones number seven, more than half of which are southern. Eight of 11 genera of the Taxodiaceae are monotypic although most of them are in the northern hemisphere. The present distribution is restricted, although most species were widespread according to fossil record. One should not speculate the geohistory based solely on the present distribution. One of the most typical examples of this is seen in the Podocarpaceae. Three of seven genera are monotypic and southern, and the other two bispecific genera are restricted to New South Wales and Tasmania or New Caledonia and Fiji, respectively. This family was therefore suggested to have had its origin in the south. The recent information including that of palynology, however, shows that this family was formerly widespread in Asia, Europe, and North America as well as the southern hemisphere, and present distribution has been traced only since later Tertiary. This does not coincide with speculation of this family as of Antarctic origin.

As to the distribution of the filmy ferns, it may be possible to refer the number of monotypic genera and sections in the south to the Antarctic origin of this family. At the same time, however, it may also be possible to consider that these diversified forms of the extreme south were specialized during southward migration. An example of this kind may be the northward migration of *Hymenophyllum levingei* in the Himalayas. The abundance of monotypic genera in the south may be due to ecology, as there are wide areas in the southern temperate zone where precipitation is extremely high. Anyway, we have no sound evidence to support Copeland's hypothesis of Antarctic origin of the filmy ferns.

Phytogeography and Speciation

Macroglena seems to be a heterogeneous assembly of unrelated species. The species are subdivided as follows:
   (i) *Macroglena* s. str. consisting of *T. meifolium* in the paleotropics from

West Africa to Polynesia, *T. gemmatum* in continental South–East Asia to Western Malesia, and *T. clathratum* in Taiwan and Northern Luzon;

(ii) *Callistopteris* and *T. schlechteri* with five species, *T. apiifolium* known in Malesia and Polynesia from Sumatra to Samoa, *T. schlechteri* in Borneo and New Guinea, the other three referred to *Callistopteris* in Pacific area, *T. bauerianum* in Norfolk and Lowd Howe Island, *T. polyanthum* in Society Islands, and *T. baldwinii* in Hawaii;

(iii) *Abrodictyum* and *T. caudatum* with four or more species, *T. cumingii* in Taiwan, the Philippines, Java, Celebes, Moluccas, and New Guinea, *A. boninense* in the Bonins, *T. caudatum* in Tahiti to Queensland, and *T. strictum* in New Zealand;

(iv) *Selenodesmium* and *T. setaceum* group consisting of some 15 species throughout the world, a variable Malesian species, *T. obscurum*, with such offshoots as *T. setaceum* in Malaya to the Philippines, *T. extravagans* in Luzon, and *T. tereticaulon* in China, *T. cupressoides* of Africa with such allies as *T. batrachoglossum* in Liberia, *T. stylosum* in East African Islands, and *T. parviflorum* in Mascarenes; Polynesian *T. dentatum* with affinities to *T. elongatum* in New Zealand and eastern Australia and *T. longicollum* in New Caledonia, and tropical American *T. rigidum* with its relative *T. mandioccanum* in Brazil. As to the dispersal of these species, we have no conclusive remarks based on sufficient evidence, although it is evident that every group is much diversified in the tropics.

Ecology and Distribution

As to the high ratio of filmy ferns compared with the total number of species, we have to remind ourselves that the number of filmy fern species is very high in tropical areas and that diversification seems to have occurred there, even though there are several specialized species without close relatives in the south. However, there are many specialized forms in the tropics, and these diversified into a number of species in moist humid places in the tropics. The present distribution of filmy ferns is influenced by their geohistory, and ecology. A recent study of *Gonocormus* or *T. minutum* (Yoroi and Iwatsuki, 1977) showed that prolifiration and apogamy seem to occur independently in warm moist conditions in the same locality.

One of the most difficult groups of the filmy ferns in species taxonomy is *Gonocormus* in which Copeland (1933) enumerated six species, two of which were obscurely distinct. Copeland (1938) listed two distinct and one variable species which may be distinguished into four. These were later united into a single species (Copeland, 1958). Contrary to these conclusions there is a general tendency to admit two species for this variable complex in most of the local revisions and cytological studies. The two forms generally accepted as two distinct species are distinguished by frond form and pinnation, with or without proliferation, and by the different reproductive forms. The form most common in the north has a small fan-shaped frond without proliferations and with a normal life-cycle, while the other is frequently found in warm wet places and has larger fronds pinnately compound in dissection, and produces spores without regular meiotic division.

Even in the fan-shaped frond which is sometimes described as dichotomously branched, the division is not dichotomous but pinnate with a short rachis. Elongation of the frond is usually caused by elongation of the rachis, and the longer frond takes an appearance of pinnate division. Moreover, we observe this proliferation in every form. The frequency of proliferation is high in the wet season, and proliferation occurs in all cultivated plants in saturated moisture. The buds of proliferation fade away in many cases, though they develop dramatically under very wet conditions. This coincides with the observation in the field, where the pinnate form with proliferation usually occurs in wet and warm places (Copeland, 1958; Bell, 1960). The reproductive feature was observed by Bell (1960) and Braithwaite (1969, 1975) who recognized a so-called apogamous type of reproduction in frequently proliferous forms. According to preliminary observations, however, the occurrence of apogamy is observed independently from the difference in phenetic features. Based on these observations we suggested that there is no distinct sign of speciation in the *T. minutum* complex, but that variation seems to be under the influence of environmental conditions, not difference in genetic features. *Gonocormus* is thus not geographically distinct, though there are records of particular local forms such as *T. alagense* in Mindoro and *T. latilabiatum* in the Marquesas.

We may develop our speculation that the filmy ferns, generally adapted to moist condition, are more at home in warm moist conditions and are going to diversify in such habitats. The southern localities where the specialized filmy ferns are rich are the places with high precipitation as represented by Holloway (1923). In the very moist areas of the Himalayas in the northern hemisphere, we find such specialized forms as *H. levingei*, recognized only when the structure of the Hymenophyllaceae is recognized as to be strictly adapted to such a habitat, though they are in many cases torrelable to a temporary xerophytic condition as in most mosses and liverworts.

It is rather difficult to make experimental comparison of the ecology of the filmy ferns, for they can be cultivated only in special conditions.

To elucidate the origin, diversification and dispersal of the filmy ferns, therefore, it is necessary to make further comparative field observations especially in the tropics.

References

Bell, P. R. (1960). The morphology and cytology of cytogenesis of *Trichomanes proliferum* Bl. *New Phytol.* **59**, 53–91.

Braithwaite, A. F. (1969). The cytology of some Hymenophyllaceae from the Solomon Islands. *Br. Fern Gaz.* **10**, 81–91.

Braithwaite, A. F. (1975). Cytotaxonomic observations on some Hymenophyllaceae from the New Hebrids, Fiji and New Caledonia. *Bot. J. Linn. Soc.* **71**, 167–189.

Copeland, E. B. (1933). *Trichomanes. Phil. J. Sci.* **51**, 119–280.

Copeland, E. B. (1937). *Hymenophyllum. Phil. J. Sci.* **64**, 1–188.

Copeland, E. B. (1938). Genera Hymenophyllacearum. *Phil. J. Sci.* **67**, 1–110.

Copeland, E. B. (1939). Fern evolution in Antarctica. *Phil. J. Sci.* **70**, 157–189.

Copeland, E. B. (1947). "Genera Filicum." Chronica Botanica Co., Waltham, MA.

Copeland, E. B. (1958). "Fern Flora of the Philippines I." Institute of Science and Technology, Manila.

Holloway, J. E. (1923). Studies in the New Zealand Hymenophyllaceae. Part I. The distribution of the species in Westland, and their growth-forms. *Trans. N.Z. Inst.* **54**, 577–618.

Holttum, R. E. (1955). "Revised Flora of Malaya II. Ferns of Malaya." Botanic Gardens, Singapore.

Iwatsuki, K. (1958). Taxonomic studies of Pteridophyta II. *Acta Phytotax. Geobot.* **17**, 161–166.

Iwatsuki, K. (1977). Studies in the systematics of filmy ferns II. A note on *Meringium* and the taxa allied to this. *Gard. Bull. Sing.* **30**, 63–74.

Morton, C. V. (1942). Recent fern literature. *Am. Fern J.* **32**, 30–31.

Morton, C. V. (1947). The American species of *Hymenophyllum* section *Sphaerocionium. Contr. U.S. Nat. Herb.* **29**, 139–201.

Morton, C. V. (1968). The genera, subgenera, and sections of the Hymenophyllaceae. *Contr. U.S. Nat. Herb.* **38**, 153–214.

Yoroi, R. and K. Iwatsuki (1977). An observation on the variation of *Trichomanes minutum* and allied species. *Acta Phytotax. Geobot.* **28**, 152–159.

The New Caledonian Genera of Araliaceae and their Relationships with those of Oceania and Indonesia

L. BERNARDI

Conservatoire et Jardin botaniques, Geneva, Switzerland

Viguier (1910–1913) had an interesting view of the importance of Araliaceae in New Caledonia's flora. He wrote: "Cette famille occupe en effet le sixième rang comme importance dans la flore de l'île . . . avec 14 genres et près de 90 espèces". Viguier did not go in to the matter deeply, but I will try to do so now.

First it is necessary to make three observations:

(i) The number of species recognized by Viguier is similar to my own observations, but the number of genera is only nine.

(ii) New Caledonia is a medium sized island (19 000 sq km) with approximately 3000 phanerogam species.

(iii) I follow the idea, proposed by Eyde and Tseng (1971), of splitting the family into two groups: those with palmately compound or lobate and palmately veined leaves, and those with pinnate leaves.

The total number of species and the division between the two main groups are given in Table 1 although the data are unfortunately incomplete. I would have liked to include the data for Indo-China, Borneo, Philippines and New Guinea, but data available to me were old and incomplete. It is clear, however, that only Madagascar and Australia have more pinnate-leaved Araliaceae than digitate or palmate. It is also remarkable that America has so few pinnate-leaved Araliaceae. New Caledonia certainly has the greatest number of species of Araliaceae in the world in comparison with its size. Fig. 1 is taken from Smith (1970), and deals with the cradle of 60 early flowering plant families, and the probable routes of migration. The Araliaceae are not included in those 60, so if we modify Fig. 2 to include New Caledonia, reducing the importance of Australia and lessening the number of

Table 1. The number of Araliaceae species in 11 parts of the world.

	Area (sq km)	Approximate no. of phanerogams	Araliaceae	
			Pinnate	Digitate
New Caledonia	19 000	3000–3500	44	44
New Hebrides, Fiji, Samoa and Tonga	36 800	2500 (?)	17	17
Hawaii	16 700	1800	8	6
New Zealand	268 000	1500	0	19
Australia	7 690 000	12 000	16	6
Java	132 000	6100	13	20
Madagascar	587 000	16 000–20 000	44	18
Taiwan	36 000	2500	3	10
Japan	372 000	3300	3	12
USSR	22 000 000	15 000–20 000	5	11
America	42 000 000	40 000 (?)	13	170–200

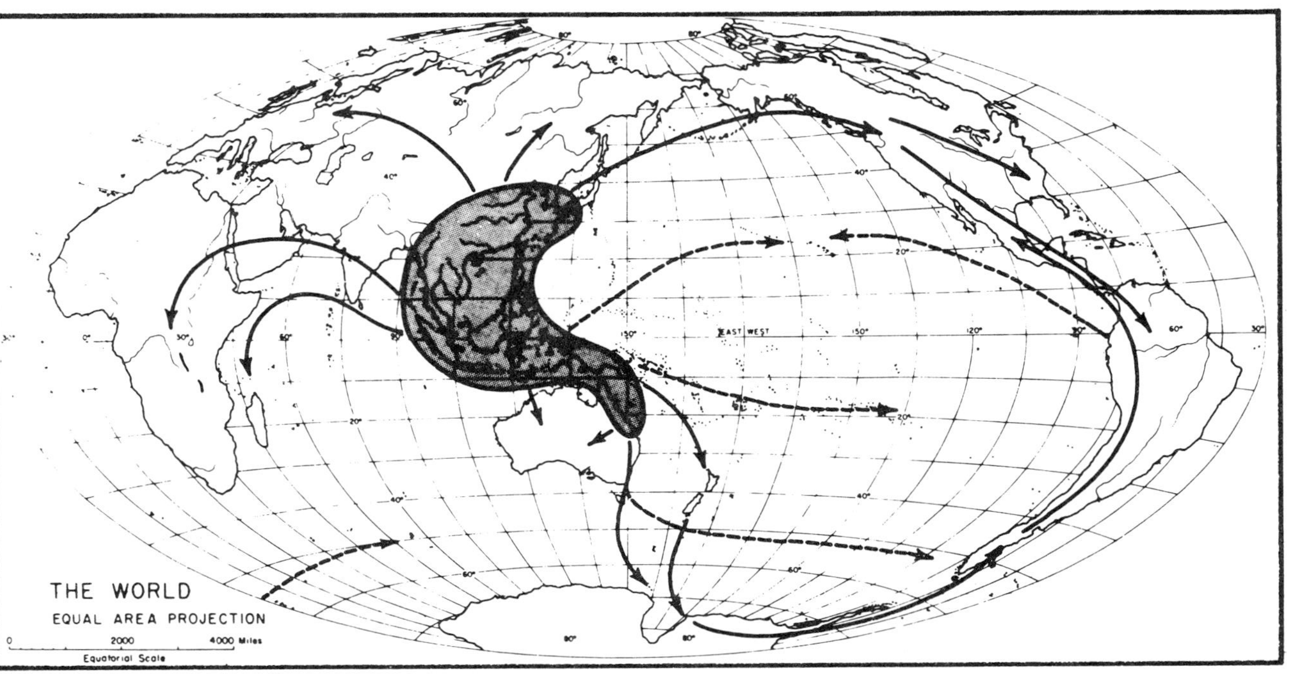

Fig. 1. A generalized model of angiosperm distribution indicating the probable routes of migration followed by early flowering plants (from Smith, 1970).

	L. *Bernardi*

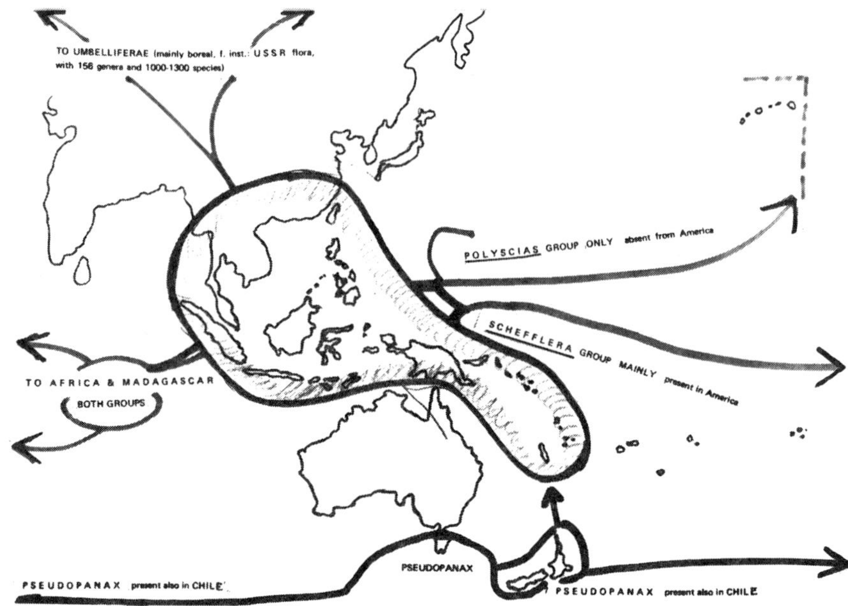

Fig. 2. General view of the distribution of *Schefflera* and *Polyscias* from the hypothetic centre of the family.

migration routes, we can visualize the Araliaceae at their beginning. I shall concentrate on two genera of New Caledonian Araliaceae: *Schefflera* for the digitate and *Polyscias* for the pinnate-leaved taxa, but first a few words about the other genera following Viguier's arrangement.

Myodocarpus is endemic, with simple and pinnate leaves. Baumann-Bodenheim (1946) made a fuss about this genus because it was already considered by Baillon (1879) and by Bentham and Hooker (1867) as true Araliaceae, but converging in fruit morphology with Umbelliferae. I do not think that *Myodocarpus* constitutes a step towards the Umbelliferae. It is only a dead end by-product of the Caledonian isolation. Its pollen morphology shows links with Cornaceae (Cranwell, 1960).

Delarbrea also has one species (*D. collina*) which is found in the New Hebrides, the Solomon Islands and islands of Indonesia. A few others are endemic to our island, one of which (*D. paradoxa*) can be merged with *D. collina*; but another undescribed species is sharply differentiated from all others and was once collected by Dr Mackee, in Dome de la Tiebaghi.

Pseudosciadum is monotypic, strictly endemic and very rare. Until recently its fruits were unknown, and it was thought by Baillon (1879) to be akin to the Queensland *Mackinlaya*. I consider it as very near to *Delarbrea*. *Pseudosciadum* like *Myodocapus* is another example of similarity to the Umbelliferae, which is convergence rather than relationship.

Apiopetalum contains three species with simple leaves and it is endemic. I have the strong feeling that this genus is not Araliaceae. I think that *Apiopetalum* stands near *Mastixia* and (or) *Corokia* (New Zealand) or *Curtisia* (South Africa), Cornaceae sensu antiquorum.

Tieghemopanax was placed in *Polyscias* by myself (Bernardi, 1971). The New Caledonian species have two-carpelled ovaries. Viguier defined *Tieghemopanax* on this character, leaving *Polyscias* for the species with three to five or more carpels. He described *Polyscias gigantea* as endemic. It is a mixed collection of a leaf of *Delarbrea harmsii* and fruits of a species of *Cissus*.

Eremopanax is endemic with simple leaves at the apex of the branches and pinnate leaves otherwise. The genus has six to eight species, and could be considered as a section or subgenus of the Indo-Malaysian *Arthrophyllum*. The only anatomical difference in the seeds stressed by Viguier (1906, p. 164) was ruminate in *Arthrophyllum* and smooth in *Eremopanax*. After studying the Madagascan Araliaceae some years ago, I concluded that ruminate versus smooth seeds is a character of very little weight, and is certainly not useful for generic delimitation.

Schefflera and its Allies

Schefflera, in preliminary works of Frodin (1977), was over enlarged by merging it with *Dizygotheca* and *Plerandra*, therefore taking *Schefflera* in a conservative sense (i.e. Harms, 1894) we have the largest and most widespread genus of the family. Leaving aside the inflorescence, and considering just the styles and stilopodia, I can recognize three subgenera: (i) *Heptapleurum* centred in Indonesia and Indo-China, (ii) *Agalma* abundant on continental Asia and New Guinea, and to a lesser extent in New Caledonia, and (iii) *Schefflera* with free styles and a range far beyond the hypothetical core of the family, from New Caledonia, to the Pacific Islands and America (in Geneva I only had scanty material of New World *Schefflera*: America does not have subgenus *Heptapleurum*) to Madagascar and Africa. In spite of the incomplete material studied we have a geographic trend of true systematic character (Fig. 3).

Dizygotheca, *Plerandra* and *Meryta* are considered "vicariant genera" of *Schefflera* in New Caledonia and nearby islands. The first has a peculiar form of the stamens which are duplicate, i.e. with double the

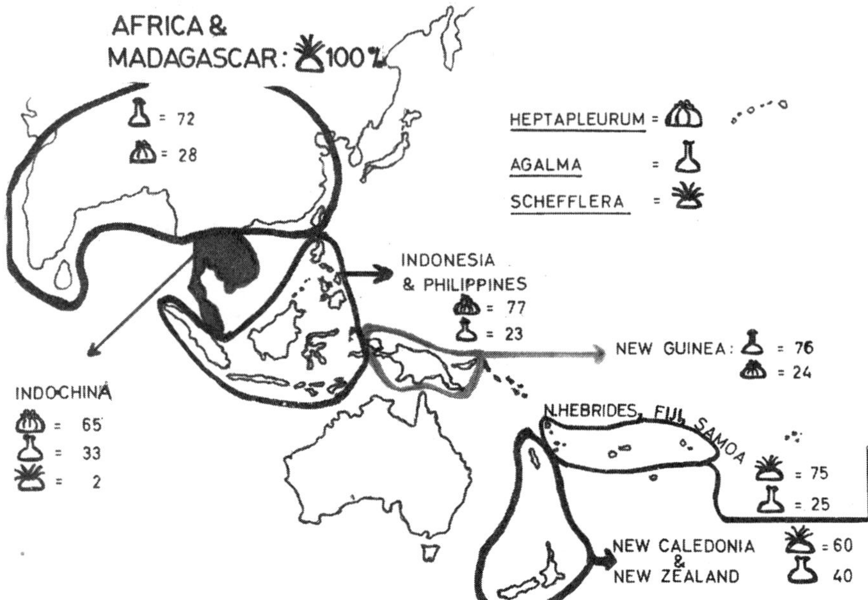

Fig. 3. Geographical trend of styles and stilopodia (in percentages) for *Schefflera*.

usual number of thecae of the flowering plants. We are confronted with a spectacular instance of the by-products of island life by Carlquist (1969). *Dizygotheca* has an isomerous androecium, with one exception: one species has 15 stamens and a more or less 15 loculed ovary. Viguier (1906, p.135) made this species the type of the genus *Octotheca*, placing it in a different tribe—Plerandreae (he belonged to the school of Van Tieghem, one of the most famous splitters of the last century). This species is now called *Dizygotheca plerandroides* since it is the living link between *Schefflera* and *Plerandra*, a beautiful genus of New Guinea, the Solomons and Fiji. I think that *Dizygotheca* and *Plerandra* are not primitive, but are taxa derived from *Schefflera* stock.

Meryta, with simple leaves, often very large, is a well-defined and morphologically rich genus, described by several authors as dioecous, but it also has polygamous–monecious species. Before trying to make any conclusions it is necessary to undertake a revision of the New Guinea material. At present, however, I do not think that *Anakasia simplicifolia* (Philipson, 1973) from western New Guinea (Vogelkop Peninsula) is worth generic status in view of *Meryta*, a taxon with diverse forms, especially in inflorescences. Its distribution, however, is interesting. It occurs from New Guinea, to New Caledonia, touching also the northern shore of New Zealand, absent in Fiji but present in Samoa, Tonga, the Society Islands and the Marquesas (Fig. 4). New

Fig. 4. *Meryta*: its distribution and more peculiar forms of inflorescence and fruit.

Caledonia is the richest island in *Meryta* species with a wide range of forms of flowers and inflorescences. Some small genera were created around *Meryta* such as *Strobilopanax* (Viguier, 1906, p. 148), *Schizomeryta* (Viguier, 1906, p. 149) and *Botryomeryta* (Viguier, 1910–1913, p. 84) which are not worth that status.

New Caledonia shares another Araliaceae taxon with New Zealand, the extremely rare *Pseudopanax scopoliae* (Baill.) Philipson, which has been collected only once this century. This species has simple leaves and a bilocular ovary which becomes a compressed fruit and appears at first sight to be a very peculiar New Caledonian *Polyscias*; but it is clearly akin to *Pseudopanax edgerley* (Hook.f.) C. Koch from the Northern Island of New Zealand.

Concerning the styles of the *Schefflera* group, I observed that while *Dizygotheca* has more but not all, species with free styles, *Meryta* always has separated styles as does the *Schefflera* subgenus *Schefflera*. Therefore the increase of floral parts (*Dizygotheca* and *Plerandra*) or the enriched inflorescence morphology (*Meryta*) are not primitive, but are derived. Fig.5 takes for granted the fact that the cradle of the Araliaceae and Umbelliferae was the spot depicted here, and that the eastward arrows (strictly intended for the *Schefflera* group) means an outgrowth or

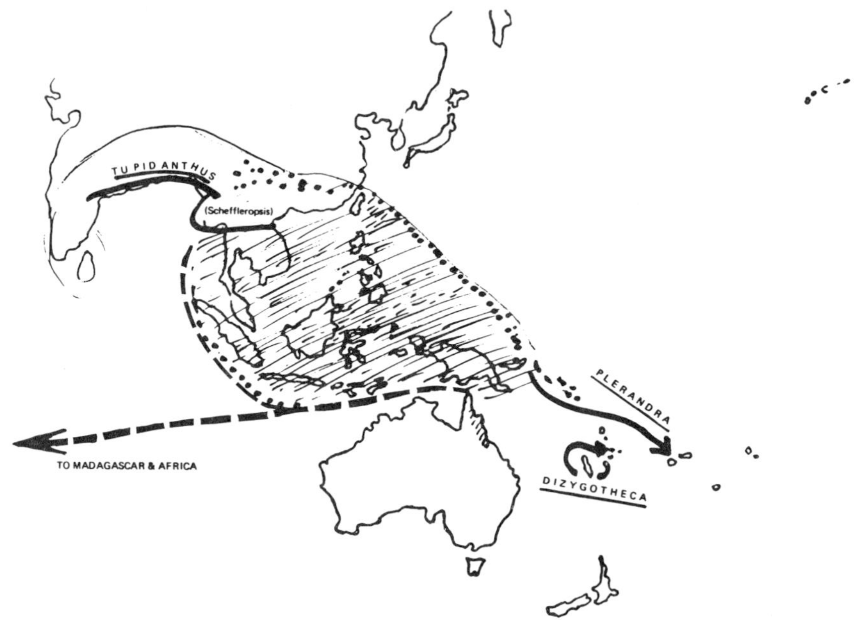

Fig. 5. Floral morphology in *Schefflera* (——— = enrichment of inflorescence, – – – = reduction of inflorescence).

enrichment of form, whereas the westward arrows, pointing to Madagascar and Africa and to a lesser extent America, signify a reduction or impoverishment of the morphogenetic power.

Polyscias

The *Polyscias* group in New Caledonia and the surrounding area has a completely different morphogenetic trend. This group is abundant in New Caledonia and has a few species in eastern Australia, the New Hebrides, Fiji as far as Samoa and the Society Islands (*Bonnierella*). This genus always has two-carpelled fruits, which are often flattened as in many Umbelliferae. Viguier assigned 19 species, to his *Tieghemopanax*, and later included *Panax scopoliae* Baill. Däniker and eventually Guillaumin both added two species, and the above taxa can probably be reduced to 16 even though two or three new species, from recent collections are undescribed.

In contrast to morphological variation of the *Schefflera* group, the New Caledonian *Polyscias* species are uniform. They resemble each other more than do the species within *Schefflera* and *Meryta*. The *Schefflera* group are mainly a forest group, whilst *Polyscias* are found on the forest margins or in the open. The type species of the genus was collected during Cook's second voyage to Tanna, New Hebrides, but *Polyscias pinnata* Forst. (= *P. scutellaria* (Burm. f.) Fosb.), a species with a three to five-carpellate ovary, is definitively not a New Hebridean species, but was introduced by the natives from Indonesia.

From New Guinea to the West as far as Madagascar and Africa and to the North through Indonesia to the Philippines, *Polyscias* species with more than two carpels occur. It seems strange that this genus does not reach the Asiatic mainland and that it is not distributed outside the Northern tropical belt. It is present in Luzon, but absent in Formosa.

Polyscias in New Caledonia seems to be a reduced or narrow-form-making taxon encountered at its southern limit (it is not present in New Zealand or Tasmania). On the other hand it is from the Indonesian–New Guinean front toward the west (Madagascar and the

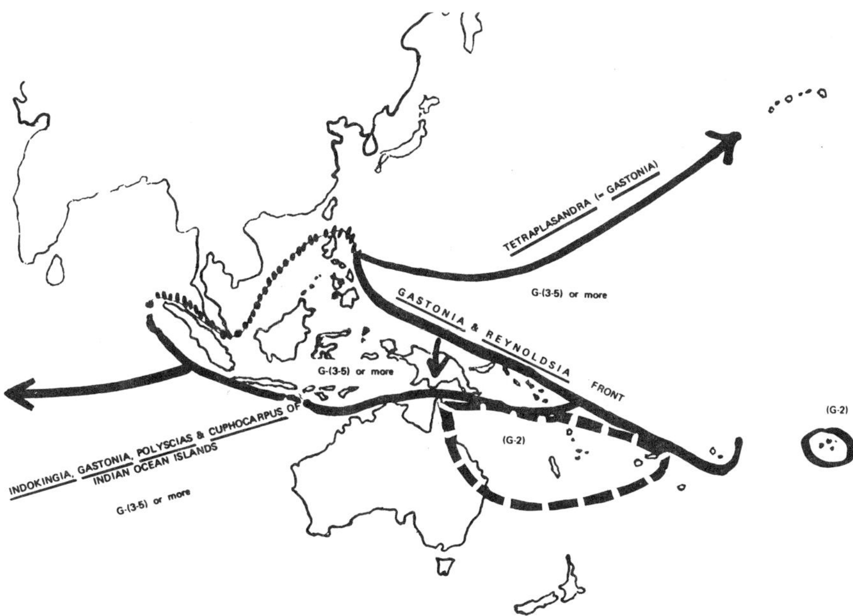

Fig. 6. Floral morphology in *Polyscias* (——— = enrichment of inflorescence, – – – = reduction of inflorescence).

Indian Ocean Islands) and toward the east (as far as Hawaii) that the *Polyscias* group started to diversify, e.g. the *Gastonia*-branch: *Indokingia* in the Seychelles, *Gastonia* in Mascarenes, Madagascar, Indonesia and also Queensland, *Tetraplasandra* and *Reynoldsia* from the Solomons to Hawaii (but lacking in Fiji).

Figure 6 shows that the *Polyscias* group has its arrows in a different direction from the *Schefflera* group. This could be viewed as supporting the Eyde and Tseng (1971) idea of an old split of the family. Nevertheless I do not agree with them that: "the ancestral Araliaceae had pinnately compound leaves and at least some of the inflorescence were paniculate". On the contrary, the ancetral Araliaceae were very like a modern *Schefflera* with compound or simple umbellae, the paniculate and racemose inflorescence being a derived character in all the taxa of Araliaceae. Eyde and Tseng (1971) also agree with others who have suggested an evolutionary link with Burseraceae and Rutaceae. The ancestral pinnate-leaved Araliaceae, show some resemblance to *Protium*, *Canarium* or *Zanthoxylon*. Finally, if the angiosperms are truly monophyletic, all of them have some mutual connections, but thinking in a more morphological way, I prefer this affinity: Hammamelidaceae, Saxifragaceae sensu antiquorum (with Escalloniaceae, Bruniaceae, Curtisiaceae, Cunoniaceae, etc.) going to Cornaceae sensu antiquorum and Caprifoliaceae–Rubiaceae.

References

Baillon, H. (1879). Rechérches nouvelles sur les Araliées et sur la famille des Ombelliféres en general. *Adansonia* **12**, 125–178.

Baumann-Bondenheim, M. G. (1946). Myodocarpus und die Phylogenie der Umbelliferen-Frucht. *Ber. Schweiz Bot. Ges.* **56**, 13–112.

Bentham, G. and J. D. Hooker (1867). "Genera Plantarum" Vol. 1, 935 pp. Reeve, London.

Bernardi, L. (1971). Araliaceae Madagascariae et Comores propositum. 2 Revisio et taxa nova Polysciadum. *Candollea* **26**, 13–89.

Brongniart, A. and A. Gris (1861). Note sur un genre nouveau d'Ombelliféres de la Nouvelle Calédonie. *Bull. Soc. Bot. Fr.* **8**, 121–124.

Carlquist, S. (1965). "Island Life." Natural History Press, Garden City, New York.

Cranwell, L. (1960). In "Principic Botanica" (L. Croizat, ed.) Vols 1–2, p. 1–31. Caracas.

Eyde, R. H. and C. C. Tseng (1971). What is the primitive floral structure of Araliaceae? *J. Arn. Arb.* **52**, 205–239.

Frodin, D. (1977). Cited by B. Stone. Notes on the system of Malayan phanerogams 25. Araliaceae. *Gard. Bull. Sing.* **30**, 275–291.

Harms, H. (1894). Araliaceae in Engler et Prantl. *Nat. Pflanzenf.* **3**, 1–62.

Philipson, W. R. (1973). *Anakasia*, a new genus of Araliaceae from West New Guinea. *Blumea* **21**, 87–89.

Smith, A. C. (1970). The Pacific as a key to flowering plant history. University of Hawaii Harold L. Lyon Arboretum Lecture No. 1.

Viguier, R. (1905). Sur les Araliacées du groupe des Polyscias. *Bull. Soc. Bot. Fr.* **52**, 285–314.

Viguier, R. (1910–13). Contribution à l'ètude de la flore de la Nouvelle Calédonie: Araliacées. *J. Bot. (Morot)* Sér. 2 No. 3, 38–101.

Willis, J. C. (1949). The birth and spread of plants. *Boissiera* **8**, 1–561.

Chromosomes and Distribution of Monocotyledons in the Eastern Himalayas

A. K. SHARMA

University of Calcutta, India

Introduction

The Eastern Himalayas cover an area of nearly 1200 square miles (Biswas, 1967) and are flanked on the east side by the Donkya and on the west by the Singalila ranges. On the ranges running from north to south are located Darjeeling, Kurseong, Gangtok and Kalimpông. The river Tista, which runs through the valleys in between, originates from the great Himalayas and ultimately flows down to the plains of North · Bengal. The zone of perpetual snow is at 15 000 ft.

The tropical to the temperate zones of the Eastern Himalayas are characterized by high humidity, very heavy annual rainfall and low temperature. The forest is tropical to temperate evergreen. The tropical foothill regions harbour a deciduous vegetation which to some extent is a mixture of flora of the Indo-Gangetic belt and a few emigrants from the temperate zone. In view of the arctic zone up to the limit of perpetual snow on the top and hot climate lower down in the plains, the Himalayas contain a high percentage of endemics (Chatterjee, 1939).

The total number of monocotyledonous species, according to Hooker (1849), occurring in the Eastern Himalayas, from the tropical to alpine zones, is approximately 1500 (Prain, 1903). Unfortunately, monocotyledons have not cytologically been so well explored as the dicotyledons. Hydrophytic vegetation is relatively scarce due to the rapidly flowing nature of the streams, except in the tropical belt where hydrophytes grow in stagnant pools caused by silting of the beds. The monocotyledons, in general, are characterized by bulbous, rhizomatous, epiphytic and climbing habit and vegetative propagation is profuse. Such asexual reproduction has resulted in the accumulation of somatic mutations due to diminished pressure of selection and consequently in the origin of cytotypes which aids in the origin of new species (Sharma, 1956, 1974).

One of the characteristic features of the vegetation of the Eastern Himalayas is the rapid succession of flora from the tropical to subtropical, temperate and alpine regions. Of the large number of species of monocotyledons, the most widely occurring ones belong to Gramineae, Cyperaceae, Orchidaceae, Scitamineae, Liliaceae, Aroideae and Commelinaceae.

Observations and Discussion

Aquatic representatives of families like Hydrocharitaceae, Alismataceae and Naiadaceae are rather scarce though a few species of *Blyxa* Noronha (2n = 16), *Sagittaria* L. (2n = 20, 22), *Tenagocharis* Hoch. (2n = 14), *Vallisneria* L. (2n = 30, 40), *Lagarosiphon* Harv. (2n = 14–100) and *Ottelia* Pers. (2n = 22 to 72) are not uncommon. These are restricted mostly to the tropical zone, within which they also have polyploid representatives.

The orchids are widely distributed in the substropical to temperate zones, as are three species of *Burmannia* L. in the tropical zone of the Khasi Hills. According to Hooker (1849), there are nearly 450 species of orchids, of which Hara (1966, 1971) recently collected 155. Amongst the 82 genera of this family, those which have the largest number of species and on which chromosome studies have been carried out (Sharma and Chatterjee, 1966; Roy and Sharma, 1972; Sharma and Sau, unpublished) are discussed here with their distribution (Table 1). Excepting a few species of a limited number of genera like *Habenaria* Willd., *Cypripedium* L., *Coelogyne* Lindl., *Calanthe* R. Br., *Liparis* Richard, *Herminium* L. and *Pholidota* Lindl. all others are chiefly restricted to subtropical to temperate regions, since the dense, evergreen forests in this zone with heavy rainfall and high humidity allow a profusion of epiphytic growth. The stunted nature of the plants, scrubby vegetation and intense sunlight behave as limiting factors against such epiphytic growth in the alphine zone. The extent to which the possible scarcity of insects and birds necessary for cross-pollination contributes to this scarcity, needs investigation.

In this family a remarkable feature of the chromosome complements is the gross uniformity in the general morphology of the chromosomes,

Table 1. Family Orchidaceae Benth. & Hook.f.

Genus	No. of species	Range of 2n chromosome number	Altitude (ft)
Oberonia Lindl.	15	30	2000–6000
Liparis Richard	27	28–80	3000–11 000
Dendrobium Swartz.	51	20–80	1000–7000
Bulbophyllum Thouars.	29	38–80	3000–8000
Cirrhopetalum Lindl.	13	30–48	3000–7000
Eria Lindl.	29	36–66	2000–6000
Coelogyne Lindl.	24	38, 40	3000–11 000
Pholidota Lindl.	10	40	6000–11 000
Calantha Br.	15	40, 44	7000–11 000
Cymbidium Swartz	11	40–80	3000–7000
Saccolabium Blum.	16	36–44	3000–7000
Goodyera Br.	11	20–44	2000–6000
Herminium L.	7	40	4000–14 000
Habenaria Willd.	41	28–64	3000–12 000
Cypripedium L.	7	20–30	5000–11 000

with mostly median contrictions and deep seated numbers like 19, 20, 21, possibly derived from x = 10 and its derivatives by hybridization. Chromosomal cytotypes have been reported in nearly all of them, but the aneuploid and polyploid numbers are mostly derivatives of the basic sets. The general morphology of the chromosomes does not show marked differences, and speciation has been shown to be affected by minor karyotypic changes. This feature is also true for orchids throughout the world (Jones, 1963). Such deep seated numbers may possibly represent genes or gene clusters in different chromosomes, all of which jointly contribute to the epiphytic adaptation of the genera. This suggestion is validated by the fact that certain species with unusual numbers for this family, viz. n = 7 as in *Orchis drudei* M. Sch. or n = 9 as in *Pogonia ophioglossoides* (L.) Ker. or n = 14 noted as in several species of *Habenaria* Willd., n = 13 in *Paphiopedilum bellatula* (Reichenb. f.) Pfltz. (Chaudhuri and Sharma, unpublished) are not epiphytic but terrestrial in habit. Cytological data is meagre and further investigation may provide confirmation of such correlation of base number with epiphytic adaptation. The species growing in the alpine zone do not show any interpopulation differences in number—another characteristic of this family.

The family Araceae, essentially purely tropical, has nearly 73 species distributed in the Himalayas, of which 34 have been reported by Hara (1966, 1971). There are altogether 19 genera, most of which are represented by one or two species each, excepting *Arisaema* Mart., *Alocasia* Neck., *Colocasia* Schott. and *Rhaphidophora* Hassk. *Alocasia* Neck. and *Colocasia* Schott. do not go beyond the marshy subtropical areas and their distribution merges into that of the Indo-Gangetic plains. The eight representative species of *Rhaphidophora* occur as huge climbers in subtropical to temperate rain forests. The most widely distributed genus, starting from the subtemperate to alpine zones, is *Arisaema* with 18 species recorded by Hooker (1849) and 17 by Hara (1966, 1971). Chromosome studies so far carried out in the Himalayan species of this

Table 2. Genus *Arisaema* Mart.

Species	2n chromosome number	Altitude (ft)
A. concinnum Schott.	28	6000–7000
A. curvatum Kunth.	28	3000–4000
A. decipiens Schott.	28	8000–8500
A. erubescens Schott.	28	5000
A. griffithii Schott.	28	6500–7000
A. helliborifolium Schott.	26	4000–5000
A. intermedium Bl.	28	6000–7000
A. nepenthoides Mart.	26+1B	9500–10500
A. sikkimense Stapf ex Chatterjee	36+1B	10 000–11 000
A. speciosum Mart.	28	6500–7000
A. tortuosum Schott.	26	6000–7000
A. utile Hook. f. ex Eng.	28	11 000–12 000
A. wallichianum Hook.	26+4B	11 000–12 000
A. ostiolatum Hara.	28	10 000–10 500
A. jacquemontii	28	7000–8000

family show a range of numbers (Sharma and Bhattacharya, 1966; Hara, 1971; Kurosawa, 1966; Mookerjea, 1955a) from 14 to 70 with aneuploidy and polyploidy and considerable difference in the size of the chromosomes even at an interspecific level. Diploids and aneuploids near the diploid range are rather common as compared to polyploids. The number n = 14 derived from n = 7 has been inferred as deep seated, from which all others have been derived. No correlation so far could be established between polyploidy, distribution and ecological preference as both diploids and polyploids occur under the same ecological conditions. Three species have been seen to contain B chromosomes, viz. *A. nepenthioides* Mart. (2n = 26 + 1B), *A. sikkimense*

Stapf. ex Chatterjee (2n = 26 + 1B) and *A. wallichiana* Hook. (2n = 26 + 4B), all of which occur approximately beyond 10 000 ft in the alpine zone. The remaining species, growing in the temperate zone, contain 2n = 26 and 28 chromosomes. These species of *Arisaema* Mart. containing B chromosomes are also characterized by the large size of the A chromosomes as compared to those of other species. It is likely that the accessories confer certain selective advantage in the alpine climate, which may involve control over chromosome coiling as well (Table 2).

In the order Scitamineae *sensu lato* 74 species have been recorded in Hooker of which Hara's record indicates widespread occurrence of 17 species distributed under 12 genera (see Table 3). In this group of

Table 3. Order Scitamineae Benth. & Hook. f. (*sensu lato*).

Genus	No. of species	Range of 2n chromosome number	Altitude (ft)
Globba L.	7	44, 48	2000–6000
Roscoea Smith	2	24, 26	8000–11 000
Cautleya Royle	4	27–36	5500–6000
Kaempferia L.	5	22, 24, 55	2000–6000
Curcuma L.	10	32–64	3000–7000
Hedychium Koenig.	17	24–66	1500–7000
Amomum L.	8	48, 52	2000–9000
Zingiber Boehm	6	22, 23, 55 (+Bs)	1000–5500
Alpinia L.	5	26, 33, 46, 48	3000–6000
Musa L.	3	22, 23	1500–4000
Costus L.	2	18, 27, 36, 44	1000–3000
Maranta Plum. ex L.	2	8–52	1000–4000

rhizomatous plants, excepting *Roscoea alpina* Royle restricted only in temperate and arctic zones, the rest are mainly confined to tropical and temperate belts in swampy areas. Excepting this species, with a characteristic chromosome number of 2n = 24, all the other genera show variation in chromosome number at intra- and interspecific levels (Bhattacharyya, unpublished; Chakravarty, 1951; Sharma and Bhattacharyya, 1959). In *Roscoea purpurea* Smith, growing at lower altitudes, the chromosome number is 2n = 26. The chromosomes of *Hedychium* Koen. species are high in those growing in comparatively higher ranges. The deep seated numbers in this group as a whole are n = 11 and 12, the former being most predominant in Musaceae and the latter in the remaining genera. All the species of *Musa* L. are restricted to the tropical and subtropical zones, growing widely in association with

332 A. K. Sharma

species of *Pandanus* Rumph ex Linn.f. on the hill slopes. The entire group is characterized by apomictic reproduction, wide occurrence of cytotypes and changes in structure and number of chromosomes involving both aneuploidy and polyploidy.

In the Eastern Himalayas, 82 species of Liliaceae, distributed under 26 genera, were reported by Hooker (1849). Hara (1966, 1971), through a series of expeditions, recorded 77 species from this area. The rest are mostly represented by one or two species occurring sporadically. Here, Liliaceae has been considered in a wider sense, including the genus *Allium* L. as well. The distributions of the genera referred to,

Table 4. Family Liliaceae Benth. & Hook. f. (*sensu lato*).

Genus	No. of species	Range of 2n chromosome number	Altitude (ft)
Smilax	20	26–60	2000–11 000
Polygonatum Tourn.	9	26, 30–88	7000–13 000
Smilacina Desf.	3	36–72	6000–12 000
Chlorophytum Ker.	4	28, 42, 56	2000–7000
Streptopus Michaux.	3	16	1000–13 000
Allium L.	7	16–28, 14+Bs	4000–13 000
Lilium L.	1	24, 33	5000–11 000
Fritillaria L.	2	24	5000–11 000
Disporum Salisb.	2	14, 16, 30	6000–13 000
Clintonia Rafin.	1	28	12 000–14 000
Trillium L.	1	20	7000–12 000
Paris L.	1 (2 var.)	10, 20	6000–11 000
Ophiopogon	3	36, 72	3000–9000

with their chromosome number, are shown in Table 4. Extensive works on Eastern Himalayan species have been carried out by Sen (1975) and of Western Himalayan by Kumar (1970). In *Smilax* L., both diploid and polyploid forms occur in a comparatively low altitude, whereas 2n = 32 chromosomes have been recorded mainly from the alpine zone. The Himalayan forms of *Chlorophytum* Ker.-Gawl. show polyploidy with structural changes of chromosomes (2n = 42, 56 in *C. arundinacean* Baker and *C. nepalense* Baker) (Sharma and Raju, 1967). In *Kniphofia aloides* Moench., growing between 5000–6000 ft (2n = 14), the Eastern Himalayan populations have been found to show distinct differences in karyotype from those growing in the North-Western and Central

Himalayas. *Convallaria majalis* L. (2n = 32, 36, 38) is restricted only to the alpine zone and none of the representatives have been found below 8000 ft. Its distribution is quite distinct.

A single species, *Clintonia alpina* Kunth. (*Clintonia udensis* Trautv. et. Meyer var. *alpina* Hara) (2n = 28), has been found not only in the arctic zone of the Eastern Himalayas, but also has a wide distribution in the arctics of Central and North-Western Himalayas. This arctic species does not show any difference in number or morphology of the chromosomes in any of the populations. Apparently the extreme conditions in the high alpine region have resulted in the adaptation of a specialized genotype, the variants of which do not stand selection in unfavourable conditions. The species of *Disporum* Salisb. form a complex, merging with one another with n = 6 chromosomes in the basic set, and with karyotypic alterations occurring mostly at the diploid level. Extensive structural alterations in complements with long chromosomes provide enough scope for reassortment of genes which have enabled them to acquire marked adaptive capacity. In the genus *Polygonatum*, the *verticillatum* and *oppositifolium* groups are represented in the Himalayas, their distribution ranging from temperate to alpine zones. It is remarkable that each population differs from the other in certain karyotypic alterations in an otherwise asymmetric karyotype. In this genus n = 9 or 10 has been assumed to be the basic set, from which other numbers have gradually been derived. *P. cirrhifolium* has a distribution range from 7000–12000 ft with chromosome numbers 2n = 26, 28, 30, 32 and 88. In general, the populations growing in extreme alpine conditions maintain a constancy in the karyotype. In *Smilacina*, most of the species show multiples of 18 chromosomes. In this genus, though occasional polyploids and aneuploids have been found, structural heterozygosity is very common, their survival possibly aided by a moist, humid, evergreen forest.

All the three genera of Polygonatae have thrown off cytotypes which are almost endemic to the Outer and Middle Himalayas. The snow line in the arctic and hot climate in the plains have provided their natural barriers. The fossil history of *Disporum* Salisb. is traced back to Pleistocene and that of *Smilacina* Desf. to tertiary, prior to continental drift. The constancy of the chromosome number of the latter genus, both in the Old and New Worlds, is considered as an evidence of their origin prior to the continental drift.

In the genus *Lilium* L., except *L. henryi* Baker, all other species collected from the Eastern Himalayas show minor karyotypic differences between populations and do not strictly follow Stewart's (1947) classification. The non-nucleolar chromosomes are also involved in structural alterations in alpine populations of *L. wallichianum*. The species of *Frittilaria* L. (2n = 24) are also restricted to the alpine zone, whereas *Paris* L. and *Trillium* L. have a distribution from temperate to subalpine regions. As compared to other parts of the Himalayas, the tetraploid species of *Paris* L. (2n = 20) mostly prevail in the Eastern Himalayas. Both *Trillium* L. and *Paris* L. have the same haploid set of 10 chromosomes.

Allium L. extends from temperate to alpine zones and most of the Eastern Himalayan representatives have a basic set of n = 8 chromosomes. The extremely alpine species *A. wallichiana* (2n = 48) does not have any chromosome variant. This confirms once more that the alpine species are characterized by a constancy in the karyotype, and as in *Chlorophytum* Ker.-Gawl., in *Allium* L. as well, species occurring at higher ranges are characterized by polyploidy. In *Allium stracheyi* Baker (n = 7) (Sharma and Aiyanger, 1961) diploids and polyploids occur in temperate regions (7000 ft) but the former is characterized by the presence of B's as well. In the alpine zone only polyploid populations occur.

In the genus *Ornithogalum* L. innumerable cytotypes have been found in the Eastern Himalayas between 5000–7000 ft. The basic set in this species is supposed to be n = 9, though in *O. thyrsoides* Jacq. 2n = 12 chromosomes have been recorded. Of the tribe *Dianelleae*, the populations of the same species *D. tasmanica* L. growing in the Khasia Hills at 5000–6000 ft have been shown to contain n = 20 chromosomes, whereas in those of the Eastern Himalayas it is n = 17 chromosomes.

In the allied family, Iridaceae, several species of *Iris* L., namely *I. clarkei* Baker, *I. nepalensis* D. Don, *I. japonica* Thunb. and *I. decora* Wall. are common throughout the Eastern Himalayas. The former two alpine species are characterized by 2n = 38 and 30 chromosomes (9000–10000 ft), whereas 2n = 36 chromosomes characterize the latter two (Banerjee and Sharma, 1971, Sharma and Talukdar, 1959).

In Amaryllidaceae, *Zephyranthes* and *Habranthus* occur in subtropical to temperate regions, the basic set being n = 12 with populations ranging to polyploid levels (Mookerjea, 1955). Summarizing the distribution of chromosome characteristics of Liliaceae and Iridaceae, in its entirety, it is clear that the temperate zones of the Himalayas have thrown off innumerable cytotypes of the different genera. Though a clear correlation between the altitude and chromosome number could not be established in certain genera such as *Chlorophytum*, *Paris*, etc., the Himalayan forms harbour polyploid types. The alpine conditions as a whole do not allow the survival of variants.

Another family which deserves mention is Dioscoreaceae, with eight species of *Dioscorea* L., distributed in tropical, subtropical and subtemperate zones. The subtropical zone is congenial for the growth of *D. deltoidea* Wall. and *D. prazeri*, the two species yielding diosgenin. The chromosome number ranges from 2n = 20 to 86 in the genus but the diploid forms (2n = 20) are mostly prevalent in colder regions.

In the tropical and temperate zones of the Himalayas, the family Commelinaceae is also well represented. Of 76 species of this family, 16 are reported to occur in the Himalayas (Hara, 1966, 1971; Rao, 1963)

Table 5. Family Commelinaceae Benth. & Hook. f. (*sensu lato*).

Genus	No. of species	Range of 2n chromosome number	Altitude (ft)
Amiscophacelus L.	1	20	1000–3000
Aneilema (Blume) Kunth.	2	38, 30, 58	1000–5000
Commelina L.	4	20–70	1000–6000
Cyanotis (L.) D. Don	2	20, 22, 24	1000–7000
Floscopa Lour.	1	24	4000–6000
Murdannia Royle	4	30–40	1000–4500
Streptolirion Edgew.	1	10, 12	6000–8000

(see Table 5). It has been inferred that there are two series of chromosome numbers noted in the family, one starting with six and the other with 10 chromosomes (Sharma, 1971). Most of the Himalayan species fall under the latter series except *Floscopa* Lour. and *Streptolirion* Edgew. which belong to the former. Extensive nomenclatural changes and redistribution of the genera have been made in the family by several authors (Brenan, 1966; Rao, 1963). High degrees of polyploidy and aneuploidy have been noted in the series with 10 chromosomes in the basic set such as *Commelina* L., *Aneilima* R. Br. and *Murdannia* Royle. Except *Streptolirion* Edgew. and *Floscopa* Lour., all other genera are distributed in the tropical to temperate zones of Himalayas. Polyploids occupy territories distinct from the diploids as specifically noted in *Murdannia spicatum*, *Commelina diffusa* and *C. sikkimense*.

Of the 175 species of Cyperaceae recorded from the Himalayas (Table 6), only about 45 have been cytologically studied. In *Cyperus* L., the high chromosome numbers are mostly from the high altitudinal zones. The chromosomes of *Fimbristylis* Vahl. show a basic set of five, with clear evidence of polyploidy, specially in the temperate areas. *Scirpus* L. is restricted to comparatively marshy areas and next to *Cyperus* L. has a series of chromosomal cytotypes. In *Eleocharis* R. Br. a

Table 6. Family Cyperaceae Benth. & Hook. f. (*sensu lato*).

Genus	No. of species	Range of 2n chromosome number	Altitude (ft)
Cyperus L.	29 (14)	16–208	1000–12 000
Eleocharis R. Br.	12 (7)	10–54	1000–2000
Fimbristylis Vahl.	29 (14)	10–30	2000–8000
Bulbostylis Kunth.	2	—	1000–12 000
Scirpus L.	14 (6)	28–80	1000–7000
Kobresia Willd.	9	—	8000–14 000
Carex L.	75	—	2000–14 000
Kyllinga Roltb.	(1)	120	1000–3000
Pycreus Nees.	(2)	54, 96	1000–2000
Lipocarpha R. Br.	(1)	26	1000–2000

diffuse centromere has been recorded (Sharma and Bal, 1956; Sanyal, 1972; Sanyal and Sharma, 1972). In absence of extensive data in this family, it is difficult to correlate the distributional pattern with the chromosome characteristics.

The allied family Juncaceae, too, is almost unexplored though a large number of species of *Juncus* L. and a few of *Luzula* DC. occur. A similar situation prevails in Gramineae, of which 179 species have been recorded from the Himalayas by Hara (1966, 1971). Very few records of chromosome numbers of this family are available. Several species, such as those of *Poa* L., *Panicum* L., *Paspalum* L., *Bromus* L., *Isachne* R. Br., *Festuca* L. occur at different altitudinal zones.

Conclusion

This analysis indicates that, of the vast number of monocotyledonous species occurring in the Eastern Himalayas, not even 20% have been cytologically explored. The subtropical to temperate zones harbour the largest number of species with the maximum concentration in the former. Polyploidy, aneuploidy and structural changes of chromosomes occur both at intra- and interspecific levels, leading to the formation of cytotypes, the survival of which is assured by vegetative reproduction. Though a clear correlation cannot be established between polyploidy and alpine distribution, the majority of alpine species are polyploids. In contrast to tropical and temperate zones, populations in the alpine areas do not differ in karyotype characteristics—due possibly to excessive selection pressure.

Acknowledgements

I would like to thank Dr Shyama Baksi (Sanyal) for her help in the preparation of the manuscript. Thanks are also due to Dr Rachel Thomas and Miss Arati Raychoudhury for their ungrudging assistance.

References

Banerjee, M. and A. K. Sharma (1971). A cytotaxonomical analysis of several genera of family Iridaceae. *Plant Sci.* **2**, 14–29.

Biswas, K. (1967). "Plants of Darjeeling and the Sikkim Himalayas" Vol. I. West Bengal Government, Calcutta.

Brenan, J. P. M. (1966). The classification of Commelinaceae. *J. Linn. Soc. (Bot.)* **59**, 349–370.

Chakravarty, A. K. (1951). Origin of cultivated bananas of South East Asia. *Ind. J. Genet. Pl. Breed.* **11**, 34–46.

Chatterjee, D. (1939). Studies on the endemic flora of India and Burma. *J. R. Asia. Soc. Beng.* **5**.

Hara, H. (1966). "The Flora of Eastern Himalaya." University of Tokyo Press, Tokyo.

Hara, H. (1971). "The Flora of Eastern Himalaya" (second report). University of Tokyo Press, Tokyo.

Hooker, J. D. (1849). Notes, chiefly botanical, made during an excursion from Darjeeling to Tonglo. *J. Asia. Soc. Beng.* **18**, 419–446.

Jones, K. (1963). Chromosomes of *Dendrobium*. *Bull. Am. Orchid Soc.* **32**, 634–640.

Kumar, V. (1970). Cytological studies in Eastern Himalayan species of *Smilacina*. Proc. 47th Ind. Sci. Congr. 360–361.

Kurosawa, S. (1966). Cytological studies on some Eastern Himalayan plants. *In* "The Flora of Eastern Himalayas" (H. Hara, ed.) p. 658–670. University of Tokyo Press, Tokyo.

Mookerjea, A. (1955). Cytology of different species of Aroids with a view to trace the basis of their evolution. *Caryologia* **7**, 221–290.

Prain, D. (1903). "Bengal Plants" Vols 1 and 2. Botanical Survey of India, Calcutta.

Rao, R. S. (1963). A botanical tour in the Sikkim State, Eastern Himalayas. *Bull. Bot. Surv. Ind.* **5**, 165–205.

Roy, S. C. and A. K. Sharma (1972). Cytological studies of Indian orchids. *Proc. Ind. Natl. Sci. Acad.* **38B**, 72–86.

Sanyal, B. (1972). Cytological studies on Indian Cyperaceae. II. Tribe Cypereae. *Cytologia* **37**, 33–42.

Sanyal, B. and A. Sharma (1972). Cytological studies in Indian Cyperaceae. I. Tribe Scirpeae. *Cytologia* **37**, 13–32.

Sen, S. (1975). Cytotaxonomy of Liliales. *Fed. Rep.* **86**, 255–305.

Sharma, A. (1971). Chromosome evolution in Commelinaceae from Eastern India. *J. Cytol. Genet.* (Suppl.) 19–25.

Sharma, A. K. (1956). A new concept of a means of speciation in plants. *Caryologia* **9**, 93–103.

Sharma, A. K. (1974). Plant chromosomes. *In* "The Cell Nucleus" (H. Busch, ed.) Vol. 2, p. 264–276. Academic Press, London and New York.

Sharma, A. K. and H. R. Aiyanger (1961). Occurrence of B chromosomes in diploid *Allium stracheyii* Baker and their elimination in polyploids. *Chromosoma* **12**, 310–317.

Sharma, A. K. and A. K. Bal (1956). A cytological investigation of some members of the family Cyperaceae. *Øyton* **6**, 7–22.

Sharma, A. K. and G. N. Bhattacharya (1966). A cytotaxonomic study on some taxa of Araceae. *Genet. Iber.* **18**, 237–262.

Sharma, A. K. and N. K. Bhattacharya (1959). Cytology of several members of Zingiberaceae and a study of the inconstancy of their chromosome complements. *La Cellule* **59**, 299–346.

Sharma, A. K. and A. K. Chatterjee (1966). Cytological studies on orchids with respect to their evolution and affinities. *Nucleus* **9**, 177–203.

Sharma, A. K. and D. T. Raju (1967). Cytological analysis of six species of *Chlorophytum. Bull. Bot. Soc. Beng.* **21**, 37–46.

Sharma, A. K. and C. Talukdar (1959). Cytotaxonomical studies on some members of the Iridaceae with special reference to the structural heterozygosity. *Nucleus* **2**, 63–84.

Stewart, R. N. (1947). The morphology of somatic chromosomes in *Lilium. Am. J. Bot.* **34**, 9–26.

Distribution Patterns of Neotropical Bignoniaceae: some Phytogeographic Implications

A. H. GENTRY

Missouri Botanical Garden, Saint Louis, USA

Although the family Bignoniaceae is pantropical in distribution and even has a few temperate zone representatives, it is predominantly (620, or 78%, of its world total 800 species) neotropical. In the process of monographing the family for Flora Neotropica, I have been able to study Bignoniaceae extensively in the field over most of Latin America; as a result its distribution is perhaps better known in several countries than that of any other plant family. If extrapolatable to other elements of the neotropical flora, the distribution patterns shown by neotropical Bignoniaceae would appear to have broad phytogeographical implications.

It is the purpose of this paper to summarize the distributions of the genera and species of New World Bignoniaceae, extrapolate from these data several patterns of biogeographic (and taxonomic) interest, and finally attempt to correlate large scale phytogeographic patterns with local diversity and community ecology in an overall synthesis of Bignoniaceae distribution.

The summary of Bignoniaceae distributions given here includes in its taxonomic data base a number of generic mergers which I have already proposed or am about to propose (Gentry, 1973, 1977c, in prep.) The species recognized include several undescribed ones as well as unpublished specific mergers and generic reassignments. The status of all species accepted here have been critically evaluated preparatory to my familial treatment for Flora Neotropica except for the Cuban and Hispaniolan species of *Tabebuia* and its segregate genera *Spirotecoma* and *Ekmanianthe*. An understanding of the complex taxonomy of West

     A. H. Gentry

Indian *Tabebuia* must await intensive field study. I have opted to accept here nearly all the species recognized by León and Alain (1957) though I suspect that many of these will be reduced to synonymy when critically studied.

Dividing the family into tribes gives a clearer picture of its distributional patterns (Fig. 1). For the family as a whole Brazil is clearly the centre of distribution. Bignoniaceae and Tecomeae, with dehiscent fruits and wind or water-dispersed seeds, are centred in Brazil since nearly all the West Indian Tecomeae represent the extensive adaptive radiation of a single section of *Tabebuia*. Crescentieae, with indehiscent fruits, are Central American; *Schlegelia*, *Gibsoniothamnus* and *Synapsis* are best segregated together as Schlegelieae and might belong to Scrophulariaceae. Eccremocarpeae and Tourrettieae are small monogeneric montane groups centred in the Peruvian Andes.

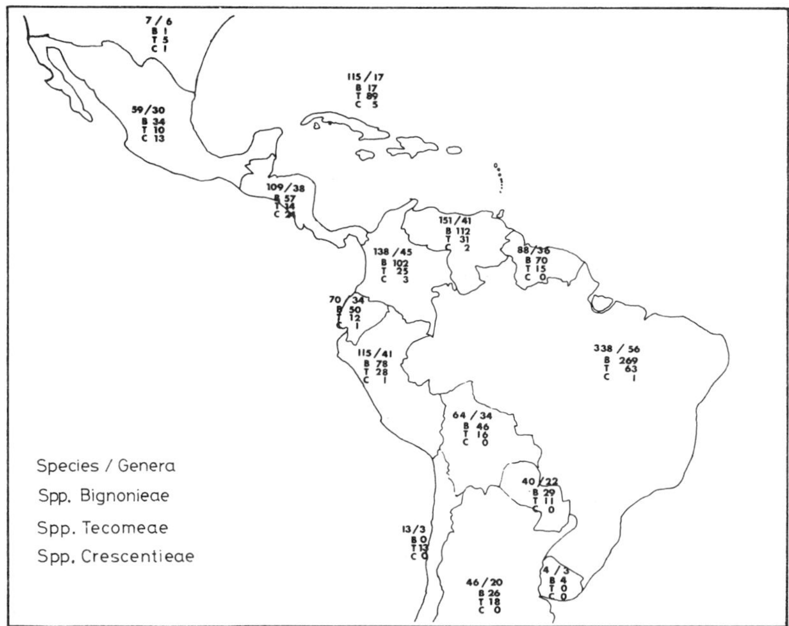

Fig. 1. Distribution of New World Bignoniaceae by country. Central American countries, West Indies (coastal islands excluded), and Guianas treated as single units. Total species/total genera for each country plus number of species per tribe for Bignonieae (B), Tecomeae (T), and Crescentieae (C). Species breakdowns for Schlegelieae with 23 species centred in Colombia and Panama, Eccremocarpeae with six species mostly in Peru, and Tourrettieae with one species from Mexico to Argentina are not listed separately. Many species of West Indian Tecomeae have not yet been critically evaluated.

One of the most significant generalities about neotropical Bignoniaceae distributions, especially from a taxonomic viewpoint, is that most of the wind or water-dispersed species have geographically ample but ecologically restricted distributions. The Central American species of Bignoniaceae provide a documented example. 94% of the dry forest species of Costa Rica and Panama reach Venezuela and 94% reach Guatemala and southern Mexico; for moist forest species these figures are 97% and 80%, for wet forest species 94% and 73%. Yet these same species show little overlap between different life zones within Panama and Costa Rica (Gentry, 1974a, 1976a). Even about half of the species from faraway Bolivia reach Central America. Unfortunately this biogeographic continuity has not been reflected by the prevalent taxonomy. Typically a different name has been applied to these widespread species in almost every country in which they occur. *Tabebuia impetiginosa* may be used to illustrate this point (Fig. 2). As I interpret the species, it occurs throughout continental tropical America from north-western Mexico to northern Argentina. It has been known under

Fig. 2. Distribution of *Tabebuia impetiginosa* (Mart. ex DC.) Standl., a typical wide-ranging species of Bignoniaceae. Known as *T. palmeri* Rose in Mexico, *T. nicaraguensis* Blake in Nicaragua, *T. dugandii* Standl. in Colombia, *T. avellanedae* Lorentz ex Griseb. in Argentina and *T. schunkevigoi* Simpson in Peru.

                    A. H. Gentry

six names in as many countries. In my experience this example is typical of our incredibly poor floristic data base for the neotropics. The result both of taxonomic parochialism and a rich and poorly collected flora, this taxonomic failure makes any kind of broad spectrum evaluation of distributional patterns, floristic affinities, or species diversity based on compilations from herbarium determinations or literature citations extremely difficult. The problem of multiple names for wide-ranging species is not peculiar to Bignoniaceae (Gentry, 1977a, 1978a). We may generalize that wide-ranging species are the rule rather than the exception for much of the neotropical flora, especially in wind-dispersed groups.

In general the tropical American flora is very poorly collected (e.g. Gentry, 1978a, Prance, 1978). Thus a corollary of the general pattern of wide distributions is that the second collection of a new species of Bignoniaceae is as likely as not to come from the opposite side of South America as the first (Table 1). *Memora juliae* A. Gentry is a particularly striking example. The first two collections were made by Spruce and Ducke in lower Amazonia a century and half century ago, respectively; last year I made the third collection on the Rio Yaguasyacu in Peru. I have previously noted similar examples from other families (Gentry, 1978b); the precautionary note for would-be describers of new neotropical species is self-evident.

At the other extreme, the species of Crescentieae have indehiscent fruits with mammal-dispersed seeds and are extremely prone to local endemism. Nearly all the exclusively mammal-dispersed species occur in only one or two adjacent countries and several of them are apparently restricted to single mountains (Fig. 3). Only a few secondarily water-dispersed species are widespread. I have suggested that the restricted ranges of Crescentieae species are a direct result of more localized seed dispersal patterns associated with use of terrestrial mammals as dispersal agents (Gentry, 1974c, 1977b). This pattern is repeated by the indehiscent-fruited Madagascar Bignoniaceae of tribe Coleae (Gentry, 1976b). If bignon distributional patterns can be extrapolated, we might predict generally higher rates of endemism for mammal-dispersed plant groups.

Water-dispersed seeds have evolved independently in many predominantly wind-dispersed genera. Most water-dispersed species tend to be as wide-ranging as their wind-dispersed congeners, several ranging from Mexico or Guatemala to Argentina and Bolivia. Water-dispersed species of Bignoniaceae have been misunderstood taxonomically, much unwarranted generic splitting having resulted from considering presence or absence of seed wings an absolute generic criterion

Table 1. Some species of Bignoniaceae with the second known collection at least halfway across Amazonia from the first collection (or localized population).

Species	First collection (or population)	Second collection (or population)
Adenocalymma prancei A. Gentry	Manaus, Brazil	Tingo Maria, Peru
Anemopaegma brevipes S. Moore	Mato Grosso, Brazil	Roraima, Brazil
Anemopaegma insculptum (Sandw.) A. Gentry	Río Caqueta, Colombia	Rondonia, Brazil
Arrabidaea cinerea K. Schum.	Bahia, Brazil	Kanuku Mountains, Guyana
Arrabidaea nicotianiflora Kränzl.	Acre, Brazil	Putumayo, Colombia
Cuspidaria subincana A. Gentry	Manaus, Brazil	Amazonas, Venezuela
Distictella laevis (Sandw.) A. Gentry	Amazonas, Venezuela	Rio Tapajós, Brazil
Haplolophium rodriguesii A. Gentry	Manaus, Brazil	Leticia, Colombia
Leucocalanthe aromatica Barb. Rodr.	Manaus, Brazil	Rondonia, Brazil
Manaosella cordifolia (DC.) A. Gentry	Bahia, Brazil	Manaus, Brazil
Mansoa onohualcoides A. Gentry	Ceara, Brazil	Amazonas, Venezuela
Memora juliae A. Gentry	Santarem, Brazil	Río Yaguasyacu, Peru
Memora longilinea A. Samp.	Manaus, Brazil	Guanai, Bolivia
Memora racemosa A. Gentry	Maranhão/Pará border, Brazil	Lely Mountains, Surinam
Memora tanaeciicarpa A. Gentry	Amazonas, Venezuela	Manaus, Brazil
Periarrabidaea truncata A. Samp.	Manaus, Brazil	Goias, Brazil
Spathicalyx duckei (A. Samp.) A. Gentry	Obidos, Brazil	Rio de Janeiro, Brazil
Tynnanthus pubescens A. Gentry	Rondonia, Brazil	Mazaruni River, Guyana
Tabebuia incana A. Gentry	Manaus, Brazil	Río Ucayali, Peru

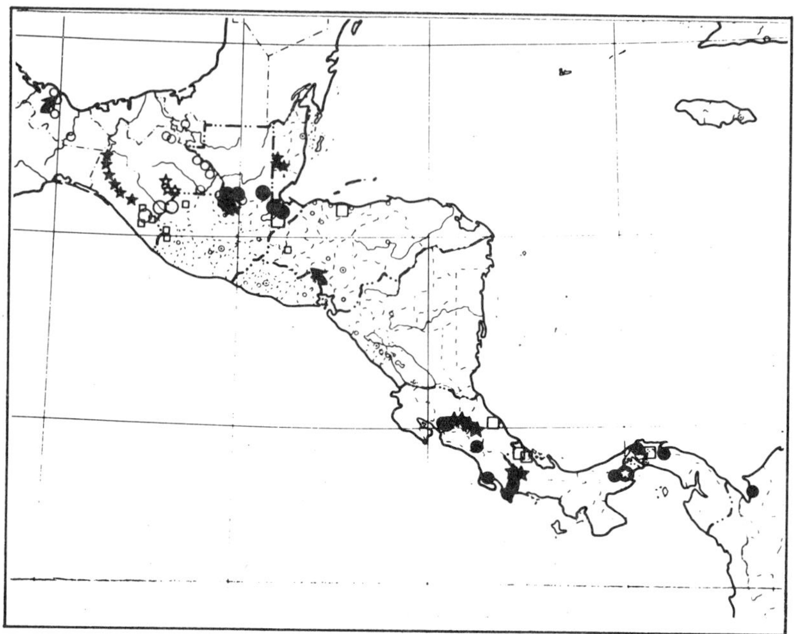

Fig. 3. Localized distributions in Crescentieae. Distributions of some animal-dispersed species of *Amphitecna*. *A. apiculata* A. Gentry, small open circles; *A. breedlovei* A. Gentry, small closed stars; *A. costata* A. Gentry, large square, *A. donnell—smithii* (Sprague) L. Wms., large closed circles; *A. isthmica* (A. Gentry) A. Gentry, small closed circles; *A. kennedyi* (A. Gentry) A. Gentry, small squares; *A. molinae* L. Wms., left-pointing arrow; *A. montana* L. Wms., tiny squares; *A. sessilifolia* (Donn. Sm.) L. Wms., large closed stars; *A. silvicola* L. Wms., open stars; *A. spathicalyx* (A. Gentry) A. Gentry, star-in-circle; *A. steyermarkii* (A. Gentry) A. Gentry, large open circles; *A. tuxtlensis* A. Gentry, right-pointing arrow. All species have rather restricted distributions, see text.

(Gentry, 1973). A modern evolutionary perspective makes it obvious that loss of seed wings in water-dispersed species is a frequent and relatively minor adaptive shift. Far from mandating different genera, the difference between wind-dispersed and water-dispersed seeds may not merit even specific recognition in every case (Hunt, 1972; Gentry, 1973). An interesting general pattern is the tendency for Amazonian populations of widespread otherwise wind-dispersed, species to have wingless or thick-winged water-dispersed seeds. An example is *Arrabidaea revillae* A. Gentry, a water-dispersed upper Amazonian relative of wind-dispersed *A. corallina* (Jacq.) Sandw. (Fig. 4). Several of these Amazonian variants are indistinguishable in flowers and leaves from their wind-dispersed relatives and their taxonomic status is not

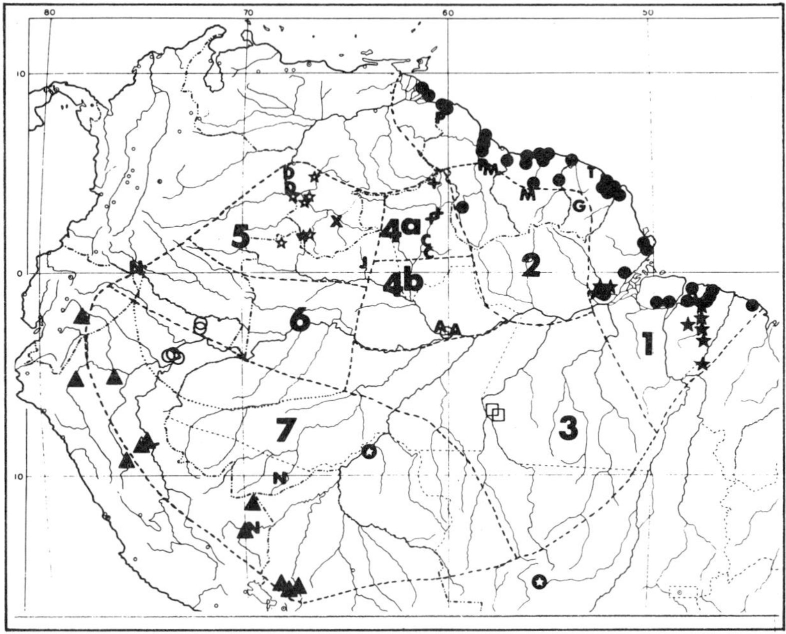

Fig. 4. Distributions of selected species of Bignoniaceae correlated with the Amazonian phytogeographic regions proposed by Prance (1977). 1. Atlantic coastal region: closed circles, *Tabebuia aquatilis* (E. Mey.) Sprague and Sandw.; closed stars, *Arrabidaea ornithopila* A. Gentry; P, *Pleonotoma echitidea* Sprague and Sandw.; T, *Tynnanthus sastrei* A. Gentry; 2. Jari–Trombetas region: G, *Anemopaegma granvillei* A. Gentry; M, *Anemopaegma maguirei* Sandw.; 3. Xingu–Madeira region: open squares, *Jacaranda egleri* Sandw.; 4. Roraima–Manaus region: crosses, *Anemopaegma jucundum* Bur. and K. Schum.; C, *Arrabidaea monophylla* A. Gentry; A, *Distictella reticulata* A. Gentry; 5. D, *Distictella arenaria* A. Gentry; open stars, *Pleonotoma exserta* A. Gentry; X, *Anemopaegma patelliforme* A. Gentry, J, *Jacaranda bullata* A. Gentry; 6. Solimões–Amazonas west region: open circles, *Arrabidaea revillae* A. Gentry (one collection also from region 4); 7. Southwest region: triangles, *Arrabidaea pearcei* (Rusby) K. Schum. ex Urb., star-in-circle, *Pleonotoma pavettiflora* Sandw., N, *Arrabidaea nicotianiflora* Kränzl. Dotted line indicates suggested modification of boundary of Solimoes–Amazonas west region to coincide with distribution of white sand substrate (cf. Kinzey and Gentry, 1978).

always clear. Even their existence was mostly unsuspected until recently. No doubt discovery of similar water-dispersed Amazonian variants may be expected in other plant families as well.

A number of species of Bignoniaceae, especially in Amazonia, are specialists for various other edaphic situations (Table 2). For example,

Table 2. Some species with edaphically determined distributions.

Tahuampa or varzea (seasonally inundated forest) species	*Anemopaegma paraense* Bur. & K. Schum. *Arrabidaea bilabiata* (Sprague) Sandw. *Arrabidaea bracteolata* (DC.) Sandw. *Arrabidaea revillae* A. Gentry *Arrabidaea spicata* Bur. & K. Schum. *Clytostoma binatum* (Thunb.) Sandw. *Crescentia amazonica* Ducke *Memora pseudopatula* A. Gentry *Macfadyena uncata* (Andr.) Sprague & Sandw. *Memora schomburgkii* (DC.) Miers *Pachyptera alliacea* (Lam.) A. Gentry *Pachyptera kerere* (Aubl.) Sandw. var. *kerere* *Phryganocydia uliginosa* Dugand *Tabebuia aquatilis* (G. Mey.) Spr. & Sandw. *Tabebuia barbata* (E. Mey.) Sandw. *Tanaecium jaroba* Sw.
White sand species	*Anemopaegma foetidum* Bur. & K. Schum. *Anemopaegma cf. oligoneuron* (Spr. & Sandw.) A. Gentry *Distictella arenaria* A. Gentry *Distictis pulverulenta* (Sandw.) A. Gentry (*Jacaranda obtusifolia* H&B. in Amazonia) *Martinella iquitosensis* A. Samp. *Pleonotoma dendrotricha* Sandw. *Pleonotoma exserta* A. Gentry *Schlegelia scandens* (Briq. & Spruce) Sandw. *Tabebuia obscura* (Bur. & K. Sch.) Sandw.
Limestone-outcrop species	*Parmentiera cereifera* Seem.
Seashore species	*Amphitecna latifolia* (Mill.) A. Gentry
Mangrove species	*Phryganocydia phellosperma* (Hemsl.) Sandw. *Tabebuia cassinoides* (Lam.) DC *Tabebuia obtusifolia* (Cham.) Bur. *Tabebuia palustris* Hemsl.

Parmentiera cereifera Seem., the widely cultivated "candle tree", is endemic to a single small limestone outcrop area in central Panama. *Tabebuia palustris* and *Phryganocydia phellosperma* are restricted to mangrove swamps and *Amphitecna latifolia* mostly to seashores. Most of the water-dispersed Amazonian species are restricted to seasonally inundated tahuampa or varzea habitats. A number of other Amazonian bignon species are restricted to white sand substrates. An interesting related distribution pattern is shown by widespread dry area species like *Jacaranda obtusifolia* which are restricted to edaphically drier white sand habitats in humid upper Amazonia. Nearly all of the bignon species restricted to these specialized, floristically relatively impoverished "marginal" habitats are obvious derivatives from extant more-widespread core-habitat species. Even though there are bignon species with edaphically determined distributions, the distributional limits of the great majority of neotropical species seem climatically determined.

Although the species of Bignoniaceae are typically wide-ranging, they do show several fairly distinct and repeated distributional patterns. Five major regions, each with appreciable Bignoniaceae endemism, might be recognized:

(i) Central America and western South America form a single rather homogeneous unit which extends south-eastward in Venezuela to the Orinoco River and south along the Andes to Bolivia and nothern Argentina.

(ii) Lowland Amazonia from near the Ecuador–Peru border to near the mouth of the Amazon has a number of distinctive species though sharing many species with surrounding areas.

(iii) Coastal Brazil, very rich in bignons and high in endemism.

(iv) The Guayana region which might better be considered a subset of Amazonia from the Bignoniaceae point of view (Gentry, 1978b).

(v) The Brazilian dry areas (cerrado and caatinga) with a rather rich and distinctive representation of Bignoniaceae.

Several of these regions are rather sharply demarcated from each other as is the case in Venezuela where the bignon flora north of the Orinoco is central American in affinities and completely different from the Amazonian bignon flora south of that river. Although its exact boundary is not yet clear, the Andean lowland/lowland Amazonian demarcation is also rather clearly defined and separates such allopatric species pairs as western *Arrabidaea verrucosa* (Standl.) A. Gentry from Amazonian *A. japurensis* (DC.) Bur. and K. Schum. and western *Tabebuia chrysantha* (Jacq.) Nichols. from Amazonian *T. capitata* (Bur. and K. Schum.) Sandw. There are only seven species endemic to the Guayana highlands, five species strictly endemic to lowland Guayana, and this list is shrinking with recent discoveries of several Guayana-centred species in coastal Brazil and Amazonia. Distributional limits of the Brazilian dry area species are even more open-ended and interdigitate with those of both coastal forest and Amazonian species which enter Central Brazil along gallery forests.

Generic distributional patterns in Bignoniaceae are consistent with those recently discussed by Kubitzki (1975, 1977) with the range of a single widespread species typically spanning that of an entire genus and concentration of species in eastern Brazil. However, Bignoniaceae differ from Kubitzki's examples in that the wide-ranging species are not demonstrably less primitive than their more restricted congeners. In fact the reverse is at least sometimes the case as some species of restricted distribution are clearly derived from widespread ones.

Any modern analysis of neotropical biogeography should consider the potential applicability of the attractive theory of Pleistocene refugia in explaining the observed distributional patterns of the group under study. Unfortunately Bignoniaceae distributions are so poorly known over much of Amazonia and its periphery that attempts to fit known bignon distributions to proposed refugia are still very tentative. Selected species of Bignoniaceae clearly fit some of the refugium-determined patterns predicted by an emerging concensus of biogeographers (e.g. Haffer, 1969, 1974; Prance, 1974; Brown, 1976; Kinzey and Gentry, 1978). Thus bignon species are known which seem restricted to each of the refugium-based phytogeographic regions proposed by Prance (Fig. 4), if the southern limit of the Solimoes–Amazonas west region is shifted south to include the Iquitos area (cf. Kinzey and Gentry, 1978).

However, a disconcerting number of Bignoniaceae species which once seemed to fit this pattern were subsequently discovered to occur also in distant regions. For example when I visited Manaus in 1974 10 species of Bignoniaceae were known to me only from the immediate area of Manaus, a distribution supporting the controversial Manaus refugium proposed by Prance (1974) but not supported by zoogeographers. However, I have subsequently seen collections of most of the so-called "Manaus endemic" bignon species from distant parts of Amazonia and only two of them now appear restricted even to the Manaus–Roraima phytogeographic region. *Tabebuia incana* A. Gentry provides an instructive example—while it was in press, being described as endemic to Manaus, I discovered it more than halfway across Amazonia at Genaro Herrera on the Peruvian Río Ucayali. The substrate of grey sand mixed with clay at Genaro Herrera is similar to some soils in the Manaus region and apparently not known at any intermediate site. Significantly some of the other species occurring with *T. incana* at the Peruvian site were new to Peru but common around Manaus. Since many of the bignon species whose distributions seem restricted to specific refugium-correlated phytogeographic regions are known from only one or two collections or localities (Fig. 3), it seems likely that many of them will also turn out to be more widespread, especially when a broader spectrum of habitats has been explored in diverse parts of Amazonia. It may well prove that Bignoniaceae (and perhaps wind-dispersed canopy lianas and trees in general) are too easily and widely dispersed, as a rule, to have retained many present day distributional patterns which can be correlated with hypothetical refugia. It is no doubt significant that most of the bignon species which are restricted to specific refugium-correlated phytogeographic regions seem to be edaphic specialists. Of the species whose distributions are mapped in Fig. 4, for example, two are restricted to periodically inundated sites, and at least seven to white sand substrates; only three of these 18 species are definitely found in core habitats.

Another key aspect of Bignoniaceae distribution is the prevalence of the family in dry areas. This is carried to such an extreme that some tropical dry forest communities are completely dominated by Bignoniaceae. For example, seven out of the 10 commonest species in one 500 sq m vegetation sample were Bignoniaceae (Gentry, 1976a). In most neotropical dry forest plant communities, at least in Central America and north-western South America, Bignoniaceae are the second most important component of the vegetation (after Leguminosae). However, the apparent increase of Bignoniaceae diversity in dry areas is largely a result of decreased diversity of other plant families: although Bignoniaceae density and dominance increase in dry forest communities, there are no more species than in lowland moist or wet forest communities.

The restriction of many bignon species to dry forest vegetation types suggests an interesting corollary to the Pleistocene refugium theory. If during glacial advances the tropics were so dry that moist and wet forest vegetations and species could persist only in the most favourable "refugia", the converse should be true today for dry forest plants. The return of moist conditions with the advent of the current interglacial must have isolated populations of dry forest species in dry "reverse refugia" protected from excess humidity by rain shadow effects or other local conditions. If "reverse refugia" for dry forest species really exist then we might expect them to be living laboratories, providing a chance to actually study the evolutionary processes hypothesized to have occurred in their full-glacial counterparts. Dry forest Bignoniaceae seem to provide evidence for the existence of such "reverse refugia". Many bignon species are restricted to the geographically isolated patches of dry forest scattered around the fringes of Amazonia, especially in the major interAndean valleys of Colombia and Peru, and along the length of Central America. Some of these dry forest species, like *Tabebuia impetiginosa* (Fig. 2) show little differentiation between their various isolated populations; others like *Tabebuia ochracea* (Cham.) Standl. seem to be actively differentiating at present with the variants I have treated as taxonomic subspecies correlated with different dry forest refugia.

Tecoma provides an example of an intermediate level of evolutionary divergence. Most species of *Tecoma* are confined to the dry inter-Andean valleys of Peru and adjacent countries. Two species—one orange-flowered and hummingbird-pollinated and one yellow-flowered and bee-pollinated—co-occur in most major valleys. The populations of each of these in each valley are perceptibly different from the related populations of adjacent valleys. Some of these isolates have been formally recognized as species or varieties but their taxonomic circumscriptions and status remain murky—exactly the kind of situation which would be predicted if these valleys are acting as the postulated "reverse refugia". *Delostoma* and the *Jacaranda acutifolia* complex show similar variational patterns. If analysed intensively, I anticipate that other, apparently more homogeneous dry forest-restricted disjuncts such as *Tabebuia impetiginosa*, *Cybistax antisyphilitica*, *Godmania aesculifolia*, *Cydista diversifolia*, *C. heterophylla*, and *C. decora* will also prove to contain genetically diverging population isolates. Study of these and similar dry forest disjuncts from other families from the refugium perspective should prove a fruitful new avenue for investigating evolutionary mechanisms in tropical plants.

To this point we have surveyed the broad distributional patterns of Bignoniaceae from a classic biogeographic perspective. I now propose to go to the opposite extreme and review local distribution patterns of Bignoniaceae from the viewpoint of community ecology. Finally I will attempt to correlate the local and large scale patterns to suggest a kind of ecoevolutionary synthesis for the neotropical Bignoniaceae.

Bignoniaceae have undergone a tremendous coevolutionary adaptive radiation with their pollination agents, and local diversity of Bignoniaceae is intimately correlated with the resultant pollination strategies. Accordingly a brief review of Bignoniaceae pollination strategies is in order. Although all Bignoniaceae flowers share the common theme of the sympetalous Tubiflorae flower, different species and genera show much diversification in floral morphology, correlated with utilization of different pollen vectors. Most species of Bignoniaceae are pollinated by large and medium-sized bees but some are pollinated by hummingbirds, bats, hawk moths, or butterflies and small bees. Moreover different bignon species pollinated by the same assemblage of large and middle-sized bees may have different pollination strategies through phenological specializations (Gentry, 1974b, c, 1976a). "Steady state" species put out one or two flowers a day over a long time period and are pollinated by traplining bees with fixed foraging routes. "Big bang" species put out all their flowers for a given year in a few days with all individuals of the species blooming at the same time; they attract a varied assemblage of opportunistic pollinators, apparently including some species which normally follow a fixed trapline. "Multiple-bang" flowerers are like "big bang" species in putting out all their flowers at once for a very short blooming period with synchronized flowering of different individuals of a species; they differ in having flowering bursts scattered sporadically through the year. Most "Multiple bang" species are non-nectar-producing mimics which receive only occasional visits from bees opportunistically searching for new nectar sources. "Cornucopia" flowerers have an extended blooming period with a distinct peak in flower production and different individuals are synchronized only as to the same general flowering season (Gentry, 1974b, c).

Curiously, about 20 strictly sympatric species of Bignoniaceae occur together in most lowland neotropical plant communities. Each of the sympatric bignon species of a given plant community has a unique pollination niche determined by pollen vector, phenological strategy (in the case of bee-pollinated species) and seasonality (in the case of "cornucopia" species). In general, in a given plant community, only a single bignon species using a given phenological strategy is in flower at any one time. I have suggested that competition for pollinators is the critical factor regulating intra-community Bignoniaceae diversity. For Bignoniaceae, at least, most tropical plant communities seem essentially closed equilibrium systems whose diversity appears to be limited to 20 species by saturation of the available pollination niches (Gentry, 1976a).

Can this local picture aid in formulating general hypotheses to explain Bignoniaceae distribution patterns? One relevant factor is that the same pattern of assortment of sympatric Bignoniaceae species into different pollination niches which is apparent in a local plant community is operative on a regional scale (Gentry, 1976a). Very few bignon species of a given habitat or life zone share the same pollination niche. The narrow ecologic amplitude noted earlier correlates with numerous pairs (and a few triplets) of closely related species replacing each other in the same pollination niche in dry, moist, or west forest communities. This seems a natural result of the combination of the ease of dispersal typical of Bignoniaceae (cf. above) with the regulated diversity of the local plant community. Any more successful competitor for one of the possible pollination niches which enters any part of a region can quickly spread throughout the area. Fine-tuning the system makes possible regional coexistence of sibling species adapted to different habitats; if each pollination niche were filled by a different species in dry, moist, and wet forest core habitats, a regional maximum of 60 (3 x 20) core habitat bignon species might be predicted. That this figure is approached by the 56 species of core habitat Panamanian Bignoniaceae which remain when edaphic specialists and local-endemic Crescentieae are excluded, suggests that such a prediction may be meaningful.

To summarize, mammal-dispersed species show very different distribution patterns from the rest of the family with high local endemism and a Central American distributional centre. For the wind and water-dispersed species of Tecomeae and Bignonieae, Brazil is the centre of diversity. Isolation of dry forest species populations in interglacial "reverse refugia" as well as mesic species in full glacial refugia may have been important in Bignoniaceae speciation and evolution but has left relatively little evidence in modern distribution patterns. The prevalent extra-Brazilian distributional pattern of wide-ranging ecologically restricted core habitat species may reflect cyclical range extensions of easily dispersed species, dry forest ones alternating with mesic ones as their respective habitats expanded and contracted, with capacity to partition local pollinator resources the critical factor through evolutionary time in limiting intracommunity diversity and determining which new arrivals would successfully enter existing communities. The significance of large scale regional distributional patterns is unclear but appears largely to correlate with present day ecological barriers (cf. Gentry, 1978c). For the Americas as a whole, the intricate diverse biotically regulated Bignoniacease community ends abruptly near the tropics of Cancer and Capricorn and the few temperate zone species seem by comparison almost random components of their communities. In the rich equilibrated core habitat tropical communities, evolutionary change may largely involve replacement, throughout a region and habitat, of species occupying particular pollination niches by superior competitors. Successful speciation leading to overall increases in number of species of neotropical Bignoniaceae probably most frequently involves entering floristically impoverished marginal habitats. Active diversification of populations isolated in refugia may lead largely to evolutionary dead ends unless the end products of this evolution are able to enter and specialize in peripheral habitats (or occasionally develop new or more efficient pollination strategies).

The coincidence of this empirically derived ecologically-determined scenario for Bignoniaceae evolutionary and distributional patterns with such theoretical ones as those of Terborgh (1972) suggests the general applicability of the model. These somewhat speculative points also lead to a readily testable hypothesis: the as yet largely unstudied concentration of bignon species in Brazil's coastal forest should prove correlated with more species restricted to marginal and peripheral habitats rather than with more diverse local bignon communities, persistence of "archaic" taxa, or greater specificity to local and subregional core habitat conditions.

References

Brown, K. S., Jr. (1976): Geographical patterns of evolution in neotropical forest Lepidoptera (Nymphalidae: Ithomiinae and Nymphalinae–Heliconiini). *In* "Biogeographie et Evolution en Amerique Tropicale" (H. Descimon, ed.). Laboratoire de Zoologie de l'Ecole Normale Superieure, Paris.

Dodson, C. and A. Gentry (1978). Flora of the Rio Palenque Science Center. *Selbyana* **4**, 1–628, i–xxx.

Gentry A. (1973). Generic delimitations of Central American Bignoniaceae. *Brittonia* **25**, 226–242.

Gentry, A. (1974a). Bignoniaceae. In Flora of Panama. *Ann. Mo. Bot. Gard.* **60**, 781–997.

Gentry, A. (1974b). Flowering phenology and diversity in tropical Bignoniaceae. *Biotropica* **6**, 64–68.

Gentry, A. (1974c). Coevolutionary patterns in Central American Bignoniaceae. *Ann. Mo. Bot. Gard.* **61**, 728–769.

Gentry, A. (1976a). Bignoniaceae of southern Central America: Distribution and ecological specificity. *Biotropica* **8**, 117–131.

Gentry, A. (1976b). Relationships of the Madagascar Bignoniaceae: a striking case of convergent evolution. *Plant Syst. Evol.* **126**, 255–266.

Gentry, A. (1977a). Endangered plant species and habitats of Ecuador and Amazonian Peru. *In* "Extinction Is Forever" (G. Prance and T. Elias, eds) p. 136–149. New York Botanical Garden, New York.

Gentry, A. (1977b). Notes on Middle American Bignoniaceae. *Rhodora* **79**, 430–444.

Gentry, A. (1977c). Generic mergers and new species of South American Bignoniaceae. *Ann. Mo. Bot. Gard.* **63**, 46–80.

Gentry, A. (1978a). Floristic knowledge and needs in Pacific Tropical America. *Brittonia* **30**, 134–153.

Gentry, A. (1978b). Bignoniaceae. In Botany of the Guayana Highland. *Mem. N. Y. Bot. Gard.* **29**, 245–283.

Gentry, A. (1978c). Extinction and conservation of plant species in tropical America: a phytogeographical perspective. *In* "Systematic Botany, Plant

Utilization and Biosphere Conservation" (I. Hedberg, ed.) p. 110–126. Almqvist and Wiksell, Stockholm.

Haffer, J. (1969). Speciation in Amazonian forest birds. *Science* **165**, 131–137.

Haffer, J. (1974). "Avian Speciation in Tropical South America." Nuttall Ornithological Club, Cambridge, MA.

Hunt, D. (1972). A note on the typification of *Adenocalymma marginatum* (Bignoniaceae). *Kew Bull.* **27**, 335–336.

Kinzey, W. and A. Gentry (1978). Habitat utilization in two species of *Callicebus. In* "Primate Ecology: Problem-oriented Field Studies" (R. Sussman, ed.) p. 89–100. Wiley, New York.

Kubitzki, K. (1975) Relationships between distribution and evolution in some heterobathmic tropical groups. *Bot. Jahrb.* **96**, 212–230.

Kubitzki, K. (1977). The problem of rare and of frequent species: the monographer's view. *In* "Extinction Is Forever" p. 331–336. Botanical Gardens, New York.

León, H. and H. Alain (1957). "Flora de Cuba" Vol. 4, p. 414–446. P. Fernandez, Havana, Cuba.

Prance, G. (1974). Phytogeographic support for the theory of Pleistocene forest refuges in the Amazon Basin, based on evidence from distribution patterns in Caryocaraceae, Chrysobalanaceae, Dichapetalaceae, and Lecythidaceae. *Acta Amaz.* **3**, 5–28.

Prance, G. (1977). The phytogeographic subdivisions of Amazonia and their influence on the selection of natural reserves. *In* "Extinction Is Forever" (G. G. Prance and T. Elias, eds). Botanical Gardens, New York.

Prance, G. (1978). Floristic inventory of the tropics: Where do we stand? *Ann. Mo. Bot. Gard.* **64**, 659–684.

Terborgh, J. (1973). The notion of favourableness in plant ecology. *Am. Nat.* **107**, 481–501.

The Solanaceae in the Neotropics:
A Critical Appraisal

A. T. HUNZIKER

National University of Córdoba, Argentina

Introduction

It is well known that the Solanaceae are predominantly American. Notwithstanding its importance from the floristic and economic points of view, the taxonomy of this family has not been significantly improved in the present century. Since 1895, when Wettstein published his outstanding monograph, the system therein devised by him, has been followed by the authors without hesitation, and practically nothing of importance has been proposed for its amelioration. The attempt of Baehni (1946) was not successful: not only on account of its artificiality derived of over-emphasizing the importance of a single character (aestivation), but fundamentally because his data on the remaining characters were taken from bibliographical sources, without a critical evaluation from his own personal experience on the plants and structures under discussion. After many years of careful and time-consuming observations on the morphology of the flower of Americans representatives, I was able to arrive at a number of generalizations, that led me to propose some substantial preliminary changes (Hunziker 1977; 1979) to the system of Wettstein. The present article is intended to explain briefly two of the main lines of evidence for its support: one in relation to the seeds and embryos, the other concerning the broad structure of the stamens.

Some Preliminary Statistical Data

Taking in consideration geographical and taxonomical features, the 88 genera of Solanaceae may be grouped in four categories: American endemics: 64, Old World endemics: 16, widespread genera (*Solanum* L., *Lycium* L., *Physalis* L. and *Nicotiana* L.): four, doubtful genera (*Heteranthia* Nees. et Mart.: Brazil; *Atrichodendron* Gagn.: Vietnam; *Atropanthe* Pascher: China; *Pauia* Deb. et Dutta: India): four. This makes a total of 88. In other words 69 genera are indigenous in America, i.e. almost 78·5%; out of these 59 are endemic to the tropics, or reach tropical regions.

A Sketch of a Revised System

These 69 American genera are arranged, for the time being, in two subfamilies (Solanoideae and Cestroideae), the first with seven tribes, and the second with five according to the following: Subfamily Solanoideae: Tribe Solaneae (*Solanum* L., *Cyphomandra* Sendt. *Lycopersicon* Miller, *Jaltomata* Schlecht., *Cuatresia* A. T. Hunz., *Witheringia* L'Herit., *Brachistus* Miers, *Larnax* Miers, *Physalis* L., *Margaranthus* Schlecht., *Leucophysalis* Rydb., *Deprea* Raf., *Chamaesaracha* A. Gray, *Capsicum* L., *Athenaea* Sendt., *Vassobia* Rusby, *Acnistus* Schott, *Iochroma* Benth., *Dunalia* H. B. K., *Saracha* R. et P., *Exodeconus* Raf., *Oryctes* S. Wats., Nov. Gen., *Physalis spruceana* A. T. Hunz.) Tribe Datureae (*Datura* L., *Brugmansia* Pers.) Tribe Jaboroseae (*Jaborosa* Juss., *Trechonaetes* Miers, *Salpichroa* Miers, *Nectouxia* H. B. K.) Tribe Lycieae (*Lycium* L., *Grabowskia* Schlecht., *Phrodus* Miers) Tribe Nicandreae (*Nicandra* Adans.) Tribe Solandreae (*Solandra* Sw., *Trianaea* Pl. et Lind.) Tribe Juanulloeae (*Juanulloa* R. et P., *Ectozoma* Miers, *Merinthopodium* Donn. Sm., *Rahowardiana* D'Arcy, *Markea* L. C. Rich., *Schultesianthus* A. T. Hunz., *Hawkesiophyton* A. T. Hunz., *Dyssochroma* Miers); Subfamily Cestroideae: Tribe Cestreae (*Cestrum* L., *Sessea* R. et P., *Vestia* Willd., *Metternichia* Mik.) Tribe Nicotianeae (*Nicotiana* L., *Petunia* Juss., *Fabiana* R. et P., *Nierembergia* R. et P., *Bouchetia* DC., *Latua* Phil., *Combera* Sandw., *Pantacantha* Speg., *Benthamiella* Speg.) Tribe Schwenckieae (*Schwenckia* L., *Protoschwenckia* Soler., *Melananthus* Walp.) Tribe Salpiglossideae (*Salpiglossis* R. et P., *Leptoglossis* Benth., *Hunzikeria* D'Arcy, *Reyesia* Clos, *Browallia* L., *Streptosolen* Miers, *Brunfelsia* Sw., *Schizanthus* R. et P.) Tribe Parabouchetieae (*Parabouchetia* Baill.) Incertae sedis (*Heteranthia* Nees et Mart.).

Seeds and Embryos

Two basic types of organization are encountered in the family concerning the morphology of the seeds and their enclosed embryos. The subfamily Solanoideae shows, in general, discoidal flat seeds, more or less compressed; these external features are correlated with linear circinnate or annular embryos i.e. clearly curved and never straight (Fig. 1). On the other hand, the subfamily Cestroideae is characterized by prismatic or reniform or subglobose (or with a slightly different form) seeds, but never discoidal nor compressed; the embryos may be straight or bent and are always immersed in a copious endosperm (Fig. 2).

An important feature, heretofore overlooked, in relation to the morphology of embryos is found in the cotyledons. When the embryo is curved, it is very important to ascertain how the cotyledons are arranged. In Cestroideae these are always incumbent or slightly oblique, but in Solanoideae two cases are encountered; the common one—in Solaneae, Datureae, Lycieae, Jaboroseae, Nicandreae and Solandreae (*Solandra*)—is when the cotyledons are incumbent or, by exception, oblique (*Trianaea*); moreover, the endosperm is usually abundant. The tribe Juanulloeae, on the contrary, shows accumbent (*Juanulloa, Schultesianthus*) or sometimes oblique (*Markea, Ectozoma*) cotyledons, and the endosperm is notoriously scanty.

Stamens

To understand the taxonomy of Solanaceae and its evolutionary relationships, usually obscured by convergence, it is of fundamental importance a thorough understanding of the general morphology of the flower and, above all, of its androecium. Hence, it may be pertinent here to discuss briefly three peculiarities of this floral whorl which are frequently ignored in the literature.

Monadelphy

It is a well known fact that the stamens of Solanaceae (Fig. 3) are usually free from each other. But in some genera, for example *Solanum*, the basal portion of the filaments forms an annular structure adnated to the corolla tube; usually it is unnoticed, except for the rare cases when its upper rim is free from the corolla (*S. argentinum* Bitter et Lillo, *S. monadelphum* Heurck et Müll., etc.). Another striking case of monadelphy is exemplified by *Parabouchetia*, where the connectives—enlarged and protruding at the upper part of the anthers—become coalescent. Also it is worthwhile to remember here the case of *Nierembergia* (Cestroideae, Nicotianeae) where the majority of the species show a slight adnation between the filaments, although a few have independent

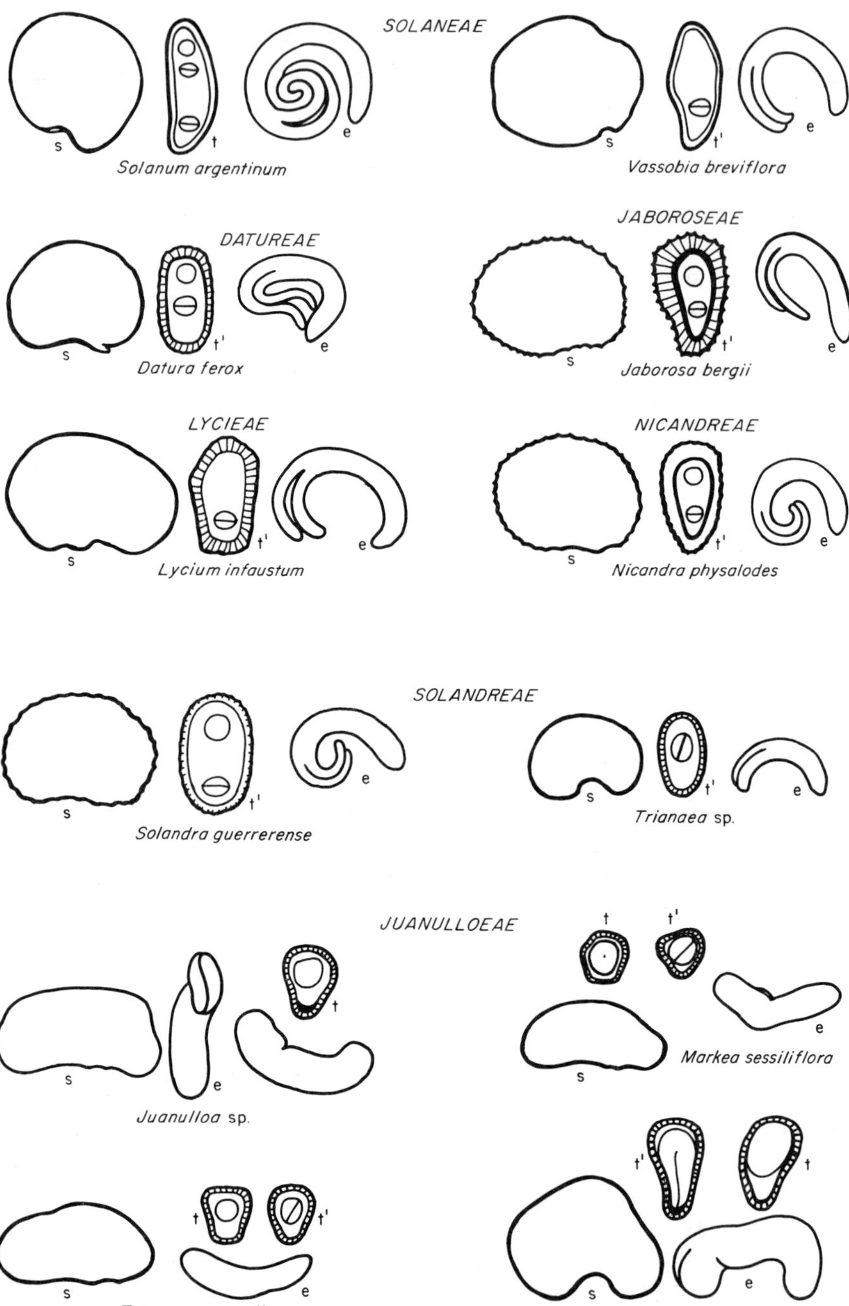

Fig. 1. Seeds and embryos (subfamily Solanoideae). Selected examples in seven tribes. Symbols: s, outline of the seed; t, transverse section near the hilum; t′, transverse section at approximately a quarter to a third of the distance from the pole opposite to the hilum, or near the centre of the seed; e, embryo. Magnifications are different in each case, therefore not comparable.

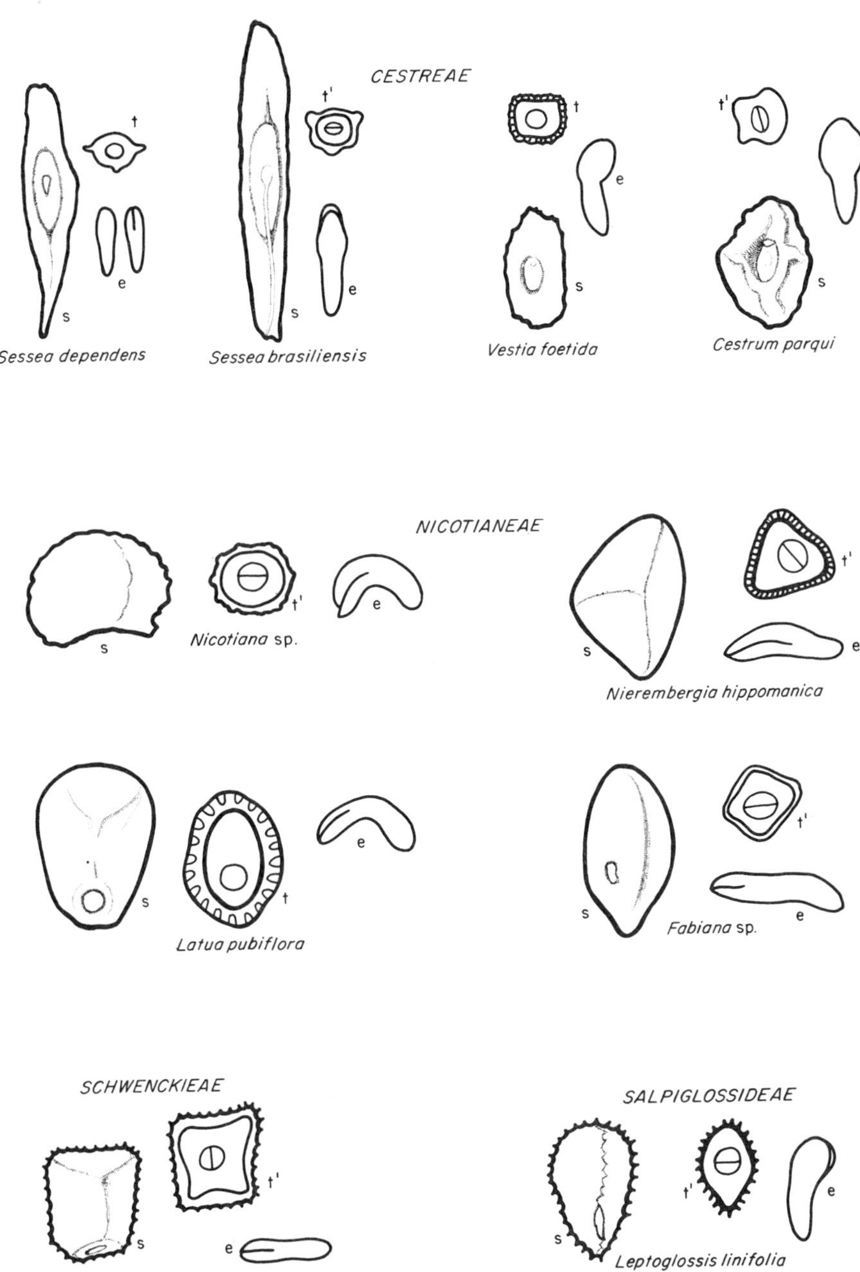

Fig. 2. Seeds and embryos (subfamily Cestroideae). Selected example in four tribes. Symbols and magnifications as in Fig. 1.

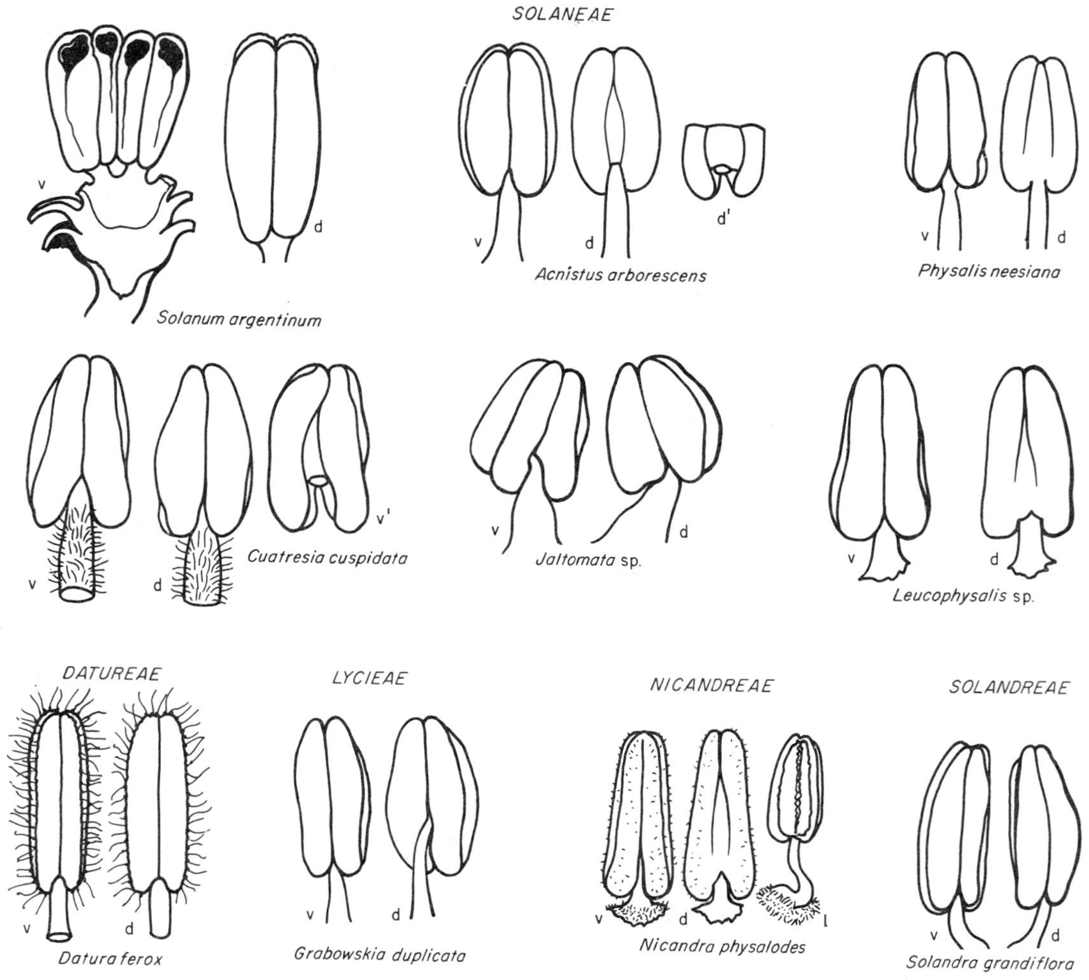

SOLANEAE
v
d
Solanum argentinum
v
d
d'
Acnistus arborescens
v
d
Physalis neesiana
v
d
v'
Cuatresia cuspidata
v
d
Jaltomata sp.
v
d
Leucophysalis sp.
DATUREAE
v
d
Datura ferox
LYCIEAE
v
d
Grabowskia duplicata
NICANDREAE
v
d
l
Nicandra physalodes
SOLANDREAE
v
d
Solandra grandiflora

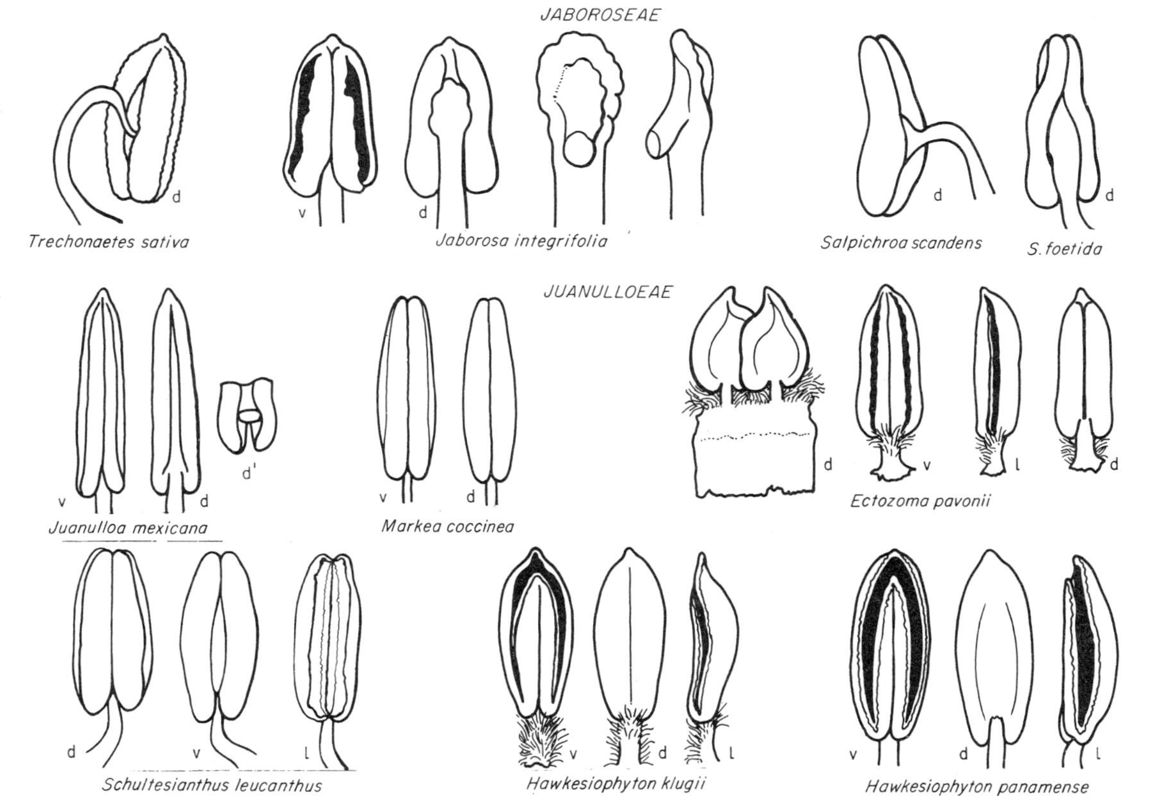

Fig. 3. Stamens (subfamily Solanoideae). Selected examples of monadelphy, adnation between the thecae, and insertion of the filament on the anther (seven tribes). Symbols: v, ventral face of anther; v′, basal ventral sector of anther, the filament having been removed; d, dorsal face of anther; d′, basal dorsal sector of anther, the filament having been removed; l, lateral view of anther; b, basal view of anther; o, staminode. Magnifications: diverse and not comparable between the genera or tribes.

stamens. Finally, in *Ectozoma* we find a well-developed ring adjacent to the insertion of the stamens to the corolla.

The Thecae and its Adnation

The tribes Solaneae, Datureae, Nicandreae, Solandreae and Juanulloeae display anthers with both thecae adnated from base to apex (Figs 3 and 4); sometimes, at the proximal pole, the thecae are separated, but the length of this sector is usually negligible. Anyway, there are exceptions as, for example, the case of *Cuatresia* (Solaneae), but at any rate the separation between both anther halves never reaches one-third of its total length. Therefore, the tribe Lycieae stands out in the Solanoideae, owing to the pronounced basal separation between the thecae: it exceeds one-third of its longitude and sometimes even equals the whole lower half.

The Insertion of the Filament on the Anther

It is surprising how this aspect of the androecium morphology in Solanaceae has escaped the attention of the botanists; in my opinion, the resulting situation is responsible for the unsurmountable difficulties to define many generic delimitations, particularly in the tribe Solaneae. In order to understand this feature, the stamens should be studied in the bud, before occurs the dehiscence.

The genera of Nicandreae, Solandreae, Datureae, and some of the Juanulloeae (*Markea, Schultesianthus*), are characterized by a typical basal insertion, i.e. the filament adheres to the lower end of the anther, in between both extremes of the thecae. Throughout the Jaboroseae and Lycieae the picture is uniform, each anther being held from its dorsal face. Finally the third type of insertion is found in some genera of Solaneae (*Cuatresia, Jaltomata*) and Nicotianeae (*Bouchetia, Nierembergia*), where the anthers are clearly ventrifixed.

It is convenient to point out that dorsifixed anthers also characterize some genera of Solaneae (v. gr. *Acnistus*) and Juanulloeae (*Juanulloa, Hawkesiophyton*, etc.), but here the insertion takes place on the back of the proximal end of the thecae; in these cases one must be careful not to mistake them with the basifixed type.

The constancy of this character is remarkable throughout the species of each genus; thereupon it should be taken always in consideration with other peculiarities to circumscribe the generic concepts.

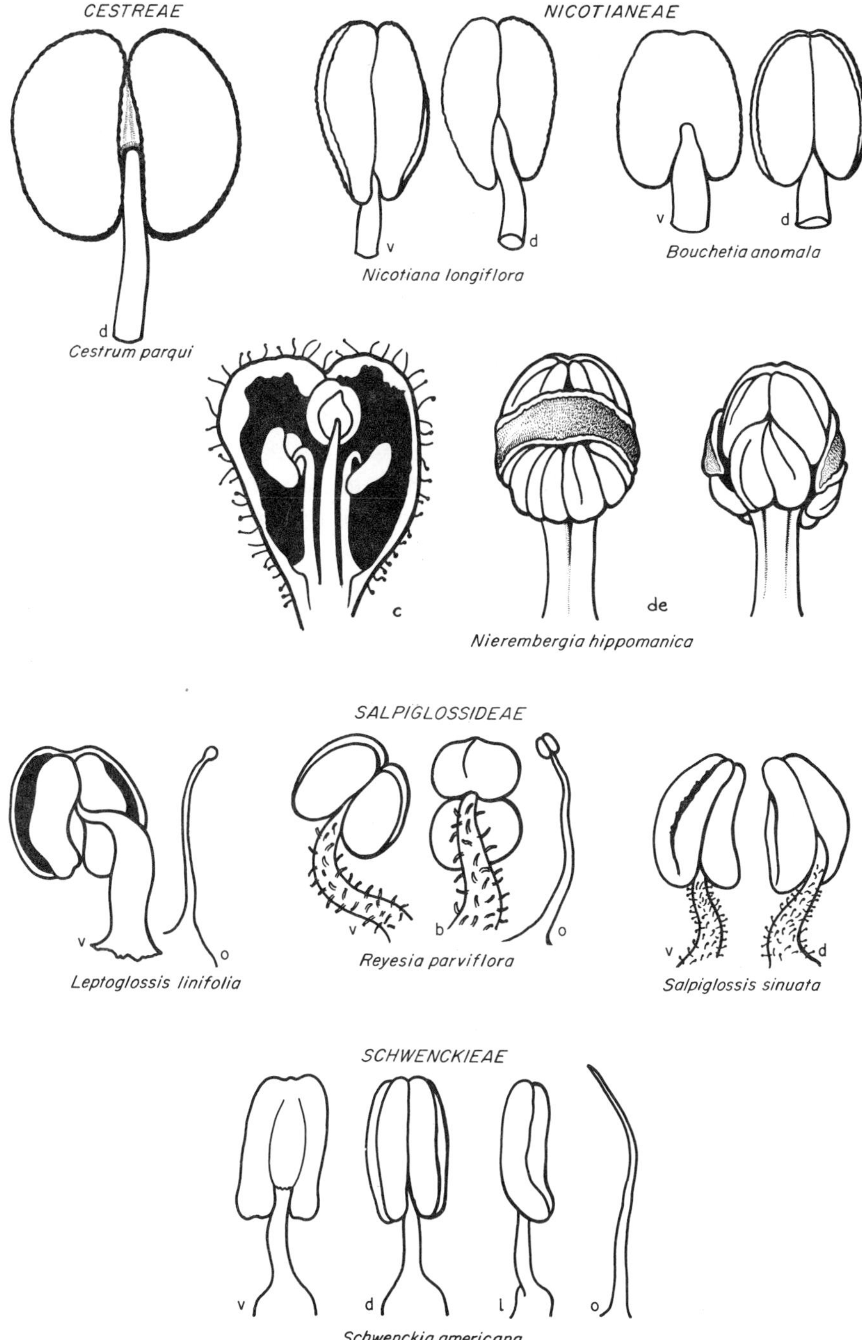

Fig. 4. Stamens (subfamily Cestroideae). Selected examples of staminal structure in four tribes Symbols: c, sector of bud with two corolla lobes and three stamens that have been forced to separate from each other; d, e, the androecium showing monadelphy and the stigma embracing the anthers; remaining symbols and abbreviations as in Fig. 3.

Conclusions

The system of Solanaceae by Wettstein (1895) has been modified according to the available information on the 69 admitted American genera (59 of these are neotropical); these changes are based mainly on characters of the flower (pedicel, aestivation, stamens, ovary, stigma), the seed and the embryo. In all, two subfamilies were recognized: Solanoideae (with seven tribes) and Cestroideae (with five tribes). For the time being, the genus *Heteranthia* has not been placed in the system; the acknowledgment of subtribal categories was postponed as well, until new and better data are at hand.

The results are undoubtedly provisional: much is still to be known from many tropical genera; for example, fully mature seeds of *Trianaea* and *Dyssochroma* were collected only once or twice, nothing is known about the chromosome numbers in the eight genera of the tribe Juanulloeae, and the monotypic genus *Parabouchetia* is based on a single floriferous collection dated 150 years ago.

Acknowledgements

I am grateful to Mrs N. M. de Flury for preparing the drawings, and to the Consejo Nacional de Investigaciones Científicas y Técnicas (Argentina) for its support.

References

Baehni, Ch. (1946). L'overture du bouton chez les fleurs de Solanées. *Candollea* **10**, 399–492.

Hunziker, A. T. (1977). Estudios sobre Solanaceae VIII. *Kurtziana* **10**, 7–50.

Hunziker, A. T. (1979). South American Solanaceae: a synoptic survey. *In* "The Biology and Taxonomy of the Solanaceae" (J. G. Hawkes and R. N. Lester eds) p. 49–85. Academic Press, London and New York.

Wettstein, R. von (1895). Solanaceae. *In* "Die Natürliche Pflanzenfamilien" (H. G. A. Engler and K. A. E. Prantl, eds) Vol. 4 p. 4–38. Engelmann, Leipzig.

Neotropical Saprophytes

P. J. M. MAAS

University of Utrecht, Netherlands

Introduction

In 1938 Jonker of the Utrecht University published "A Monograph of the Burmanniaceae", and in the same year the German botanist Giesen published a revision of the Triuridaceae in Engler's "Pflanzenreich". These two families, together with the family of Gentianaceae, are the most important saprophytic families in the neotropics. Until now the saprophytic genera of the Gentianaceae were never thoroughly investigated. Forty years later, with a team of four people, three advanced students (Mrs H. Snelders, Mr J. van Benthem, and Mr P. Ruyters) and myself, we started to revise those three families, comprising approximately 80 species, for Flora Neotropica. In the beginning we studied herbarium specimens only, but it soon became evident that that was not sufficient, and therefore I made an extensive collecting trip last year in various parts of South America during which time I restricted myself almost completely to collecting and studying saprophytes. The present paper is a summary of the first results of our investigations.

Most saprophytes are very small plants, lacking chlorophyll, their stems being white, yellow, blue, red or purple. They are often so tiny (sometimes only a few centimeters high) that they are overlooked by most botanists. Therefore the number of herbarium collections is very low. However, if one is concentrating in the field on saprophytes, they often appear not to be rare at all, but even common.

The favourite habitat of saprophytes is undisturbed lowland rain forest. In their main centre of distribution, the Guianas, they strongly prefer marshy Mora forest. This forest type is often found in periodically inundated areas like margins of streams. *More excelsa* in this forest type is strongly dominant in the upper storeys as well as in the under-growth, where their seedlings form dense thickets. The forest floor in this forest type is covered with a thick layer of leaf-litter. The thickest layer of leaf-litter is found between the buttresses of *Mora excelsa*, and it was there that we found most of our saprophytes. Another place they prefer is the margin of small forest streams. Most saprophytes occur in rain forests, but there are a few exceptions to this rule. A species like the very common Gentianacea *Voyria aphylla* prefers dry vegetations like savannas, although it also occurs in rain forests. Neotropical sap-rophytes are mostly restricted to low elevations, between sea level and *c*. 1000 m. There are only a few species reaching 2000 m, an example of this is the Burmanniacea *Gymnosphon suaveolens* (of Central America and western South America), which is mainly found at elevations between 1000 and 2300 m.

In the Guianas, a small fern, *Schizaea fluminensis*, is often found growing in association with saprophytes. As soon as we saw that fern we were bound to find saprophytic phanerogams. It is not impossible that this small fern will also turn out to be saprophytic. In Manaus I contacted two mycologists, Dr R. Singer and Dr T. St John. They identified some of the fungi of different saprophytes for me, and I hope that sometime they will find out if *Schizaea fluminensis* is indeed sap-rophytic.

How are the seeds of saprophytes dispersed? The fruits of all sap-rophytes contain numerous, very small seeds (less than 1 mm long)*. In some cases the seeds are provided with two very long hair-like outgrowths, the terminal one representing the integument and the basal one the funicle. This type of seed more or less comes into a category which is called "Fadenflieger" in German and "plumed dia-spores" in English (Ulbrich, 1928; van der Pijl, 1972). It may be compared with the seeds of the genus *Tillandsia* (Bromeliaceae) and those of most Asclepiadaceae, and there seems to be no question that they are dispersed by the wind. This type of seed is found in three species of the gentianaceous genus *Voyria* and one species of *Burmannia*. It is not surprising that the four species all have a wide distribution. The three species of *Voyria* are the only ones known in the West Indian Islands.

The seeds of most saprophytes lack the wing-like structures and it is not known exactly how they are dispersed. It is possible that as they grow along creeks (like the Burmanniacea *Campylosiphon purpurascens*)

---

* Professor U. Hamann (Ruhr-Universität Bochum, BRD) and collaborators are now intensively studying the seed structure of Burmanniaceae and Triuridaceae, their results will be incorporated in our Flora Neotropica revision.

they may be dispersed by the water. They have small, light seeds with a reticulate surface pattern, and they may well float on the water surface. Another possibility is dispersal by earthworms as already suggested by Beccari (1880) who found a saprophytic orchid in a place where the ground was full of earthworms. He suggested that the seeds might be eaten by earthworms, which are in turn eaten by birds (thrushes and ortolans), and later ejected some distance away. Beccari (1880) suggested that other saprophytes might be dispersed in the same way. I am not sure whether these suggestions about water- and worm-dispersal are really relevant, but what I hope is that they will induce tropical field workers to become more interested in studying the seed-dispersal of the various saprophytes.

Three saprophytic families will now be dealt with in some detail. I have not included the saprophytic orchids as that group still needs to be investigated (by Dr G. F. J. Pabst, Herbarium Bradeanum, Rio de Janeiro, Brazil).

Gentianaceae

There occur two genera in the neotropics:

(i) *Voyria* has about 20 species, occurring throughout the neotropics, but with its main centre of distribution in the Guianas. One species, *V. primuloides*, inhabits west tropical Africa, whereas the genus does not occur in Asia. The genus is taxonomically not highly problematic, as it has very good distinguishing characters in the calyx, corolla, stamens, nectarial glands (on ovary and calyx) and seeds. In some species the seeds have two hair-like wings.

(ii) *Voyriella* is a genus restricted to the Guianas and the Amazon Basin. They are white, very small (a few centimeters high), with a very compact, almost globose inflorescence.

Triuridaceae

This is a small (exclusively saprophytic) family with a pantropical distribution. Its centre of distribution is in Asia (three genera with 53 species), whereas in Africa two genera with five species occur, and in America four genera with 12 species. Apart from the low number of genera and species in the neotropics, the number of known herbarium collections is very low (less than 200). There are four genera in the neotropics:

(i) *Sciaphila*, a genus with a pantropical distribution. The number of species in the neotropics is seven. *Sciaphila albescens* is rather peculiar because of its very fragrant flowers with a smell like that of *Convallaria majalis* (lily of the valley). This is the only neotropical saprophyte in which I could observe a distinct smell, although some species of the Gentianacea *Voyria* also seem to have a faint smell (Ruyters, pers. comm.). *Sciaphila* has another very peculiar species, namely *S. purpurea*. This species has termite nests as its habitat; it grows there so well that it sometimes reaches a height of *c.* 1·5 m.

(ii) *Soridium* is a monotypic genus restricted to northern South America; it is closely related to *Sciaphila*.

(iii) *Triuris* is a rare, neotropical genus with two species. Its tepals possess three very long tail-like appendages (from which the genus name Tri . . . uris has been derived), which probably attract insects. So far there are no observations on pollination of *Triuris*. The male flowers of *Triuris* (see Fig. 1) have a conical outgrowth of the receptacle in which the stamens are imbedded. The female flowers consist of a flat receptacle with many ovaries. Between those ovaries there are slime ducts which according to Poulsen (1890) may play a role in seed dispersal.

(iv) *Peltophyllum* (=*Hexuris*) has two species, occurring in south-eastern Brazil, northern Paraguay and northern Argentina and is closely related to *Triuris*.

Burmanniaceae

This large family has a pantropical distribution. All neotropical genera are cited in Fig. 1. This paper only deals with some of the problematic genera.

Tribe Burmannieae

(i) The most common genus of this tribe is *Burmannia*. This pantropical genus has only one saprophytic species in the neotropics, namely *B. tenella*. The remaining 17 neotropical species all have small (to large) green leaves. These non-saprophytic species are frequently found in savannas.

(ii) *Gymnosiphon* is another pantropical genus. It is without any doubt the most problematic genus of all saprophytes, not so much from the taxonomic point of view, most species having good distinguishing characters in inflorescence and flowers, but rather because the upper part of the perianth (tepals, stamens and stigma) fall off in a very early stage, and all that is left of the flower is a naked tube or a "gymno . . . siphon" (see Fig. 1). Most herbarium material lacks com-

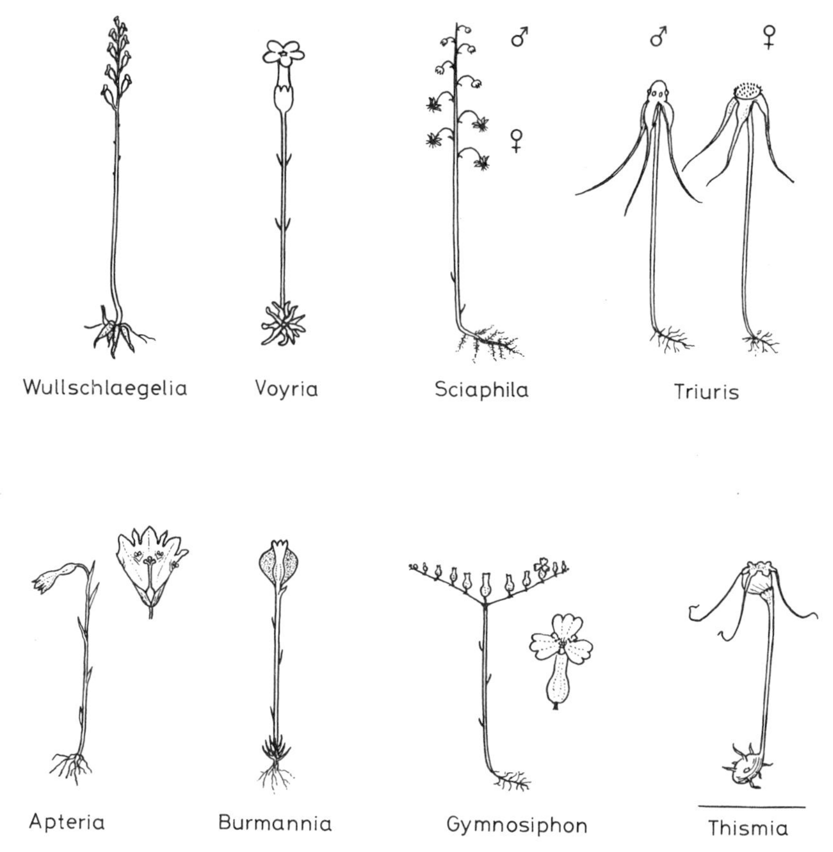

Fig. 1. Neotropical saprophytes

ORCHIDACEAE: *Pogoniopsis* (2 spp.), *Uleiorchis* (1 sp.), *Wullschlaegelia* (2 spp.). GENTIANACEAE: *Voyria* (15 spp.), *Voyriella* (2 spp.). TRIURIDACEAE: flowers mostly unisexual, many free ovaries, tepals 3–6(–8), often with tail-like appendages, *Sciaphila* (7 spp.), *Soridium* (1sp.), *Peltophyllum* (=*Hexuris*)* (1 sp.), *Triuris* (2 spp.), BURMANNIACEAE: flowers hermaphrodite, tepals six, base tubular, stamens three or six, ovary inferior, one- or three-locular, (i) Burmannieae *Apteria* (1 sp.), *Burmannia* (18 spp.), *Campylosipon* (1 sp.), *Cymbocarpa* (2 spp.), *Dictyostega* (1 sp.), *Gymnosiphon** (15 spp.), *Hexapterella* (1 sp.), *Marthella** (1 sp.), *Miersiella** (1 sp.), (ii) Thismieae *Glaziocharis** (1 sp.), *Thismia** (7 spp.), *Triscyphus** (1 sp.), *Mamorea** (1 sp.)

* means genera are insufficiently known

plete flowers, and that makes it almost impossible to identify the material (as in the *G. divaricatus*-complex). I therefore urge all collectors to conserve at least some flowers of the plants in alcohol.

(iii) *Marthella* is a monotypic genus, restricted to Mount Tucuche in northern Trinidad. It is only known from two collections, the most recent made in 1898! I hope that this genus, whose systematic position is not completely clear, will be rediscovered.

Tribe Thismieae

The representatives of this tribe are the most fascinating and attractive saprophytes, particularly their urceolate to campanulate flowers which possess intriguing, sometimes tail-like appendages (see Fig. 1) Perhaps there is no group of saprophytes in the neotropics as poorly known as the Thismieae. One of the nine species, the Panamanian *Thismia panamensis*, is rather common and well known, but of the remaining eight species there exist only 15 herbarium collections, most of which were collected about a century ago by the famous botanist Glaziou. He collected most of his specimens near Serra Macahé (Novo Friburgo), Rio de Janeiro, Brazil. About a year ago, I visited that locality, but all original vegetation was almost completely destroyed so that no saprophytes are likely to be ever expected from that place again. It is very strange, however, that no collectors since Glaziou have collected any of these species in neighbouring localities where there still exists some good vegetation. Or are those species really extinct?

Concluding, I should like to state that I sincerely hope that in the future more botanists may pay attention to the completely neglected, but very fascinating group of saprophytes. Their attention may help to come to a better understanding of the biology and the systematics of neotropical saprophytes.

References

Beccari, O. (1880). Dispersal of seeds by earthworms. *Malesia* **3**, 325.
Giesen, H. (1938). Triuridaceae. *In* "Pflanzenreich" (A. Engler, ed.) Series IV, Part 18, 1–84.
Jonker, F. P. (1938). "A Monograph of the Burmanniaceae" p. 1–279. Kemink, Utrecht.
Poulsen, V. A. (1890). *Triuris major. Bot. Tidskr.* **17**, 293.
Pijl, L. van der (1972). "Principles of Dispersal in Higher Plants." Springer Verlag, Berlin.
Ulbrich, E. (1928). "Biologie der Früchte und Samen." Springer Verlag, Berlin.

Comments on Breeding Systems in Neotropical Forests

M. T. KALIN ARROYO

University of Chile, Santiago, Chile

Introduction

Botanists have paid little attention to breeding systems, dispersal mechanisms and other facets of reproductive biology in tropical vegetation. It is evident, nevertheless, that knowledge of these is essential if patterns of population differentiation in tropical ecosystems are to be fully understood, and if the relative fragility of different kinds of tropical vegetation is to be determined. Whether neotropical forests have higher proportions of outbreeding species than paleotropical forests, whether aseasonal tropical vegetation is more closely adapted for outcrossing than seasonal types, and the extent to which adverse weather conditions in tropical cloud forests have lead to the abandonment of obligate outcrossing systems are but a few examples of questions, the answers to which would lead to the understanding of the relative migrational capacity of different vegetational types, not to mention the interpretation of vegetational changes in more recent, post-Pleistocene times. High mountain neotropical floras are recent as compared with lowland floras. To what extent have the breeding systems of migrants from lowland tropical floras been changed in response to lower pollinator diversity in high mountain areas as they have moved into these regions? Human activity has, and continues to enhance the spread of secondary vegetation in the tropics, and if secondary vegetation is the near-future for these regions, far greater attention should be devoted to the reproductive methods of colonizing and early successional species. Such information is not only of great theoretical interest, but also of practical value for future management policies and for the selection of species of potential economic value (Bawa, 1976). The floras of tropical oceanic islands derived by long-distance dispersal are a special case. The floras on adjacent mainlands, from which the progenitors of colonizers of tropical oceanic islands are derived, are likely to be characterized by highly evolved pollination arrangements. Pollinator-independent breeding systems, such as apomixis and self-compatibility might have been favoured to a greater degree among migrants to tropical islands as compared with those on temperate oceanic islands, and the breeding systems of temperate and tropical oceanic floras of similar ages might be profitably compared.

Present State of Knowledge

Breeding systems surveys entailing experimental tests for self-incompatibility and apomixis have been recently completed for a secondary, deciduous forest (Ruiz and Arroyo, 1978) and a montane cloud forest (Arroyo and Cabrera, 1978; Sobrevila and Arroyo, in prep.) in northern Venezuela. The secondary forest is situated at 110 m elevation, 10°30′ N, 66°53′ W on the foothills of the south-west side of the Valley of Caracas, Edo. Miranda, an area which receives *c.* 762 mm annual average precipitation and sustains a long dry season from December through April. The median annual temperature is 21·8°C. The secondary forest is poorly structured, comprising an emergent tree stratum of 5–8 m height and a lower stratum of shrubs and herbaceous perennials. A total of 22 tree species have been collected on a 2-hectare site. According to historical records, the area occupied by the present forest was extensively grazed and burned up until the late 1930s and the present forest, thus is probably less than 40 years old. The montane cloud forest is situated at approximately the same latitude at Altos de Pipe, 1700 m elevation along the top of an exposed ridge. The area receives *c.* 994 mm average annual precipitation and has a median annual temperature of 17·5°C. A long wet season extends from April through January, alternating with a short, little-pronounced dry season in February and March. During the wet season the forest may be covered in dense cloud for periods ranging from one to two hours in the late morning up to several days at a time. The montane forest is structurally complex and is considered to represent the original vegetation of the area. Three well defined woody strata are evident: an upper discontinuous emergent stratum, followed by a lower, continuous tree stratum and then an inferior stratum of small tree, shrubs and perennial herbs. The emergent tree may achieve 30 m height and a total of 44 tree species have been collected on the 1·6-hectare study site.

In the secondary deciduous forest pollination by small bees predominates. In the montane cloud forests, trees of the mergent stratum and upper tree stratum are principally pollinated by small and medium sized bees, whereas hummingbirds are the main pollinators of shrubs and perennial herbs.

In the secondary forest we asked whether selection has favoured pollinator-independent breeding systems during the early stages of succession, whereas in the cloud forest we set out to assess whether adverse weather conditions limit the possession of outbreeding systems, and if not, how the forest community deals with conditions unfavourable for insect pollination. The secondary forest revealed a high incidence of species adapted for obligate outbreeding (Table 1). Among the tree species, 24% are dioecious, whereas 67% are genetically self-incompatible. A total of 91% of the trees in this forest, therefore, are adapted for obligate outbreeding. Among other life-forms, self-incompatibility, sometimes associated with distyly is the predominant breeding system, and indeed, only two of a total of 22 species studied in detail were found to be genetically self-compatible to any degree. Both the latter (*Capparis flexuosa*: Capparidaceae, and *Securidaca scandens*: Polygalaceae) are frequently among the first species to appear on newly exposed sites in the area and these would appear to represent early colonizing elements of the secondary deciduous forest. The situation in the montane, cloud forest is quite different from that in the secondary deciduous forest, and indeed, also departs significantly from that described by Bawa (1974) for a mature, lowland, semi-deciduous forest in Costa Rica (Table 1). The 30% dioecism encountered among the tree species of the montane, cloud forest, is the highest frequency of dioecy recorded to date in neotropical forests (c.f. summary of data from several forests in Bawa and Opler, 1975). In the secondary deciduous forest and semi-deciduous forest only 9% and 14% respectively of the species are genetically self-compatible, whereas in the montane cloud forest well over half of the hermaphroditic tree species are self-compatible, and self-compatible species comprise 43% of the tree species in the forest. Dioecism is essentially lacking among other life-forms in the montane cloud forest, however, as among tree species, self-compatibility is strongly represented. Two shrubs (*Clidemia fendleri* and *Miconia spinulosa*) are non-pseudogamous apomicts. The total proportion of trees adapted for obligate outbreeding in the montane, cloud forest is 56% as opposed to 91% and 76% in the secondary deciduous and mature, semi-deciduous forests, respectively (Table 1).

Caution must be exercised in interpreting data on self-compatibility

Table 1. Comparison of relative proportions of different breeding systems among tree species in three neotropical forests.

Breeding system (% species)	Semi-deciduous forest, Costa Rica*	Secondary deciduous forest, Venezuela†	Montane cloud forest, Venezuela‡
Dioecious	22	24	30
Monoecious	10	0	3
Hermaphrodites in original sample %	39	43	43
Self-incompatible§	54	67	26
Self-compatible	14	9	43¶
Total % species adapted for obligate outcrossing	76	91	56

* Bawa (1974).
† Ruiz and Arroyo (1978).
‡ Sobrevila (1978); Arroyo and Cabrera (1978).
§ Based on the proportion of self-incompatible species in the sample size mentioned.
¶ Includes monoecious species.

in neotropical forests. In studies carried out to date, seed set following emasculation has been used as a test for apomixis, while seed set following artificial self-pollination, but not with emasculation, has been taken as an indication of the presence of genetic self-compatibility. These tests permit detection of non-pseudogamous apomixis, however they do not allow differentiation of pseudogamous apomicts from sexual, self-compatible species. Many of the species judged as self-compatible sexual species, therefore may reproduce apomictically, or at least do so on a facultative basis. This problem clearly is not restricted to breeding work in tropical plants, for it is a potential source of error in all breeding studies in which conventional bagging and artificial selfing techniques are used. It can only be satisfactorily resolved by recourse to elaborate embryological work although protein electrophoresis has an obvious application here. Irrespective of this limitation, present work in neotropical forests can be used to estimate the proportion of obligately outbred species in a community.

Distance Effects and Self-compatibility

A number of new questions have surfaced as recent, community-based information on breeding systems in tropical forests has come to light. Baker (1959) and Fedorov (1966) suggested that self-compatibility might be frequent in species-rich tropical forests, given that wide spacing of individuals, and the structural complexity of the forest itself, would restrict inter-tree pollinator movement. The following four questions are as yet unsatisfactorily answered:

(i) Is there any evidence that low population density in tropical forests actually limits seed production in obligately outbred species?

(ii) If so, will this automatically lead to the selection of self-compatibility?

(iii) Assuming possible prezygotic selection for self-compatibility, is the post-zygotic survival of the progeny of a self-compatible genotype automatically ensured?

(iv) What avenues are open to self-compatible lineages, should they become established?

Some insight to the first problem is obtained by consideration of reproductive efficacy (Ruiz and Arroyo, 1978). The latter is defined as percentage of fruit set following natural (open) pollination, compared with percentage of fruit set following artificial cross-pollination. Reproductive efficacy measures seed set relative to pollination conditions, and therefore it also provides a measure of pollinator efficiency in self-incompatible and dioecious species. In genetically self-compatible species it estimates natural cross-pollination, geitonogamous selfing, and in the case of autogamous species, intra-flower selfing in addition. Use of this parameter for comparisons of reproductive output eliminates the problem of variation in post-fertilization fruit-drop due to differences in the physiological state of different individuals and species in a population.

Comparing reproductive efficacy in the three neotropical forests, the values for the self-compatible species are high, and roughly of similar magnitude (0·66–0·73; Table 2). Noteworthy differences appear, however, in comparing the self-incompatible species of the three forests. The average reproductive efficacy for self-incompatible species in the semi-deciduous Costa Rican forest is 0·16 or approximately a quarter of the value registered for self-compatible species in the same forest. The average reproductive efficacy of self-incompatible species in the low-diversity, secondary deciduous forest is 0·36, or approximately double the value obtained among species of the Costa Rican forest. Reproductive efficacy has been established for only two

Table 2. Comparison of reproductive efficacy for species with different breeding systems in three neotropical forests.

	No. species studied	Reproductive efficacy Range	Mean
Semi-deciduous forest, Costa Rica*			
Self-incompatible hermaphrodites	22	0·00–1·03	0·16
Self-compatible hermaphrodites	4	0·30–1·16	0·67
Secondary deciduous forest, Venezuela†			
Self-incompatible hermaphrodites	6	0·16–0·57	0·36
Self-compatible hermaphrodites	3	0·03–1·66	0·66
Montane cloud forest, Venezuela‡			
Self-incompatible hermaphrodites	2	0·65–1·15	0·90
Self-compatible hermaphrodites	8	0·47–1·12	0·73
Apomictic hermaphrodites	2	1·18–1·24	1·15

* Bawa (1974).
† Ruiz and Arroyo (1978).
‡ Sobrevila (1978).

self-incompatible species of the montane, cloud forest, which when averaged give a value similar to that obtained for self-compatible species. The Costa Rican forest is richest in species numbers. Assuming that average density declines with increasing species richness, it would appear that wide spacing of individuals can lead to significant lowering of seed set in self-incompatible species.

Although wide spacing leads to lower seed output in self-incompatible species, species-rich forests may contain high proportions of obligately outbred species, as exemplified by the Costa Rican forest. One factor that might be instrumental in the maintenance of large numbers of outbreeding species under circumstances that must eventually lead to total reduction in seed output, is inbreeding depression following the acquisition of self-compatibility in an otherwise, highly outbred species. If individuals of an otherwise genetically self-incompatible species are prevented from setting seed because of distance effects, mutations conferring self-compatibility will be automatically selected for. Inbreeding depression in the first generation progeny of a previously outcrossed lineage might so severe as to eliminate the progeny entirely. Sorensen (1969) determined the proportion of zygotic lethals following enforced self-fertilization in the highly outcrossed *Psuedotsuga menziesii* (Pinaceae). Many trees had in excess of 90% zygotic lethals. Post-zygotic lethals and deleterious genes would further reduce survival during later phases of development. Interspecific competition with non-inbred progeny of other self-incompatible on dioecious species would also lessen the survival chances of inbred progeny. Each time an individual is prevented from setting set due to distance effects, it would either leave no progeny at all, or alternatively, if self-compatible, its progeny might be largely eliminated. Upon eventual removal of such an individual from a population, the average distances between individuals would momentarily increase once more, bringing things back into line. Individuals replacing an eliminated adult would be either outcrossed progeny of other species in the area, or, if by chance dispersal should bring diaspores of the same species into the area, other outcrossed progeny of the same species. This type of dynamic expansion/contraction of local species ranges accompanied by replacement of one species with another must be going on in all obligately outbred species in tropical forests, and the combination of the two processes would account for maintenance of high proportions of obligately outbred species in an area.

How does wide spacing affect seed set under obligate cross-pollination in genetically self-incompatible species and when do distances between conspecific individuals become critical for seed set? Earlier authors (e.g. Baker, 1959), stressed the flight capabilities of pollinators as a limiting factor for the cross-pollination of widely spaced individuals. However, it is now evident that tropical forest species are serviced by pollinators capable of flight distances exceeding the average distances separating conspecific individuals (e.g. Janzen, 1971; Heithaus *et al.*, 1975; Frankie, 1976). These considerations force the conclusion that distance must exert its effect in another manner. Heslop-Harrison (1975) has shown that callose is deposited on stigmas as part of the incompatibility reaction. Howlett *et al.* (1975) demonstrated a depression in seed set when compatible pollen grains are intermixed with incompatible pollen grains and applied to the stigma. Self-fertilization could be stimulated to some extent in self-incompatible species when wall components (containing self-incompatibility substances) of self-compatible grains were mixed with self-incompatible grains on the stigma. These findings might have direct bearing on levels of seed set in self-incompatible hermaphrodites, especially when widely spaced. As has been pointed out previously (Arroyo, 1976), intraplant geitonogamous pollen transfers will greatly exceed interplant transfers in animal-pollinated species. In *Copaifera pubiflora*, a tropical legume tree of the central savannas of Venezuela, pollen may be deposited upon almost 100% of the stigmas, yet cross-fertilization is usually in the order of 10% of total flower number even in individuals separated from cross-pollen sources by distances nowhere reaching the known flight capacities of its bee pollinators. This effect will be multiplied as individuals become more widely spaced. Reduced seed set in widely spaced individuals in tropical forests perhaps, is not insomuch a direct effect of the isolation of individuals beyond pollination range, but rather a secondary effect resulting from the accumulation of incompatible, self-pollen on stigmas due to exaggerated intra-individual pollen transference. The higher seed output of some dioecious species relative to self-incompatible species might find its explanation in lack of inhibitory effects of self-pollen on the stigmas of female plants, as such plants will bear only compatible grains.

The previous arguments provide a logical framework for results obtained in the mature semi-deciduous Costa Rican forest and secondary deciduous Venezuelan forest, yet they are not entirely consistent with the situation that has been encountered in the montane cloud forest in Venezuela. Self-compatibility is frequent among the trees and shrubs of this moderately species-rich forest. Reproductive efficacy is high in both self-compatible and self-incompatible species, and the fruit/flower ratios for dioecious species are not significantly different from those registered in self-incompatible species. In the montane cloud forest, flowering in concentrated toward the latter half of the wet season (trees) or toward the centre of the wet season (shrubs), periods in which cloud cover may extend over the forest for several hours each day. Extended periods of cloud cover, which reduce insect activity, could materially affect seed set in many species growing in the forest, and the only way out would be through autogamy or through apomixis, which permit seed set without pollen transfer. Interestingly, a number of the self-compatible species of the montane cloud forest are autogamous, while two non-pseudogamous apomictic species have been revealed (Sobrevila, 1978). If pollinator independent breeding systems have become established more frequently in cloud forests, then why is dioecism so strongly represented in the same forests? The situation in the montane cloud forest might provide an example of an often-stated viewpoint (e.g. Raven, 1973) that dioecism is the easiest way in which obligate outbreeding may be reestablished following loss of self-incompatibility, i.e. dioecism may have evolved secondarily as a means for reducing inbreeding. Nevertheless this argument rests on very slender evidence of reproductive efficacy of a small number of self-incompatible and dioecious species. The dioecious species in the montane cloud forest pertain to plant families that are frequently dioecious in other kinds of vegetation, and the real reason for the origin of dioecy might be quite distinct from that suggested by present-day reproductive dynamics of the community. Whether or not dioecism has been selected for in relation to outbreeding, another means for reducing inbreeding is through apomixis. As already mentioned, two species in the Venezuelan cloud forest are non-pseudogamous apomicts and Ashton (see p. 35) has provided evidence for apomixis in some South-East Asian Dipterocarpaceae. This mode of reproduction might be far more common in tropical forests than is usually supposed and future research efforts might be oriented towards developing manageable techniques for its detection.

Acknowledgements

Some of the data cited for Venezuelan forests in this paper are derived from undergraduate theses for the degree of Licenciatura en Biología, Universidad Central de Venezuela of T. Ruiz and C. Sobrevila, carried out in collaboration with the author. The technical assistance of Haydee Abreu is greatly appreciated. This work is supported in part by Conicit Grant No. 31.26.SI.0591.

References

Arroyo, M. T. K. (1976). Geitonogamy in animal-pollinated tropical angiosperms: a stimulus for the evolution of self-incompatibility in the angiosperms. *Taxon* **25**, 15–20.

Arroyo, M. T. K. and M. Cabrera (1978). Preliminary self-incompatibility tests for some tropical cloud forest species in Venezuela. *Incomp. Newsl.* **8**, 72–76.

Baker, H. G. (1959). Reproductive methods as factors in speciation in flowering plants. *Cold Spring. Harb. Symp. Quant. Biol.* **24**, 177–199.

Bawa, K. S. (1974). Breeding systems of tree species of a lowland tropical community. *Evolution* **28**, 85–92.

Bawa, K. S. (1976). Breeding of tropical hardwoods: an evaluation of underlying bases, current status and future prospects. *In* "Tropical Trees, Variation, Breeding and Conservation" (J. Burley and B. T. Styles, eds) p. 43–59. Academic Press, London and New York.

Bawa, K. S. and P. A. Opler (1975). Dioecism in tropical trees. *Evolution* **29**, 167–179.

Federov, A. A. (1966). The structure of tropical rainforest and speciation in the humid tropics. *J. Ecol.* **60**, 147–170.

Frankie, G. (1976). Pollination of widely dispersed trees by animals in Central America, with an emphasis on bee pollination systems. *In* "Tropical Trees,

Variation, Breeding and Conservation" (J. Burley and B. T. Styles, eds) p. 179–159. Academic Press, London and New York.

Heithaus, E. R., T. H. Fleming and P. A. Opler (1975). Foraging patterns and resource utilization of seven species of bats in a seasonal tropical forest. *Ecology* **56**, 841–853.

Heslop-Harrison, J. (1975). The physiology of the pollen grain surface. *Proc. R. Soc. Lond. B.* **190**, 275–299.

Howlett, B. J., R. B. Knox and J. D. Paxton (1975). Pollen wall proteins: Physiochemical characterization and role of self-incompatibility in *Cosmos bipinnatus*. *Proc. R. Soc. Lond. B.* **188**, 167–182.

Janzen, D. (1971). Euglossine bees as long distance pollinators of tropical plants. *Science* **171**, 203–205.

Raven, P. (1973). Evolution of subalpine and alpine plant groups in New Zealand. *N.Z. J. Bot.* **2**, 177–200.

Ruiz, T. Zapata and M. T. K. Arroyo (1978). Reproductive ecology of a secondary deciduous forest in Venezuela. *Biotropica* **10**, 221–230.

Sobrevila, C. (1978). Ecología reproductiva de un bosque montañoso simpreverde de Venezuela. Tésis de Licenciatura, Universidad Central de Venezuela.

Sorensen, F. (1969). Embyonic genetic load in coastal douglas fir, *Pseudotsuga menziesii* var. *menziesii*. *Am. Nat.* **103**, 389–398.

Lycopodium in Ecuador—Habits and Habitats*

B. ØLLGAARD

University of Aarhus, Denmark

Introduction

The genus *Lycopodium* s. lat. consists of approximately 500 species, according to Herter (1949, 1950). A little fewer than half of these species are from tropical America. At this preliminary stage of my revision of the Ecuadorean *Lycopodium* I recognize approximately 90 taxa some of which are of infraspecific rank. This is a slightly greater number than Herter recognized, largely because of new discoveries. The total number of species is likely to exceed 100, considering the rate at which new species are being discovered as previously unknown, remote mountains are explored. The occurrence of so large a proportion of the world's *Lycopodium* species in such a small area, suggests that Ecuador is an important area for the evolution of this genus. Probably Ecuador has the highest concentration of *Lycopodium* species of any area in the world. However, it must be remembered that from the point of *Lycopodium* biogeography Ecuador is not a natural entity, but is part of the whole north Andean region. *Lycopodium* is a genus of very ancient origin. However, as shown by van der Hammen (1974), it was only during late phases of the Andean uplift that the northern Andes were sufficiently elevated for the páramos to develop.

As the majority of the Ecuadorean species occur in the páramo, these species must be considered to be the result of a modern evolutionary burst, when suitable *Lycopodium* habitats became available. Because of circumstances described below, I believe that the evolution is still very active. The reasons why the upland habitats are particularly suitable for *Lycopodium* are also discussed below.

* AAU Ecuador project, contribution No. 11.

The literature dealing with the ecology of *Lycopodium* is scarce and usually of a general nature. The accounts given by Spring (1850), Pritzel (1900), and Herter (1909) are still valid in their outline. More detailed information on *Lycopodium* habitats is mostly scattered in regional vegetation descriptions. No attempt is made to gather this information here.

Thanks to a grant from the Danish Natural Science Research Council I had the opportunity to study *Lycopodium* in the field in Ecuador in 1976. In the present paper I wish to summarize my observations of more or less obvious adaptations of *Lycopodium* species to the conditions of a variety of habitat types. My field observations have been supplemented by the study of ample herbarium material, which I am using in revising *Lycopodium* for the Flora of Ecuador.

Subgeneric Characteristics and Adaptive Features of Habit

In *Lycopodium* I recognize three subgenera which are distinct in habit: subgenus *Urostachya* Pritzel, subgenus *Lycopodium*, and subgenus *Lycopodiella* (Holub). The latter name is not yet formally established at subgeneric rank. The species of subgenus *Urostachya* (Figs 1–3, 6–10) are always isotomously branched, ascending, erect or pendulous plants. Except for a few species they lack stems specialized for short distance migration, and therefore the single sporophytes are restricted to their place of origin. They can do very little vegetatively to adapt to unfavourable changes in their immediate environment. As other species of *Lycopodium* they disperse by means of air borne spores, and none of the Ecuadorean species propagate by means of bulbils, as do several species of temperate regions. Their reproduction is entirely sexual or at least gametophytic. Both gametophyte maturation and sexual development of embryos is belived to be slow in the genus, except in subgenus *Lycopodiella*. In all subgenera it is important to recognize the role that gametophyte biology may have in the distribution of species. Unfortunately our knowledge of the gametophyte biology is practically non-existent, and my own observation of *Lycopodium* habitat relations are based on sporophytes alone.

In subgenus *Urostachya* the stems are practically devoid of mechanical tissue and depend on turgidity to keep erect, or they are lax and pendulous. The sporangia are situated in the axils of leaves with or without specialized structure. These so-called sporophylls may protect or expose the sporangia, but there are no mechanisms which serve to protect the developing sporangia and to help release the spores at

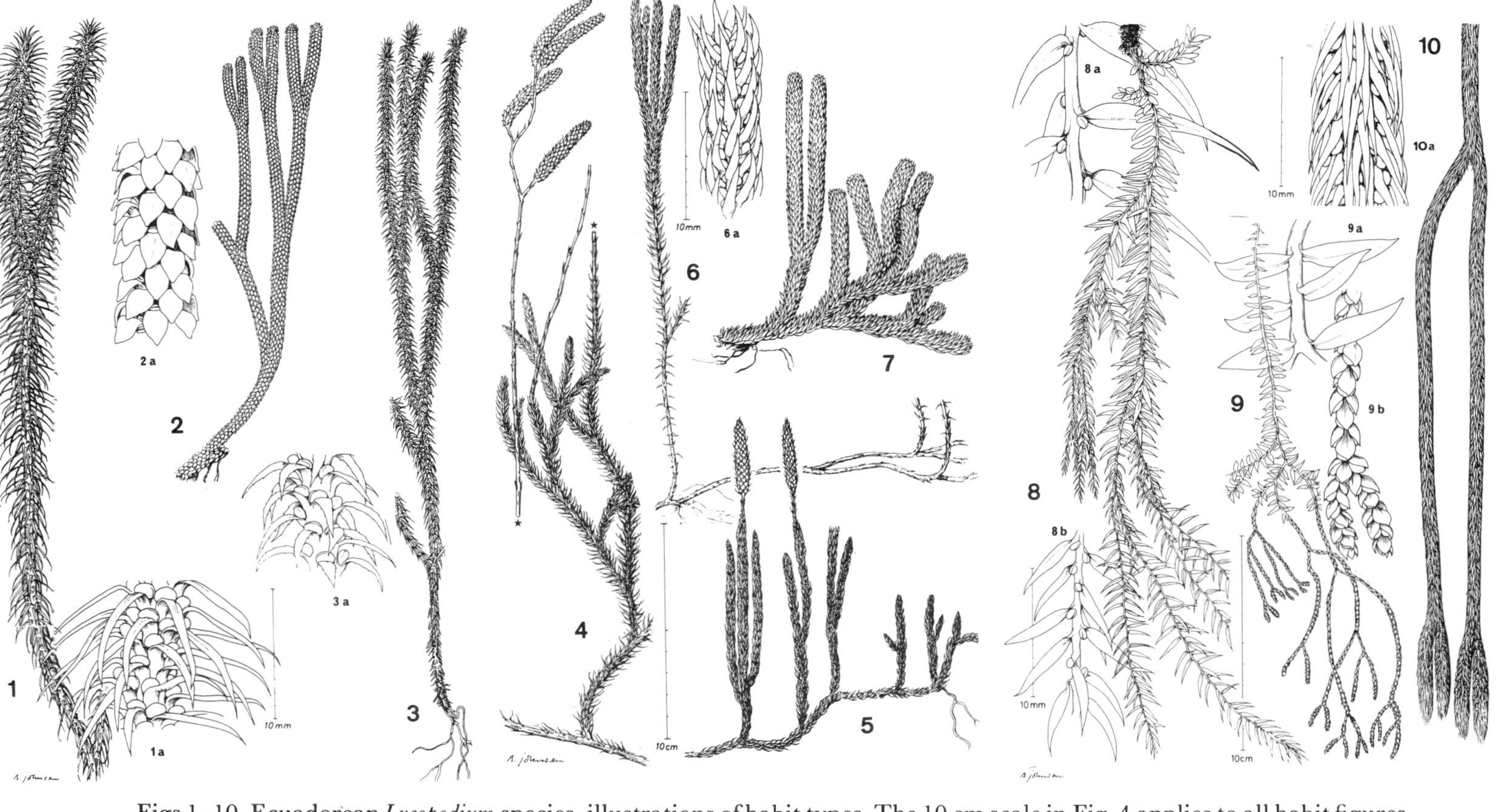

Figs 1–10. Ecuadorean *Lycopodium* species, illustrations of habit types. The 10 cm scale in Fig. 4 applies to all habit figures. The 10 mm scales apply to all detail figures. 1 = *L. hippurideum* Christ, 2 = *L.* sp. aff. *rufescens* Hook., 3 = *L. reflexum* Lam., 4 = *L. clavatum* L. spp. *clavatum*, 5 = *L. clavatum* L. spp. *contiguum* (Klotzsch), 6 = *L.* sp. with subterranean basal stems, group of *L. crassum* H. and B. ex Willd., 7 = *L.* sp. aff *L. crassum* H. and B. ex Willd., 8 = *L. linifolium* L., 9 = *L. myrsinites* Lam. spp. *phylicifolium* (Desv.), 10 = *L. funiforme* Cham. ex Spring.

maturity. In many species the young developing sporangia are first protected and and later uncovered, but the uncovering is not directly correlated with dehiscence.

The subgenera *Lycopodium* (Figs 4, 5) and *Lycopodiella* are both anisotomously branched and develop mechanical tissue and specialized stems, which serve for short distance migration. Also both subgenera have specialized strobili, in which the sporophylls protect the developing sporangia, and dry out and become reflexed at spore maturity. These features indicate a better adaptation to relatively dry conditions than the features of subgenus *Urostachya*. There is one important habit difference between the two subgenera, namely the origin and development of the upright branches (Øllgaard, 1979). In subgenus *Lycopodium* the branching is bilateral throughout, and the upright branch systems arise laterally on the creeping or scandent main stem. In subgenus *Lycopodiella* the upright branches arise from the upper side of the creeping, scandent or looping horizontal stems and in most species are much taller than in subgenus *Lycopodium*. I believe that this difference in a habit implies a difference in the capacity to compete for light.

The overall distribution of *Lycopodium* in Ecuador seems to be determined by its requirements of constantly humid conditions. The *Lycopodium* species are devoid of efficient xeric adaptations. Accordingly their number is greatly reduced, or they are absent in areas with a dry or pronounced seasonal climate. Within the humid areas, however, the *Lycopodium* species show a variety of adaptations which are correlated with different degrees of exposure and humidity of a wide range of habitats. Under humid and protected conditions the stems tend to be amply branched (Figs 4, 8, 9) without mechanical tissue (subgenus *Urostachya*) and erect or ascending (Figs 1–3), the leaves are large (Figs 1, 8), patent-reflexed (Figs 1, 3, 8) with thin cuticle, the developing sporangia are exposed (Figs 1, 3, 8). Under relatively xeric or exposed conditions the stems tend to be sparsely branched (Figs 5, 10) with mechanical tissue (subgenera *Lycopodium* and *Lycopodiella*) and creeping or pendulous (Figs 4–10), the leaves are small (Figs 2, 4, 5, 10) appressed (Figs 7, 10) with thick cuticle, and the sporangia are permanently protected or exposed only at maturity (Figs 4, 5). Table 1 gives examples of the occurrence of some of these characters in correlation with habitats.

Habitats

In the following I will discuss some characteristics of a number of habitat type with regard to the occurrence of *Lycopodium* species and

Table 1. A preliminary survey of the altitudinal ranges, habits and habitats of *Lycopodium* in Ecuador. Taxa arranged according to assumed affinity, bracketed numbers indicate number of collections used to determine the altitudinal range, inadequately known taxa and material with inadequate data are excluded. I: M, mature forests habitats; O, open immature vegetation in the forest zone. II: 1, grass páramo; 2, moderately exposed cushion páramo; 3, strongly exposed cushion páramo; 4, wet páramo depressions; 5, immature, regenerating vegetation in the páramo zone. III (Ephiphyte habit): E, erect; R, recurved; P, pendulous. IV (Terrestrial habit): E, erect or ascending; EC, erect with creeping basal stems; CE, creeping–scandent, epiterranean; CS, creeping, with usually subterranean rhizome. V (Leaf habit): P, patent—reflexed; A, appressed; EA, basally expanded, apically appressed and reduced (*L. phlegmaria*-habit); L, large, long or both; S, small, short or both.

Name	Altitude (m)	I	II	III	IV	V
Subgenus *Urostachya*						
L. hippurideum (16)	2800–3600	M			E	P, L
L. bolivianum (5)	3000–3800				E	P, L
L. sp. aff. *hippurideum* (9)	2750–3600	M			E	P, L
L. sp. aff. *sieberianum* (8)	1800–3200	O			E	P, S
L. urbanii (7)	2800–3400	O			E	P, S
L. reflexum (41)	950–3000	O			E	P (L)
L. blepharodes (6)	2400–3600	O			E	P (L)
L. ecuadoricum (42)	2000–3600	O			E	P, S
L. weddellii vel. aff. (7)	2700–3400	O			E	P, L
L. sp. aff. *ulixis* (4)	3200–3500		(1)2		E	A, L
L. sp. aff. *ulixis* (3)	3000–3450		2		E	A, (S)
L. sp. (2)	2750–3800				E	A, (S)
L. weberbauerii (3)	3000–3400		(1)2		E	(A), L
L. hystrix (9)	3200–4100		5		E	(A), L
L. sp. aff. *hystrix* (2)	3700–3800				E	(A), S
L. cumingii (16)	3600–4100		1, 2, 5		E (C)	A, (L)
L. capellae (12)	3000–4100		1		E	A-P, L
L. crassum (48)	3600–4300		2, 5		EC	A, L
L. sp. aff. *crassum* (5)	4100–4300		3		EC	A, L
L. sp. aff. *crassum* (25)	3600–4200		2, 4, 5		E (C)	A, L
L. sp. aff. *crassum* (9)	3950–4050		5		EC	A, L
L. sp. aff. *crassum* (54)	2750–4200		2, 4, 5		CS	A (L)
L. attenuatum (9)	3100–3750	O	2, 5		EC	A (S)
L. tetragonum (35)	2850–3950	O	2, 5		E (C)	A, S
L. sp. (8)	3600–4200		2, 5		E	A (S)
L. arthurii (5)	2700–3500		2		E	(A) S
L. sp. aff. *hohenackerii* (1)	3950		5		E	P, S
L. sp. aff. *hohenackerii* (1)	3800–3900				E	A-P, S

Name	Altitude (m)	I	II	III	IV	V
Subgenus *Urostachtya*—continued						
L. sp. (1)	3950		5		E	(P), S
L. rufescens (12)	3500–4150		2, 5		E	P, S
L. sp. aff. *rufescens* (9)	3350–4000		2, 5		E	(P) S
L. compactum (14)	3000–3500		2		E	P, S
L. trencilla (2)	3700–3800				E	P, L
L. molongense (4)	2600–3400	M		P		EA
L. heteroclitum (3)	2400–3400	M		P		EA
L. sp. aff. *robustum* (8)	2400–3400	M		P		EA
L. myrsinites spp. *phylicifolium* (16)	3000–3800	M		P		EA
L. myrsinites var. (2)	2300–2950	M		P		EA
L. subulatum (10)	2700–3500	M		P		EA
L. callitrichifolium (3)	2700–3000	M		P		EA
L. ericifolium (5)	950–1800	M		P		EA
L. sp. aff. *aqualupianum* (3)	150–1000	M		P		EA
L. hartwegianum (6)	2400–3500			R, P		(A)
L. rosenstockianum (10)	3000–3800	M		P		P, L
L. lindenii (18)	3500–4400	M		P		P-(EA)
L. passerinoides (5)	1000–2800?	M		P		P, L-(EA)
L. funiforme (5)	100–1200	M		P		A(S)
L. linifolium (11)	30–1600	M		P		P, L
L. sp. aff. *linifolium* (3)	2300–3000	M		P		P, L
L. sp. aff. *linifolium* (18)	600–2000	M		P		P, L
L. sarmentosum (6)	2400–2900	M		P		P, L
L. dichotomum (3)	30–1050	M		R		P, (L)
L. trichodendron aggr. (15)	50–2700	M		E		P, L
L. setaceum (2)	50–1200			P		P-(A), S
L. tenue (14)	2700–3400	M		P		P-(A), S
L. curvifolium (11)	400–1800	M		P		P-(EA), S
Subgenus *Lycopodium*						
L. spurium (25)	3000–4100	O	1, 2		CE, CS	S
L. jussieui (39)	1700–3700	O			CE	S
L. clavatum ssp. *clavatum* (42)	1500–3100	O			CE	S
L. clavatum ssp. *contiguum* (61)	2600–3800	O	1, 2, 5		CE	S
L. vestitum (19)	2700–3800	O	1, 2, 5		CE	S
L. complanatum s.l. (52)	2000–3400	O			CE	S
Subgenus Lycopodiella						
L. cernuum (43)	0–2800	O			CE	S

Name	Altitude (m)	I	II	III	IV	V
Subgenus *Lycopodiella*—continued						
L. sp. aff. *cernuum* (12)	1000–1500	O			CE	S
L. *trianae* (1)	0–100	O			CE	S
L. *pensum* (2)	1700–2600	O			CE	S, A
L. *glaucescens* (12)	1600–3000	O			CE	S
L. *pendulinum* (12)	2500–3500	O			CE	S
L. *alopecuroides* (17)	1100–3000	O	4		CE	(S)

their adaptations in habit to each habitat. One factor of obvious importance in the grouping of the habitat types is the presence or absence of a forest cover. The forested habitats form one group, while habitats without forest cover, i.e. páramos and land slides, roadbanks etc. in early stages of regeneration of vegetation, form another group.

Forest Habitats

The most important feature which separates the forested from the non-forested habitats is the strong shading of the terrestrial habitat. This seems to exclude all terrestrial species of *Lycopodium*, except from the forests close to the timber-line, which have lower and less dense canopies. Otherwise in the forest region the terrestrial species are only found in, or near, clearings. In this respect *Lycopodium* is remarkably different from the genus *Selaginella* for instance. The forest species are almost exclusively epiphytic, and they are usually not abundant. Nevertheless, in my material of epiphytes there are almost 30 taxa. Almost all of these are pendulous forms (Figs 8–10), a few being erect or recurved. According to Madison (1977) pendulous epiphytes are also well represented in other families (ferns, Gesneriaceae and Ericaceae). He indicates that the advantage of this habit is that a pendant plant will not be blown or knocked off the branch as easily as an erect plant on top of the branch. The epiphytic *Lycopdium* species all belong to subgenus *Urostachya*. This subgenus is characterized by the lack of mechanical tissue apart from the long internal cortical roots. These may improve the tensile strength of the stem, but hardly its rigidity. Accordingly an erect species of this subgenus must largely depend on turgidity to keep in upright position. The epiphytic habitats are subject to frequent, rapid, and extreme fluctuations in the availability of water, even in very wet places. Under such conditions the *Lycopodium* species are likely to be almost desiccated occasionally, due to their lack of efficient protection against evaporation, and to desiccation of the substrate. The lax pendulous habit seems to be an adequate adaptation to avoid mechanical damage under these conditions. In other words I regard the pendulous habit as a xeric adaptation in this group, although at the same time the pendulous plants also often possess features which are regarded to be hygromorphic.

The relatively low competition in the niche occupied by the pendulous epiphytes may also have been an advantage for the evolution of this habit. However, this niche is usually more shaded than the niche above the branch. It is uncertain whether the critical juvenile stage of sporophyte establishment is subject to the same competition pressure in both erect and pendulous epiphytes. A study of the position of the juvenile plants or the rooting position in mature plants in relation to their substrate and to light is needed to elucidate the significance of competition for the evolution of the pendulous habit. The number of epiphytic *Lycopodium* species, and their quantity appears to increase with ascending altitude. This distribution may reflect partly uneven collection, since the collection of epiphytes in the tall canopy of low altitude forests present greater difficulties than in the relatively low canopies of high altitude forests. De la Sota's (1972) study of the Costa Rican pteridophytes includes only a few species of *Lycopodium*, and they do not indicate an altitudinal distribution similar to the Ecuadorean distribution.

There are, however, environmental features which could explain the trend as it appears in Ecuador. First, there is a greater risk of total and fatal desiccation resulting from the higher temperatures of low altitudes, than there is in cooler habitats. Second, the growth rate is rapid and hence the competition is stronger at low altitudes, while the growth in elfin forests is known to be almost stagnant (Howard, 1970). As both drought and competition are strong delimiting factors in *Lycopodium* distribution there are reasons to expect increasing frequency with ascending altitude. A wide range of variety in leaf morphology and arrangement is found in the group of pendulous epiphytes. Some have spreading, relatively ample leaves which might represent adaptations to protected, constantly humid conditions (Fig. 8), while the species with reduced or clasping and imbricate leaves might be adapted to more exposed and occasionally xeric conditions (Figs 9, 10). Some of my observations indicate such a correlation, while others do not. The *L. phlegmaria* habit, i.e. with large expanded basal leaves, and reduced, appressed leaves in apical portions of the plants, as in *L. myrsinites* (Fig.

9), has not been found to correlate with a particular type of habitat. Much more detailed information of epiphytic habitats is necessary to make a qualified correlation. Also the relations between epiphyte and host tree are practically unstudied.

Very few species occur normally on the forest floor, and those which do are restricted to near the timberline. These are *L. hippurideum* (Fig. 1) and an undescribed close relative, and rarely other species. Both these species are remarkable because of their very long spreading leaves, well adapted to the protected, semi-shaded habitat. Epiphytic species may occasionally fall to the forest floor with broken branches, and in some cases they continue to grow. In such cases they sometime assume an ascending recurved habit, different from their original habit, and presenting a pitfall for taxonomists without field experience.

Non-Forest Habitats

More than 50 of the Ecuadorean *Lycopodium* species occur in rather exposed terrestrial habitats. Some of these species occupy rather distinct niches, while others are less specific. I will discuss the non-forest habitats under six headings.

Bunch grass páramo

In typical bunch grass páramo the ground between the grass tussocks is usually shaded and somewhat protected from wind by the grass. On the other hand the grass is frequently burned off or grazed by animals. Few species tolerate these conditions and none of these are conspicuous elements. *Lycopodium clavatum* spp. *contiguum* (Fig. 5) *L. spurium* and *L. cumingii* all tolerate some shading and superficial fires, but they all seem to thrive better under more exposed conditions. *Lycopodium capellae* is most often found in the shade of grasses. It has rather tender, unprotected tissue, and when the upper branches eventually protrude from the grass cover they are almost invariably deformed or damaged by cold and desiccating winds. If the apical meristems are scorched by fire, the plant may regenerate by means of basal adventitious buds, a feature which is not common among other terrestrial species.

Cushion—plant communities, moderately exposed

One often finds patches with low mature vegetation with few bunch grasses intermixed with, and often forming an intimate mosaic with, the proper grass páramo in sloping or undulating terrain. These habitats are more exposed and less or even uninfluenced by páramo fires. Several *Lycopodium* species can be found in this type of habitat. Many of them are, however, more frequent in other habitats. *L. crassum* var. *crassum* is characteristic and abundant here, usually above 3900 m elevation. It is worth mentioning the following species, which are less specifically bound to this habitat type: *L. tetragonum*, *L. attenuatum*, *L. spurium*, *L. clavatum* spp. *contiguum* (Fig. 5), members of the *L. rufescens* group (Fig. 2), and other varieties of *L. crassum*. Grazing is as common here as in the proper bunch-grass páramo, but the animals evidently avoid biting *Lycopodium*. Some *Lycopodium* species are known as violent purgatives in folk medicine. They contain alkaloids which may be the toxic compounds.

Strongly exposed cushion plant páramos

At altitudes well above 4000 m, often on small mountain tops the vegetation is often reduced to a dense cover of cushion plants (*Werneria*, *Plantago rigida*, *Distichia*, *Azorella* etc.) A dark red, robust varity of *L. crassum* is often the most protruding plant in this community. The conditions can be extremely harsh. During periods of bad weather this plant can be seen glazed by frost. Although the leaves are evidently damaged in many of the individuals, this taxon can be extremely abundant. The growth and establishment of new plants here is very slow. During a search for the gametophytes of this plant, I found that what at first appeared to be juvenile plants, with shoot apices level with the surrounding plant cushions, proved to be shoots up to 15 cm long, which had evidently been growing for a long time concurrently with the plant cushions.

Wet páramo depressions

The wet depressions in the páramo accommodate some of the most conspicuous *Lycopodium* populations in Ecuador. Two as yet unnamed species of the *L. crassum* group are particularly abundant. Such populations can usually be spotted from a long distance because of their vivid red colours. Essential features of the wet depressions are the scarcity of bunch grasses, and other shading and easily inflammable material and the fact that cold air will gather in the depressions at night. Because of the soft and often slightly unsafe bottom, these areas are not usually visited by grazing animals. At the lower altitudes (usually above 3600 m) one species (Fig. 6) with subterranean basal branches is dominating. Higher up it is gradually replaced by a slender, tufted close relative of *L. crassum*, which often is found growing together with the type variety.

A great number of species which are regularly subjected to frost exhibit various degrees of red colouring. The quality and intensity of these colours are both taxonomically and ecologically correlated, although the colours are by no means stable characters. The range of variation is often distinct, although somewhat overlapping, and in most species, single individuals completely without red colour are frequent. The intensity of the colours appears to be correlated with frost. It is significant that shaded individuals are green or pale. The species of the grass páramo are almost pure green. The species of exposed high altitude habitats and of páramo depressions which are cold at night are dark red, while the moderately exposed species on sloping or undulating terrain are usually rather light orange–brick red. During a week's measurements of temperature at two points in the Páramo de Guamaní we found constantly lower night temperatures in a wet depression at 4060 m altitude than on an extremely exposed hill top at 4260 m altitude near by. The difference was usually more than 1°C and probably the sloping terrain adjacent to the depression would have the highest night temperature of them all. The chemical nature of the red colour is unknown. In the literature it is reported to be located in the cell wall, but this information is based on dried material.

Immature regenerating vegetation of road banks, landslides and other disturbed sites above the timberline

During my field work I have repeatedly found a strikingly richer *Lycopodium* flora in immature habitats than in mature habitats in the páramo zone. Moreover the kind of diversity is different in the two habitat types. While the mature vegetation types are usually inhabited by relatively few and clear-cut species, the immature habitats usually contain a number of additional species and a confusion of intermediates and aberrants. It appears that the relatively low competition in these habitats allows the species to display a wide range of their variation potential, and to establish new forms, which may act as steps in speciation. Part of the aberrant development is possibly caused by different soil structure in comparison with the surrounding mature vegetation. In the mature vegetation several species are well characterized by the subterranean, ascending, or creeping course of their basal stems. Because the soil of immature habitats tends to be very solid and poorly aerated, these same species all tend to develop epiterranean, creeping basal stems. The species belonging to the páramo zone usually show very obvious xeric adaptations in leaf morphology, as a response to the so-called physiological dryness of the wet but cold environment. The majority of the species of exposed habitats have tightly appressed leaves with a thick cuticle on the abaxial side, thus protecting sporangia and stem, achieving maximum reduction of the evaporating surface. Further the rigid appressed leaves may compensate for the lack of mechanical tissue in the stem. They are sufficiently rigid to keep the stem erect, even when a loss of turgidity occurs. This type of adaptation is found in the *L. crassum* group (Figs 6, 7), and in *L. attenuatum*, *L. tetragonum* and other species. The same group is characterized by differentiation of branches into short creeping or ascending, vegetative branches (basal branches) and stiffly erect and ultimately fertile branches (Fig. 7). In another group of páramo species the leaves are patent or reflexed but very short and with a very thick and heavily cuticularized epidermis. This is found in the *L. rufescens* group (Fig. 2), *L. compactum* and *L. trencilla*.

Immature exposed vegetation in the forest region

The species content of immature types of habitats changes considerably through the altitudinal zones. However, the immature types of habitat are not subject to such a drastic environmental transition as found in mature vegetation in the change from forest to páramo. Correspondingly the transition of their species content is more gradual. Typical páramo species are often found on road banks well below the timber-line, whereas forest region species of subgenus *Lycopodium* may occur above the timber-line. In contrast to the páramo zone the mature forests cannot accommodate the terrestrial species found in landslides etc., because of light deficiency. These species therefore depend to some extent on the constant creation of new open soil, as their habitats become extinct by the regeneration of shading vegetation. They may find refuges on very steep slopes where the vegetation seldom becomes completely closed. It is therefore probable that the construction of roads will help to extend the available habitats of these species. In Ecuador I do not know of natural openings at low elevations comparable to savannas or campinas, which are inhabited by *Lycopodium* species. The open vegetation found in the upper forest region on exposed ridges (sometimes called paramillos) are here treated as páramo habitats.

Altogether the conditions of this zone seem to be rather unfavourable for the species of subgenus *Urostachya*. This group is represented here by approximately 10 species, mostly with long spreading or reflexed leaves, indicating a limited need of protection from drought. They may be exposed to rather strong sunlight, but rarely to strong winds. Their soil water reserve is usually ample and because of higher temperatures than in the páramos the physiological dryness is not pronounced. These features may explain their hygromorphic appearance. The species are: *L. reflexum* (Fig. 3), *L. ecuadoricum*, *L. blepharodes*, *L. urbanii* and a few rare or undescribed species. A number of species which are normally epiphytic in the adjacent forests can occasionally be found hanging on overhanging rocks or banks, but they play a minor role. The species of subgenera *Lycopodium* and *Lycopodiella* are particularly abundant in this zone, probably because they are more capable of competing for light with the surrounding vegetation. While all the terrestrial species of subgenus *Urostachya* are low and without even short distance migration capacity all species of the two other subgenera are rather long creeping or even scandent. The species of sugenus *Lycopodium* often form long festoons hanging over road banks, at middle and upper elevations. Thus they can exploit a large area, and they can persist even in high vegetation. The species of this subgenus found in the montane forest region are *L. clavatum*, *L. vestitum*, *L. complanatum* and *L. jussieui*. In the same groups of species there is a strong trend of simplification of the shoot systems with ascending altitude. The two vicarious subspecies of *L. clavatum* spp. *clavatum* and spp. *contiguum* offer a good example of this trend (Figs 4, 5).

Especially subgenus *Lycopodium*, but to some extent also subgenus *Lycopodiella*, possesses xeric adaptations such as mechanical tissue, small leaves and a definite correlation of spore dehiscence and reflexed sporophylls. For this reason the species of subgenus *Lycopodium* can often be found in areas which are too dry for any species of subgenus *Urostachya*. Some species of the subgenus *Lycopodiella* (*L. cernuum*, *L. trianae*, *L. glaucescens* and *L. pendulinum*) are even better equipped for competition for light than subgenus *Lycopodium*. They also spread by means of runners, which are usually looping or scandent. From the upper side of the runners they produce erect shoots up to at least 1 m high. In *L. glaucescens* these may become several metres long and scandent through shrubs, while *L. pendulinum* in the upper montane forests is always low and sparsely branched, producing another example of simplification correlated with ascending altitude. *L. cernuum* occurs commonly in wet forest areas up to *c*. 2000 m altitude, usually on poor soil. The reproductive biology of this species seems well adapted to lowland habitats with relatively rapid regeneration of vegetation. In other tropical areas it is known that spore germination and gametophyte maturation in this species is rapid. This applies also to other species of the subgenus, in which the gametophytes are known. The other subgenera are notorious for their slow development of a sporophyte.

L. alopecuroides, which belongs to another section of subgenus *Lycopodiella*, is a low, rather short creeping plant. It occurs on wet peaty soil in montane forest clearings and in open peat bogs up to *c*. 3000 m altitude.

Conclusion

In general terms the genus *Lycopodium* is characterized by high requirements of both light and humidity, and by a low competitive ability. The forested and the non-forested habitats are quite separate environments which offer entirely different conditions for speciation. Indeed, their respective species complements have little in common, and there are very few taxonomic relations across their border, as it is defined in this paper.

The extent to which the terrestrial species are confined to disturbed habitats with regenerating immature vegetation is surprising. With the exception of *L. cernuum*, *Lycopodium* species are not usually recognized as pioneer plant. Such features as their assumed slow growth and slow reproductive biology would seem to make them particularly unfit for quick colonization. It must, however, be remembered that the assumption of their slow growth is due mainly to observations in temperate regions with depauperate *Lycopodium* floras, where growth conditions are not uniform and may be unfavourable for long periods of the year. We know very little about the growth rate of *Lycopodium* species in the tropics, and very little about their reproductive biology. One reason for their relative success as colonizers may be found in their wind dispersed spores, and in the apparent lack of adaptation to quick colonization of the mature vegetation surrounding the disturbed habitats. In Ecuador as well as in the adjacent regions several factors contribute to the abundance of suitable terrestrial habitats. Because of the steep topography, frequent earthquakes, and constant high precipitation, landslides are frequent. Especially at higher elevations these sites provide ample light and little competition. They offer possibilities for the establishment of genetically more diverse populations and thus for perpetuation of deviating forms, which may act as stepping stones in speciation. At the same time especially the páramo region offers extensive areas with a variety of habitats which can accommodate well adapted forms. It is my impression that these features of the Andes have been of particular importance for the high diversity of the terrestrial Andean *Lycopodium* flora.

Acknowledgements

I wish to express my gratitude to the Danish Natural Science Research Council for a grant for field work in Ecuador. I wish to thank Dr G. T. Prance, Professor R. M. Tryon, canad. scient. H. Balslev for useful suggestions and criticism of the manuscript, and Bent Johnsen for preparing the drawings.

References

Hammen, T. van der (1974). The Pleistocene changes of vegetation and climate in tropical South America. *J. Biogeog.* **1**, 3–26.

Herter, W. (1909). Beiträge zur Kenntnis der Gattung Lycopodium. *Bot. Jahrb.* **43**, *Beibl.* **98**, 1–56, Tables 1–4.

Herter, G. (1949). Systema *Lycopodiorum*. *Rev. Sudamer. Bot.* **8**, 67–86.

Herter, G. (1950). Systema *Lycopodiorum*. *Rev. Sudamer. Bot.* **8**, 93–116.

Howard, R. A. (1970). The "alpine" plants of the Antilles. *Biotropica* **2**, 24–28.

Madison, M. (1977). Vascular epiphytes: Their systematic occurrence and salient features. *Selbyana* **2**, 1–13.

Øllgaard, B. (1979). Studies in Lycopodiaceae. II. Branching patterns of *Lycopodium* s. lat. and their relation to infrageneric groups *Am. Fern J.* **69**, 49–61.

Pritzel, E. (1900). Lycopodiaceae. *In* "Nat. Pflanzenfamilien 1" (A. Engler and K. Prantl, eds) Vol. 4, p. 563–606. Engelmann, Leipzig.

Sota, E. R. de la (1972). El epifitismo y las Pteridofitas en Costa Rica. *Nova Hedwigia* **21**, 401–465.

Spring, A. (1842). Monographie de la famille des Lycopodiacées, premiére partie. *Mém. Acad. R. Belg.* **15**, 1–110.

Spring, A. (1850). Monographie de la famille des Lycopodiacées, seconde partie. *Mém. Acad. R. Belg.* **24**, 1–358.

Growth Forms of the Espeletiinae and their Correlation to Vegetation Types of the High Tropical Andes

J. CUATRECASAS

Smithsonian Institution, Washington, USA

Introduction

Some of the most distinctive vegetation types of tropical America are found in the high areas of the Andean Cordilleras above the rather undefined timber-line. In the northern Andean section (western Venezuela, Colombia and Ecuador), these open areas of the Cordilleras are generally known as paramos. These represent one of the four main broad belts of vegetation that can be roughtly distinguished in a vertical section on a Colombian Cordillera as follows:

(i) The tall neotropical rain forest of the lowlands, 0–1000 m altitude, 30–22° average yearly temperature, rainfall 10000–1800 mm.

(ii) Subandean rain forest, 1000–2400 m altitude, average yearly temperature 23–16°, rainfall 5000–1000 mm, cloudiness and fog frequent particularly in the upper levels.

(iii) Andean forest 2400–3800 m altitude, average temperature 15–6°, rainfall estimated at about 3000–1000 mm, cloudiness and fog very frequent if not constant.

(iv) Paramos, from bushy zone, 3200–4700 or 5000 m with permanent snow, cold and humid, average temperature 12 to −2°, frequent precipitation (3000–700 mm), fog and humid ground.

Within the paramo we distinguish three subbelts (see Cuatrecasas, 1934, 1957, 1958; Cleef, 1977):

(i) Subparamo—the lower parts with alternation of thickets, low tree and scrub formation, at an ecotonic zone or transitional zone to the adjacent Andean forest.

(ii) Paramo proper—densely covered with meadows, sclerophyllous shrubs, grasses in bunches (tussocks) and sessile or tall caulirosuletum.

(iii) Superparamo—with more open or scattered vegetation on sandy or rocky or gravelled ground, occupies the highest zone of the mountain tops from nearly 4000–4700 m (eventually 5000 m) altitude, submitted to regular or frequent snowfalls.

The present talk focuses on the growth forms developed by the genera of the Espeletiinae (Compositae: Heliantheae) which are important members of the north-eastern Andean flora. Some *Espeletia* species have particularly called the attention of travellers, geographers, botanists and ecologists for their peculiar life form: the "caulirosula". This consists of a large rosette of usually long coriaceous rigid dense white-woolly leaves at the top of a woody, undivided, erect stem. The stem may be short, the rosettes appearing to be sessile, but in most species and in undisturbed paramos the stem reaches a height of several meters. The stems are either isolated or more commonly gregarious, and they produce a contrasting effect emerging from the green paramo meadows. Particularly impressive are the *Espeletia* communities which may cover large areas of paramo valleys and Andean slopes dominating the landscape. Caulirosulas usually share the territory with other paramo growth forms according to local ecology, and are always a relevant physiognomical indicator of the vegetative spectrum.

Caulirosulas belong to a broader concept of life-forms comprising large groups of plants, such as the palms, cycadoids, fern trees, many Liliaceae, Bromeliaceae, and some other Compositae. There is a great amount of literature on this subject but this aspect was not taken into account in earlier classical life forms classifications mostly based on northern vegetation biotypes (e.g. Raunkier, 1934). Nevertheless, several names have been given more or less formally to these life-forms, like tuft trees (Warming 1909), Schopfbäume (Drude 1890; Warming 1918, 1933; Troll, 1937), Rosettenbäume (Troll, 1937), rosette-trees (Du Rietz, 1931), megaphytes (Cotton, 1944), cabbage-trees (many authors referring to *Dendrosenecio*), pachycaul-trees and monocaul-trees (Corner, 1954), phanerophyte scapose (Braun-Blanquet, 1964; Schnell, 1970, 1971), monocaule-megaphylle (Schnell; 1970) and giant rosette plants (Hedberg, 1968). Most of these names are inadequate because the actual caulirosulas really are not trees. They impress us on account of their large, rich rosette foliage, either bearing sessile or supported by their long erect stem. These are significant morphologic characters which also produce a physiognomic effect.

Du Rietz (1931) considered two kinds of rosette-trees: simple (unbranched) rosette trees (ferns, palms, cycads, Liliaceae, Lobeliaceae and Compositae), the inflorescence being lateral, only in few cases terminal; and branched rosette trees (e.g. *Dracaena*, *Yucca*, *Cordyline*, *Vellozia*, Pandanaceae, Epacridaceae, Compositae, *Dendroseris*, *Robinsonia* and the African Giant *Senecios*). Troll (1937) divides his Rosettenbäume in two groups: hapaxanthae, monocarpic with terminal inflorescence (e.g. *Puya*, *Fourcroya*, the palms *Metroxylon* and *Coryphae*, a few *Lobeliae*); and pollakanthae with lateral inflorescences (e.g. most palms, Meliaceae, Anacardiaceae, Sapindaceae, *Clavija* and *Senecio keniodendron*).

Corner (1949, 1954) made the distinction of pachycaul from the leptocaul type of trees, and this is applicable here in a partial definition of caulirosula, with the young initial stem almost as thick and massive as in the adult stage. The anatomy of the young stem of *Espeletia hartwegiana* with its characteristic broad, flat apical meristem was studied and its anatomic development meticulously followed, described and illustrated by Weber (1958) and Rock (1972). With respect to its basic architecture, *Espeletia* caulirosula fall into the group of so-called "modèle de Corner" under the division of "arbres non ramifiés" of Hallé and Oldeman (1970).

Although I have only mentioned the *Espeletia* model of biotype as the one generally known, there are several biotypes in the Espeletiinae which have to be considered. These are listed in Table 1 and then defined.

Table 1. Synopsis of the life forms in the Espeletiinae.

Caulirosulae
    Polycarpic Caulirosula, synflorescence axillary
        Caulirosula ± tall
        Caulirosula sessile (rhizomatic, tuberose)
    Monocarpic Caulirosula, synflorescence terminal
        Monocaul
            Caulirosula ± tall
            Caulirosula sessile (rhizomatic, tuberose)
        Pauciramose, low branched bush

Trees
    Synflorescences terminal
        Monopodial basic branching
        Sympodial pseudodichotomic branching
    Synflorescences axillary, monopodial basic branching

Caulirosulae

Polycarpic Caulirosula, Monocaul, ± Tall

Stem erect from 10 cm–14 m tall, 8–12 (–14) cm diameter above base, monopodial and undivided. Live rosette 30–80 cm diameter of a compact crowd of coriaceous, apparently thick-lanate or sericeous whitish leaves; the leaves are usually long, simple, entire, arranged in close spiralling continuous circles with no visible internodes, the sheaths are very large (one-tenth–one-fifth of the blade) more or less geniculated at apex all regularly imbricate, appressed to each other and to the stem, building a multilayered tight cloak around the stem. The apical dominant meristem produces, although slowly, a continuous uninterrupted growth of the stem and distally new leaves at the same time that the proximal old leaves are progressively being lost. The marcescent blades remain, supported by their sheaths hanging around the trunk contributing to the apparent thickness of the column that supports the living rosette. The adult stem is cylindrical with a thick central white pith, a solid vascular cylinder, and a tight cortex, the thickness of which depends on species or age of specimens. Although the whole stem is apparently uniform, the lower part becomes progressively thicker than the distal section due to the increase with age in thickness of the cylinder of secondary wood, the contrast being more evident in old individuals of some height. Synflorescences are produced from axils of medial leaves of the rosette, several simultaneously in one to two apparent circles, spirally arranged; this production is rhythmic and apparently genetically controlled. The rates of periodicity remain unstudied, but general gross observations indicate that reproductive branching does not occur or is rare in drought periods. The inflorescence branchlets dry out after fructification, turn down, and either degenerate or remain persistently among the mass of the marcescent leaves. *Espeletia grandiflora* H. & B. and E. *hartwegiana* Cuatr. are here considered as the model for this type of caulirosula (Fig. 1).

This type, caulirosula polycarpica, is found in all the Colombian–Ecuadorian and most Venezuelan species of *Espeletia*, in almost all species of *Espeletiopsis* and all species of *Coespeletia*. By definition this biotype is said to be monocaul and simple. Nevertheless, there are species in which populations may be found with branched individuals; but in most of the latter cases there is clear evidence that an accidental mutilation of the shoot-apex in young individuals has produced a pleiochasial branching; these have two to six branches, growing undivided, erect and parallel and each being topped by a regular rosette, the

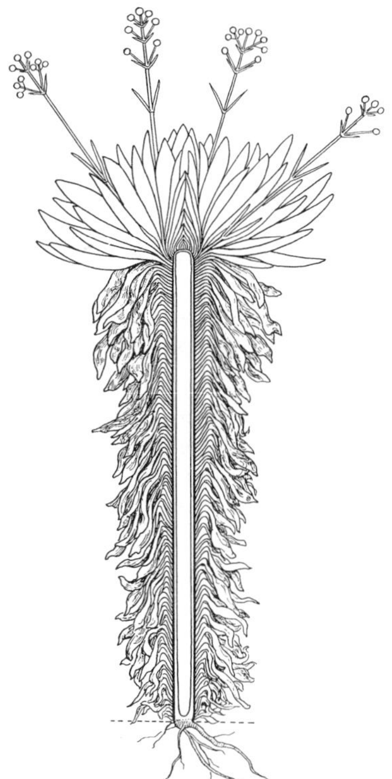

Fig. 1. Typical polycarpic caulirosula in vertical section. *Espeletia grandiflora* H. & B., from a specimen about 1·6 m high.

basal common trunk is thick and very short and usually unnoticed (e.g. *Espeletia uribei*). *Espeletiopsis pleiochasia* is of similar appearance, but the frequency of the kind of branching found in this species suggests that it is also spontaneously produced. There are also a few other species of *Espeletiopsis* normally, but sparingly, branching, departing a little from this basic biotype. A few species depart from the general rule in that the dead leaves fall off instead of remaining attached to the stem. In these cases one sees a tall, relatively narrow, naked stem topped by a large rosette, (e.g. *Espeletiopsis insignis*) (Fig. 5. 8).

This type of life-form, which is the prototype of caulirosula, is profusely extended throughout the paramo area from northern Ecuador to western Venezuela. Short-stemmed or tall-stemmed up to 10 (−14) metres, these rosettes may dominate extensive open or semi-open areas of humid meadows or mountain slopes. Tall caulirosulas with synflorescences of the dichasial and racemose types, are among the Espeletiinae occurring at higher altitudes, such as the rocky crests

402   *J. Cuatrecasas*

around Nevado del Cocuy and Sierras Nevadas of Mérida (Figs 1, 5.
10). Caulirosulas with synflorescences of monochasial corymbiform
type (Figs 5. 8, 5. 9) thrive abundantly in less humid or better drained
grounds, in open paramos in the eastern Colombian Cordillera and
Venezuela. They also grow very tall inside the Andean Forest in its
upper levels, especially on steep forested and often rocky slopes.

Caulirosula Polycarpic, Sessile, Rhizomatic

This biotype is characterized by a subterraneous thick, short, woody
rhizomatose caudex. The pachycaul apex produces a sessile rosette of
perennial life and axillary inflorescences. The subterraneous stem may
be undivided or with very few short branches. Examples are *Espeletia
marthae* Cuatr., *Espeletiopsis caldasii* Cuatr. (Figs 5. 13, 5. 14). This sessile
biotype is found in open areas of Venezuela between 3000 and 4000 m,
it is rare in Colombia.

Caulirosula Polycarpic Sessile, Tuberose

Here the caudex is subterranean and tuberose, having a thick, rather
spheroid shape. This habit has given the name to some species commonly known as "frailejón batato". The sessile rosette also produces
axillary inflorescences. Continued apical growth of the caudex may
produce a new tubercle adnate to the top of the old one resulting in an
irregularly shaped elongate caudex with an obtuse or truncate distal
end. Examples are *Espeletia weddellii* Sch. B., *E. batata* Cuatr. (Figs 2, 5.
12). This sessile tuberose biotype is common in some open paramo
areas of Venezuela, between 3000 and 4000 m altitude.

Caulirosula Monocarpic, Monocaul

Stem is erect up to several metres tall, monopodial and undivided.
Rosette is terminal with a construction similar to previous biotypes but
with a broader range of variation in leaf structure. The basic feature is
the apical meristem, which, after a few too many years of growing
vegetatively, suddenly produces a terminal, usually large, synflorescence. This effort represents the consumption of all reserves of the
cormus, with the whole plant dying after fructification, this form is
monocarpic, hapaxantic. Under the Hallé and Olderman (1970) system it may be classified among the "arbres non ramifiés" as the
"Holttum modèle".

The caulirosula monocarpic form with a terminal synflorescence

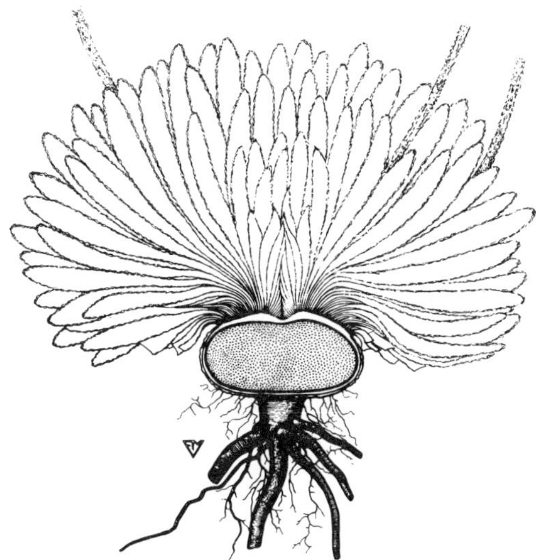

Fig. 2. Polycarpic caulirosula, tuberose form *Espeletia batata* Cuatr.

(Figs 3, 5. 5) was first detected in a few species in Venezuela on my 1969 trip (Cuatrecasas, 1971). An example is *Ruilopezia figueirasii* Cuatr. (Cuatrecasas, 1976, Fig. 2). This biotype is probably restricted to the genus *Ruilopezia*, where it occurs with three variant forms. It is almost exclusively in the Venezuelan subparamos and adjacent local paramos. The plants may grow profusely and become rather tall, especially near the summits of the lower hills of state Trujillo and in the Andes south-west of Mérida with an average altitude of 2800–3200 m. These areas are of the subparamo type, or locally of open paramo, especially in some higher hills (up to 3600 m). Some species grow well in the forest at 2500–2800 m, and a very few rise the Cordillera of Mérida to 3800 m altitude.

Caulirosula Monocarpic, Sessile, Rhizomatic

The rosula is sessile on the ground, the stem is pachycaul subterranean, rhizomatous except for the head or distal end which produces and supports the crowded rosette. The terminal synflorescence might be very large. At the base of the rosette, axillary to the lower leaves or leaf base, several surviving buds are usually present, some of which may eventually develop into new shoots. The growing of the lateral buds may take place when the apical shoot has been accidentally des-troyed—a frequent cause of mutilation of the apical meristems is

                                    J. Cuatrecasas

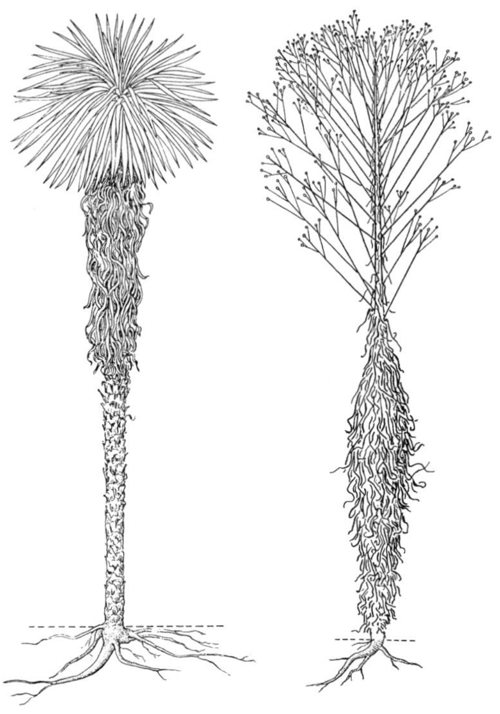

Fig. 3. Monocarpic caulirosula, leafy sterile and fructified individuals. *Ruilopezia figueirasii* Cuatr. Specimens about 3 m high.

grazing cattle. An example for this life-form is *Ruilopezia atropurpurea* (A. C. Sm.) Cuatr. (Fig. 5. 6). This biotype grows in subparamos of Trujillo and south-western Mérida, between 2600 and 3200 m. In the Sierra Nevada de Mérida it can rise to 3700 m.

Caulirosula Monocarpic, Sessile, Tuberose

The caudex grows thick and subterranean and the whole rosette dies with the fructification of the terminal synflorescence. An example is *Ruilopezia jabonensis* Cuatr. (Fig. 5. 7). This form is found between 2800 and 3400 m in the paramos and subparamos of Trujillo and in some parts of Meridan Andes.

Caulirosula Low Branched Bush

Sparsely branching from near the base, pleiochasial and monopodial, or also apparently dichotomic and sympodial. The branches end with a

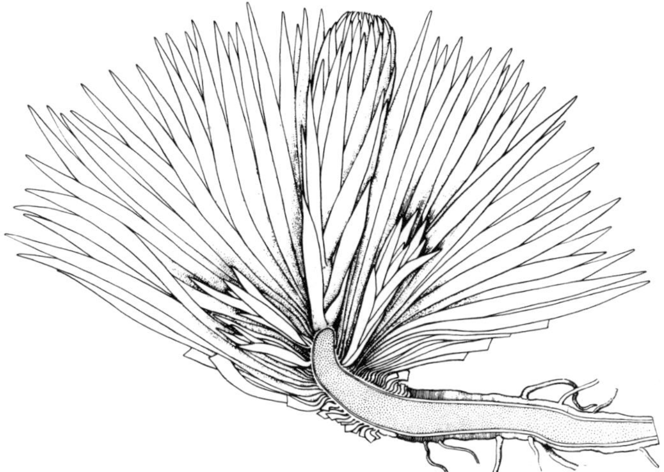

Fig. 4. Monocarpic caulirosula with initial terminal inflorescence and few lateral young shoots. *Ruilopezia bromelioides* Cuatr.

large, dense monocarpic rosette. Each rosette produces a usually large, heavy, terminal synflorescence which degenerates after fructification. The branches are covered with a thick compact cloak of the marcescent leaf bases, and they are prostrate on the ground. The plant usually forms a more or less large cluster or cushion over the ground with up to 15–20 rosettes. This number may be often much smaller, down to an unbranched short stem with a single rosette. In this last case the plant is monocarpic, otherwise only each rosette with its corresponding branch is monocarpic. An example is *Ruilopezia jahnii* (Fig. 5. 3). This architecture belongs to the group of "modèle de Leeuwenberg" of Hallé and Oldeman (1970). This life-form is found on slopes or high areas of the subparamo and local paramos in the Venezuelan "Sierras del Sur", south-west of Mérida and bordering Táchira.

Trees

Monopodial Basic Branching

Includes trees with monopodial axis and monopodial pleiochasial basic pseudoverticillate branching. Distal sympodial growth of branches is frequent and leaves along the branchlets are alternate, tending to be in bunches at the distal ends. Synflorescences terminal, with definite monochasial (alternate) branching. The mature trunk is massively

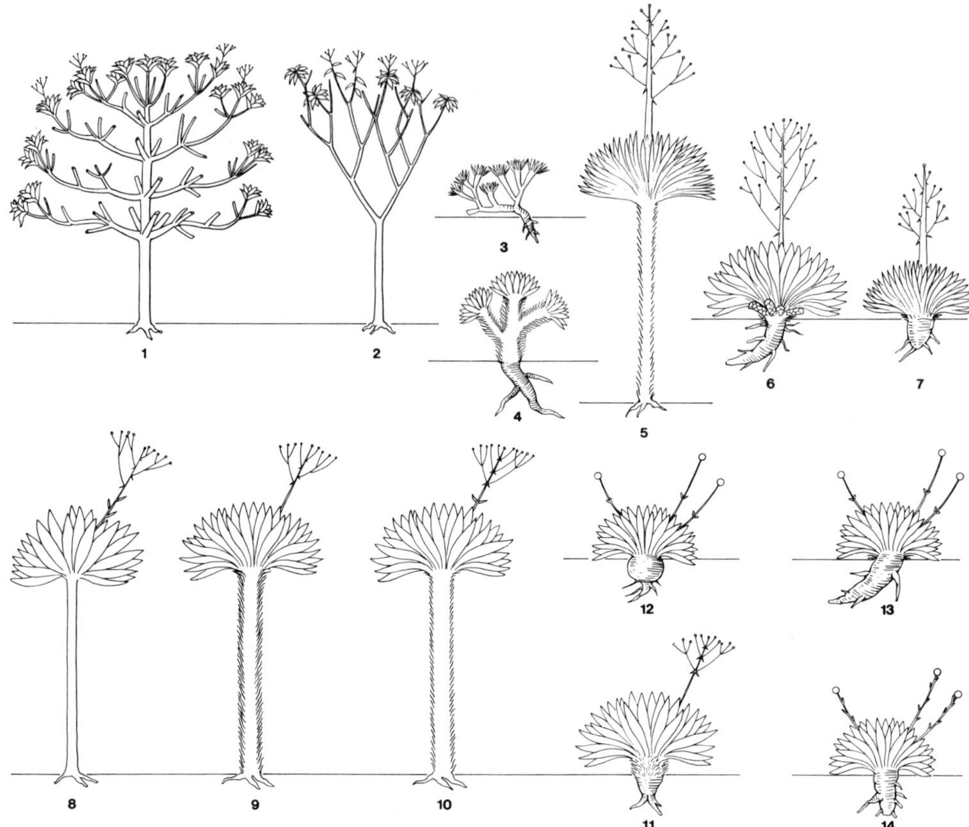

Fig. 5. Diagramatic representation of Espeletiinae main growth forms: 1, tree of *Libanothamnus* with basic monopodial branching; 2, tree of *Tamania* with pseudodichotomic sympodial branching; 3, caulirosula pauciramosa, low branching, sympodial bush, basically monocarpic like in *Ruilopezia jahnii* (St.); 4, occasional low branching caulirosula, monopodial; *R. bromelioides*, *Espeletiopsis meridensis* Cuatr.; 5, monocaul moncarpic caulirosula, tall, *Ruilopezia* spp. 6, sessile monocarpic caulirosula, rhizomatic, *Ruilopezia atropurpurea*; 7, sessile monocarpic caulirosula tuberose or subtuberose, *Ruilopezia* spp; 8, polycarpic caulirosula with naked stem, *Espeletiopsis* insignis Cuatr.; 9, polycarpic caulirosula with persistent cover of marcescent leaf bases on the stem and lateral monochasial synflorescences, *Espeletiopsis corymbosa* spp. *zipaquirana* Cuatr.; 10, same with dichasial synflorescences, *Espeletia grandiflora* H. and B.; 11, polycarpic caulirosula sessile or subsessile like in *Espeletia barclayana* (young stage); 12, sessile tuberose polycarpic caulirosula, *Espeletia weddellii* Sch. B.; 13, sessile rhizomatic polycarpic caulirosula with dichasial inflorescences, *Espeletia marthae* Cuatr.; 14, sessile rhizomatic polycarpic caulirosula with monochasial inflorescences, *Espeletiopsis caldasii* Cuatr. Reductions arbitrary.

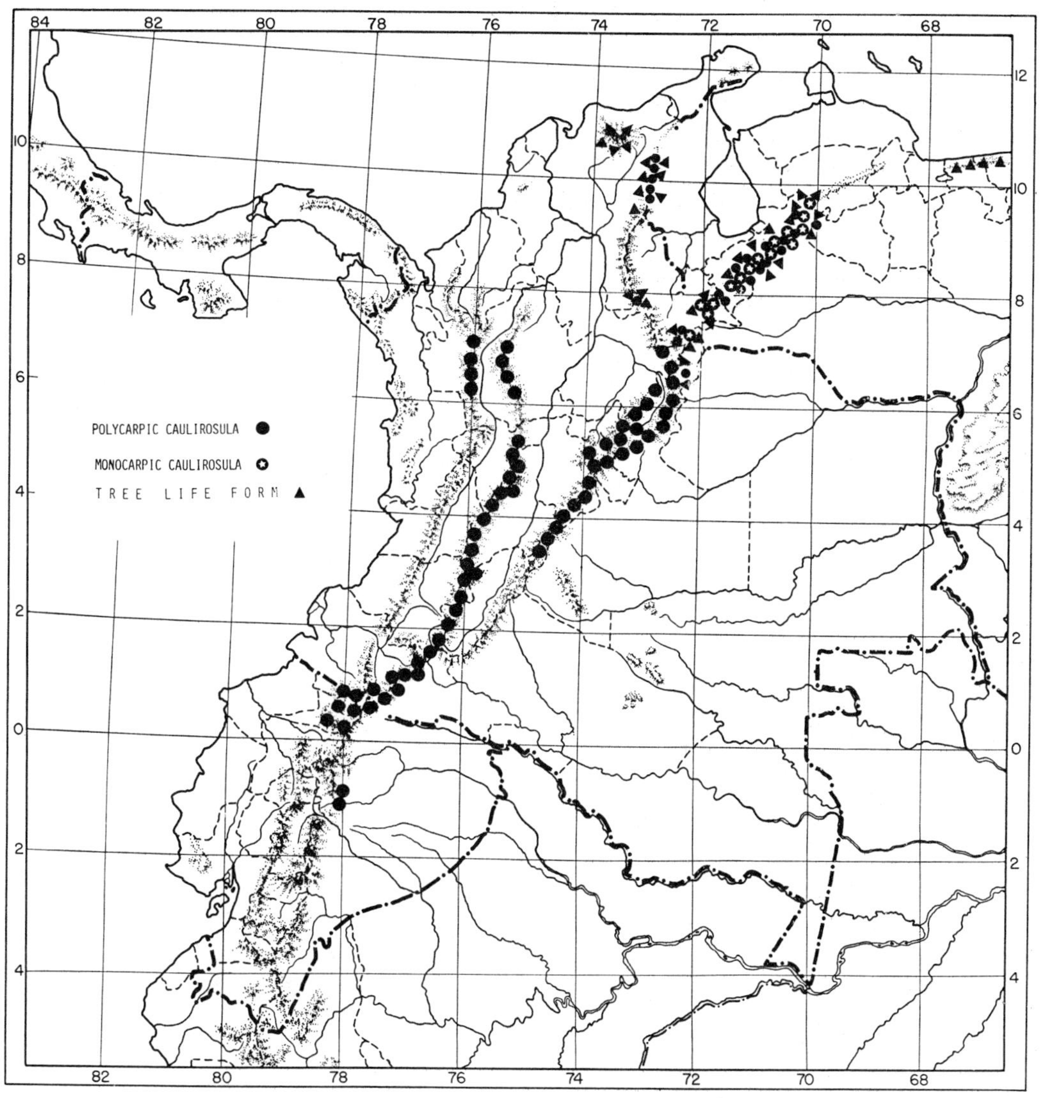

Fig. 6. Distribution of the Espeletiinae growth forms in the northern Andes, north-western section of South America.

                    J. Cuatrecasas

woody (Fig. 5. 1), an example of which is *Libanothamnus neriifolius* (B. ex H.) Ernst. (Cuatrecasas, 1976).

All tree forms of the Espeletiinae are basically members of the Andean Forest belt. They may be found in the whole respective area of the Venezuelan Andes and in the adjacent areas of Colombia (Fig. 6). Their area is disjunct, extending to the coastal mountain range of Avila-Naiguatá in Venezuela and to Sierra Nevada de Santa Marta in Colombia. They are abundant from 2000 (–1800) to 3200 m altitude, but some representatives grow very well in higher mountains up to 3000 (–3800) m in the "Sierras Nevadas" (Fig. 7).

Sympodial and Pseudodichotomic Branching

Synflorescences terminal with determinate monochasial (alternate) branching. The mature trunk is massively woody (Fig. 5. 2). An

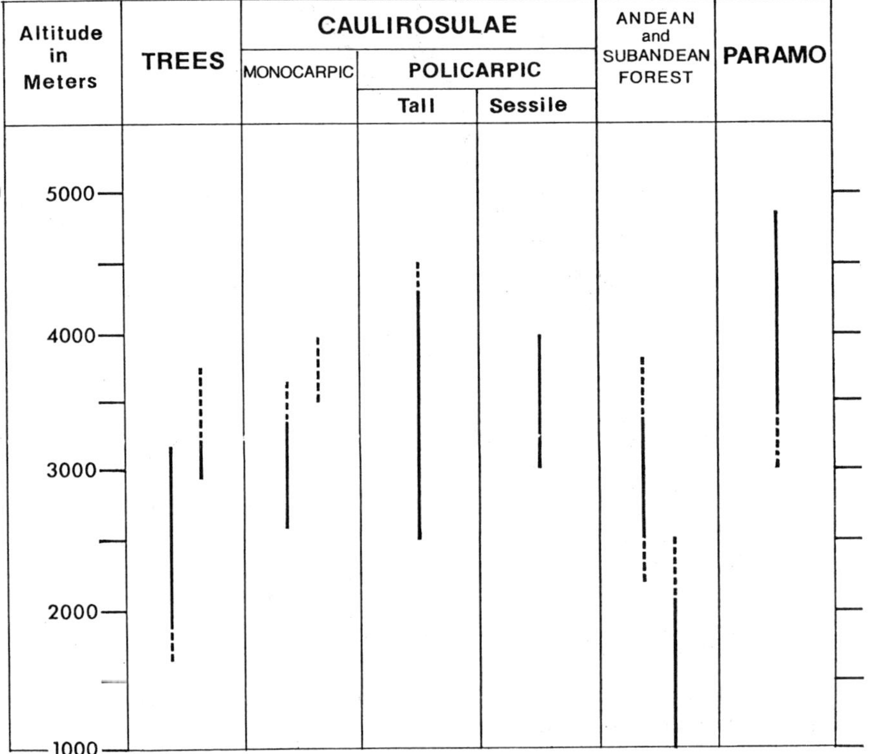

Fig. 7. Chart indicating the altitudinal ranges of the Espeletiinae growth forms.

example is *Tamania chardonii* (A.C.Sm.) Cuatr. This corresponds to the groups of "modèle de Leeuwenberg" of Hallé and Oldeman (1970).

Basic Monopodial Alternate Branching

Synflorescences axillary. The leaves are mostly large and green and the mature trunk is massively woody or with a thin pith. An example is *Carramboa pittieri*.

References

Braun-Blanquet, J. (1964). "Pflanzensoziologie. Grundzuge der Vegetations-kunde." Springer-Verlag, Berlin.

Cleef, A. M. (1977). Secuencia altitudinal de la vegetación de los Páramos de la Cordillera Oriental, Colombia. Symp. Int. Ecol. Trop. Panamá.

Corner, E. J. H. (1949). The Durian theory or the origin of the modern tree. *Ann. Bot.* **13** (52).

Corner, E. J. H. (1954). "The Evolution of Tropical Forest; Evolution as a Process" (J. S. Huxley, A. C. Hardy and E. B. Ford, eds) p. 34–46. Allen, London.

Cotton, A. D. (1944). The megaphytic habit in the tree *Senecios* and other genera. *Proc. Linn. Soc. Lond.* **156**, 158–168.

Cuatrecasas, J. (1934). Observaciones geobotánicas en Colombia. *Trab. Mus. Nac. Cienc. Nat. Ser. Bot.* **27**, 1–144.

Cuatrecasas, J. (1949). Rosette-trees, a tropical growth form. *Bull. Chicago Nat. Hist. Mus.* **20**, 6–7.

Cuatrecasas, J. (1954). Outline of vegetation types in Colombia. Cong. Int. Bot. Paris Sect. VII, 77–78.

Cuatrecasas, J. (1957). A sketch of the vegetation of the North Andean Province. Proc. 8th Pa. Sci. Congr. Bot. Vol. 4, 167–173.

Cuatrecasas, J. (1958). Aspectos de la vegetación natural de Colombia. *Rev. Acad. Colomb. Cienc.* **10**, 221–268.

Cuatrecasas, J. (1968). Paramo vegetation and its life forms. *Coll. Geog.* **9**, 163–186.

Cuatrecasas, J. (1971). Notes on neotropical flora. *Phytologia* **20**, 475.

Cuatrecasas, J. (1976). A new subtribe in the Heliantheae (Compositae): Espeletiinae. *Phytologia* **35**, 43–61.

Drude, O. (1890). "Handbuch der Pflanzengeographie." Engelhorn, Stuttgart.

Du Rietz, G. E. (1931). "Life-forms of Terrestrial Flowering Plants" p. 1–95. Almqvist and Wiksell, Stockholm.

Hallé, F. and R. A. A. Oldeman (1970). "Essai sur l'Architecture et la Dynamique de Croissance des Arbres Tropicaux." Masson, Paris.

Hedberg, O. (1968). Taxonomic and Ecological Studies on the Afroalpine Flora of Mt Kenya.—Hochgebirgsforschung Heft 1: 171–194. München.

Raunkier, C. (1934). "The Life Forms of Plants and Statistical Plant Geography." Clarendon Press, Oxford.

Rock, B. N. (1972). Vegetative anatomy of *Espeletia* (Compositae). Ph.D. Thesis, University of Maryland.

Schnell, R. (1970). "Introduction à la Phytogeographie des Pays Tropicaux" Vol. I. Gauthier-Villars, Paris.

Schnell, R. (1971). "Introduction á la Phytogeographie des Pays Tropicaux" Vol. II. Gauthier-Villars, Paris.

Troll, W. (1937). "Vergleichende Morphologie der höheren Pflanzen" Vol. 1, Part 1. Verlag Gebrüder Borntraeger, Berlin.

Warming, E. (1909). "Oecology of Plants. An Introduction of the Study of Plant Communities." University Press, Oxford.

Warming, E. and P. Graebner (1918). "Lehrbuch der Oekologischen Pflanzen-geographie" 3rd edition. Verlag Gebrüder Borntraeger, Berlin.

Warming, E. and P. Graebner (1933). "Lehrbuch der Oekologischen Pflanzen-geographie" 4th edition. Verlag Gebrüden Borntraeger; Berlin.

Weber, H. (1958). Die Paramos von Costa Rica und ihre pflanzengeographische Verkettung mit den Hochanden Suedamerikas. *Abh. Akad. Wiss. Lit. Mainz* **3**.

The Alkaloid Content of the Genus *Cinchona* in Relation to Altitude *

M. ACOSTA-SOLIS

Instituto Ecuatoriano de Ciencias Naturales, Quito, Ecuador

During World War II I worked as head botanist for some of the Cinchonan expeditions of the American Mission in Ecuador and had the opportunity to collect material for the genus *Cinchona*, both for the herbarium and for analysis of the bark content for total crystallizable alkaloid (TCA). The bark samples were sent directly from the jungle to the laboratory of the mission in order to provide quick and exact indication of the feasibility of further investment in local exploitation.

The present paper deals with analyses made of material collected in Prov. Bolivar, Ecuador. The material belongs to two altitudinally distinct varieties of *Cinchona pubescens* Vahl. (Syn. *C. succirubra* R. & P.), both occurring in Prov. Bolivar. The variety *Cascarilla serrana* occurs at altitudes between 2000 and 3000 m in undisturbed Subandean sotobosque, while the variety *Cascarilla roja* that occurs between altitudes 580 and 1980 m is now in cultivation, being propagated by means of cuttings and seeds.

Chemical analyses of the bark samples suggest that the TCA contents tend to decrease with ascending altitude. For instance samples of *Cascarilla serrana* from El Olivo (atitude 2950 m) and Descenso de Sambulona (altitude 2900 m) showed values of 1·2% and 2·1% respectively while samples of the same variety from San Pablo de Atenas (altitude 2850 m) and Chillanes (altitude 2600 – 2800 m) gave values of 3·4% and 4·8%. Similar trends are found in the *Cascarilla roja*. Unfortunately precise meteorological data for extensive periods are not available for the sites where collections were made. Therefore it is not

* Originally prepared for the 1st Latin American Congress of Botany and Pharmacognosy in Lima 1957, revised 1978.

possible to correlate the TCA contents with temperature directly.

The same type of climate rules throughout the province studied, so these differences cannot be caused by climatic differences other than those correlated with altitude. Therefore I believe it better to state that altitude is decisive as a factor for TCA content.

This thesis appears to be confirmed by data from other species and other areas, such as *Cinchona officinalis* L. in Prov. Azuay, which gave values of 3·5% at altitude 2500 m and 2·6% at altitude 2800–2900 m (data of Dr J. A. Steyermark).

Other varieties of *C. pubescens* collected by Dr Steyermark in Prov. Azuay gave a decrease in TCA contents as follows:

"Pata de gallinazo blanco"—altitude 2400 m, 3·7%; 2950 m, 1%.

Cascarilla rosada—altitude 1375 m, 3·3%; 2200 m, 2·1%.

Cascarilla roja—altitude 600 m, 5·5%; 850 m, 4·8%.

Cinchona pitayensis is the only high altitude species which produced high amounts of TCA, a general average of 5%, occasionally producing up to 7% at altitude 3000 m. Other species such as the *C. humboldtiana* Lamb. and *C. delessertiana* Standl. produce less than 0·86%. *C. micrantha* R. & P., which grows at about the same altitude as *C. pubescens* has a higher TCA production than *C. pubescens* and *C. officinalis*. There are exceptions from the general trend. Thus *C. officinalis* from South Ecuador produced 3·5% at altitude 2700 m. Consequently it cannot be claimed that only certain elevations give a good production. The production also varies with the taxon.

Even within the same individual the TCA content varies both during the life of a tree as shown by Taylor (1945) and in different parts of the tree. Taylor showed that a tree of the species *C. ledgeriana* grown in Java reached its maximum contents at an age of five years. In *C. pubescens* of Ecuador the maximum content is reached at an age of five to seven years. Analyses of trunk bark and branch bark from Peru have shown marked differences. The *C. pubescens* variety "colorada" had a three to four times higher content in trunk as opposed to branch bark. The *C. humboldtiana* variety "negra" had approximately one-third higher content in branches than in the trunk bark. The *C. micrantha* variety "Huanuco" had a slightly higher content in the branches than in the trunk, and the variety "monopol" had approximately one-third higher content in the branch bark than in the trunk bark.

In spite of the general trend that TCA content decreases with ascending altitude there seems to be no universal factor ruling the concentration of TCA. Further studies, including extensive sampling of bark and meteorological data, are needed to establish conclusive evidence about this problem.

References

Acosta-Solis, M. (1951). "Cinchonas del Ecuador." Ecuadorian Institute of Natural Sciences, Quito.
Taylor, N. (1945). "Cinchona in Java." Greenberg, New York.

V. Conclusion

Concluding Remarks

G. T. PRANCE

New York Botanical Garden, USA

First of all, as a foreign delegate, and on behalf of all of us at this symposium, I would like to congratulate the University of Aarhus on its Golden Jubilee, also the Botanical Institute on its fifteenth anniversary, and its staff on the fantastic organization of the symposium which has been an unqualified success in all aspects.

It is exciting and encouraging to see a young institute placing such a definite emphasis on tropical botany, with major programmes as far afield as Thailand and Ecuador. We have heard only about the tropical side of the Institute in this symposium, but it also has impressive temperate region programmes in Europe, and even from Greenland to Tierra del Fuego.

I have been asked to give a lecture here in Denmark in a few weeks time on the subject of "Why should we in the north temperate countries be doing botanical work in the tropics?" I think that this symposium has provided many of the answers to this question for the university authorities. Many speakers have emphasized the extraordinary gaps in our knowledge of the tropical flora, the need for greater urgency of future studies before the natural tropical ecosystems are destroyed, the extreme shortage of personnel to write Floras, the responsibility of the developed countries to train and assist those in developing areas, and to make our resources, such as libraries and herbaria, available to those who do not have such facilities.

One of the important aspects for the future must be the training of nationals of the tropical countries. Even in this aspect the young Botanical Institute of the University of Aarhus is to be congratulated. Several Thai students have already been trained here in Aarhus and one of the coordinators of this symposium, is shortly to be seconded for two years to work at an Ecuadorian University financed by the Danish developmental aid programme.

This symposium has shown that there is great shortage of botanists to produce even the tropical Floras which are under way, and that too many experts are tied up in teaching, administration and other distractions from our main goal. We will need the greater assistance of local botanists in tropical countries. Even with the lack of resources in some tropical countries, there is much that local botanists can and must contribute. Nothing replaces the day by day observation of biological phenomena. For example, dispersal of diaspores has been a frequent theme of this symposium. We have heard complaints about the lack of data on dispersal and pleas to collectors to pay more attention to this. We can stimulate local residents, who do not have the resources to write a Flora, to make and publish for themselves observations on dispersal, phenology, pollination, breeding systems etc. Many speakers in this symposium have emphasized the need for much more alpha taxonomy, and our symposium resolution also calls for this with its emphasis on Floras. This is certainly necessary or we will never complete the tropical Floras, but, we must encourage an expanded alpha taxonomy that considers dispersal, pollination, phenology and other biological phenomena, chromosome numbers etc., at the same time that it is producing the Floras. We need Floras to identify plants, but also need to understand the history and dynamics of the tropical ecosystem in order that it can be used rationally, and that we are not just producing museum-piece Floras of an extinct flora. Utilization and conservation have not been the central theme of this symposium, but we are all concerned with this. A more experimental approach could give us more tropical flora to study in the future because its results will lead more naturally to conservation.

The diverse programme of this symposium has encouraged this balanced approach by including such subjects as a chromosome study of Himalayan monocots in relation to altitude, the palynology of the South American Andes region, the dispersal of Malesian plants, the breeding systems of Venezuelan plants, intermingled with reports about local Floras from around the tropics. Surely this also gives us a hint as to where we should head in the future.

We have had a symposium whose proceedings will provide many useful data. I hope that they will stimulate us and other botanists to unite in more collaborative efforts with our tropical colleagues to intensify our studies of the tropical vegetation.

Resolution

P. BRENAN, F. R. FOSBERG, R. HOWARD and K. LARSEN

Whereas plants are the basis of all life and essential for continual human existence and as an adequate classification of plants is the only practical framework on which plant knowledge can be organized so that it can be utilized; and *whereas* accurate identifications and nomenclature are the only means for retrieval and dissemination of plant information; and *whereas* adequate systematic accounts of plants, known as Floras, are lacking for most tropical areas, wherein are the developing nations that would benefit most from such information:

We, the members of the Symposium on Tropical Botany, meeting at the University of Aarhus, Denmark, representing 17 countries, and recognizing the pressing, economic and scientific importance to developing countries of the completion of tropical floras, urge most strongly that all national governments, appropriate research councils and foundations, ministries, and other agencies, national and international, should ensure adequate provision, both in funding and staffing, to facilitate speedy completion and publications of such Floras as are at present in preparation, and for such field and herbarium work as is needed for this purpose.

We also urge that sufficient provision be made for the training by appropriate universities and herbaria, of promising taxonomists, including those from developing countries, and for residence of specialists in those countries which desire such collaborative aid.

We urge that both policy and priorities for future study of tropical plant life and the preparation of floras can be considered, and recommendations made by a representative body.

Systematic Index

D

Forsteronia portoricensis 213
Fourcryoa 399
Freziera 244
Freziera steyermarkii 219
Fritillaria 332, 334
Froesia 190, 199
Froesia crassiflora 200
Froesia tricarpa 200
Froesia venezuelensis 200, 203, 215
Fuchsia 171, 179, 190
Furcraea 172

G

Gaiadendron 177, 182
Galactophora pulcella 209
Galeopsis 91
Galium 179
Garcinia 47, 150
Garhadiolus 266
Gastonia 323, 324
Gaultheria 179
Gaultheria tatei 218
Gaylussacia 179
Gearum 299, 301
Geissanthus 177
Gentianaceae 365, 367, 369
Gentiana 173, 179
Gentianella 28, 179
Geonoma 169
Geraniaceae 173
Geranium 28, 179, 182
Geranium santanderiense 193
Gesneria onacaensis 218
Gibsoniothamnus 340
Gigantochloa 117, 119
Glaziocharis 369
Globba 331
Glochidion 277, 278, 280
Gloeospermum 189, 199
Gloeospermum sphaerocarpon 199, 217
Gloeospermum sphaerocarpum 200, 203, 217
Glomeropitcairnia 189
Glomeropitcairnia erectiflora 217
Glossocordia setosa 148
Glossonena varians 145

Glossostigma diandra 147
Gnaphalium 179
Gnaphalium caerleocanum 218
Gochnatia 243
Godmania aesculifolia 350
Gomphia serrata 115
Gomphichis 177
Gonatanthus 298, 306, 307
Gonatopus 291, 295
Gonglolepis 206
Goniopteris megalodus 218
Gonocormus 312, 313
Goodeniaceae 132
Goodyera 329
Gordonia 118
Gorgonidium 299, 301
Grabowskia 356
Grabowskia duplicata 360
Graffenrieda weddellii 199, 200, 201, 203, 217
Gramineae 28, 117, 118, 183, 328, 336
Grammadenia 177
Grammitis limula 217
Gratiola 179, 182
Greigia 177
Greigia aristeguietae 195
Grewia tenax 147
Guacamaya superba 209
Guadua 170
Guadua angustifolia 169
Guaiacum officinale 256
Guatteria venezuelana 216
Guazuma ulmifolia 169
Guettarda 243
Gunnera 28, 171
Gunnera pittieriana 216
Gurania simplicifolia 211
Gustavia 77
Guzmania confinis 193
Guzmania cylindrica 202, 215
Guzmania hedychioides 216
Guzmania membranacea 217
Guzmania sanguinea 217
Guzmania steyermarkii 211
Guzmania venamensis 211

M